INVENTIONS
THAT CHANGED
THE WORLD

To the inventors, past, present and future.

'O brave new world, that has such people in't!'

INVENTIONS
THAT CHANGED
THE WORLD

by Rodney Castleden

Futura

A *Futura* Book

First published by Futura in 2007

Copyright © Omnipress 2007

ISBN: 978-0-7088-0786-6

Produced by Omnipress, Eastbourne

Printed in the EU

Futura
An imprint of Little, Brown Book Group
Brettenham House
Lancaster Place
London WC2E 7EN

Photo credits: Getty Images

CONTENTS

Introduction 13

I: THE ANCIENT WORLD

The invention of stone tools (about 500,000 BC) 20
The invention of paint (38,000 BC) 23
The invention of arrowheads (18,000 BC) 26
The development of settled life (8000 BC) 28
The invention of trepanning (6500 BC) 31
The invention of the wheel (6400 BC) 33
The invention of maps (6200 BC) 37
The invention of irrigation (6000 BC) 40
The invention of pottery (6000 BC) 43
The invention of copper smelting (5500 BC) 46
The invention of monuments (4700 BC) 48
The invention of the ard (4500 BC) 52
The invention of papyrus (4000 BC) 54
The invention of sun-dried bricks (4000 BC) 58
The invention of writing (3000 BC) 62
The invention of soap (2800 BC) 65
The invention of silk (2640 BC) 67
The invention of currency (2200 BC) 69
The invention of the aqueduct (1700 BC) 71
The invention of coins (640 BC) 74
The invention of the stirrup (500 BC) 76
The invention of geometry (300 BC) 78
The invention of the Archimedean screw (265 BC) 82
The invention of the astrolabe (150 BC) 83
The invention of the encyclopedia (35 BC) 86
The invention of paper from pulp (AD 105) 92
The creation of the first world map (150) 94
The invention of woodblock printing (200) 97
The invention of the horseshoe (450) 100

The creation of the Christian calendar (525) 102
The invention of the windmill (600) 105

II: THE MEDIEVAL AND RENAISSANCE WORLD

The invention of gunpowder (900) 108
The invention of paper money (960) 111
The invention of the magnetic compass (1086) 112
The invention of anaesthetics (1236) 115
The invention of spectacles (1265) 117
The invention of the clock (1280) 119
The invention of the watermark (1282) 121
The invention of the movable type printing press (1440) 123
The invention of the rain gauge (1441) 126
The invention of the paddle wheel (1450) 128
The invention of the arquebus (1475) 130
The invention of the globe (before 1490–92) 134
The invention of the Gregorian calendar (1582) 138
The invention of the knitting machine (1589) 140
The invention of the flushing water closet (1589) 143
The invention of the thermometer (1593) 146
The invention of the microscope (1595) 149

III: THE ENLIGHTENED WORLD

The invention of the telescope (1608) 152
The invention of logarithms (1614) 154
The invention of the slide rule (1614) 157
The invention of the micrometer (1636) 160
The invention of the calculating machine (1642) 162
The invention of the barometer (1643) 164
The invention of the air pump (1650) 168
The invention of the pendulum clock (1656) 169
The invention of the reflecting telescope (1668) 172
The invention of calculus (1666) 175
The invention of the pressure cooker (1675) 179

The invention of the steam pump (1698) 181
The invention of the piano (1700) 183
The invention of the seed drill (1701) 186
The invention of iron smelting using coke (1709) 187
The invention of the steam engine (1712) 190
The invention of the mercury thermometer (1714) 192
The invention of inoculation against smallpox (1718) 194
The invention of the diving suit (1715) 199
The invention of the flying shuttle (1733) 202
The invention of the ship's chronometer (1735) 204
The invention of the lightning conductor (1752) 208
The invention of the ribbing machine (1758) 210
The invention of the spinning jenny (1764) 212
The invention of the condensing steam engine (1765) 214
The invention of Venetian blinds (1769) 216
The invention of Coade stone (1771) 218
The reinvention of the flushing water closet (1775) 221
The invention of the spinning mule (1779) 223
The invention of the hot-air balloon (1783) 225
The invention of the parachute (1783) 227
The invention of the power loom (1785) 229
The invention of the threshing machine (1786) 232
The invention of coal gas lighting (1792) 235
The invention of the cotton gin (1793) 238
The invention of the internal combustion engine (1794) 240
The invention of the pencil (1795) 242
The invention of lithography (1796) 244
The invention of cast-iron beams and columns (1796) 246
The invention of the top hat (1797) 249
The invention of the electric battery (1800) 251
The invention of the metal lathe (1800) 253

IV: THE NINETEENTH-CENTURY WORLD

The invention of the Jacquard loom (1801) 258
The invention of the suspension bridge (1801) 260

The invention of the steamboat (1802) 262
The invention of the steam locomotive (1804) 265
The invention of the refrigerator (1805) 267
The invention of carbon paper (1806) 270
The invention of the tin can (1810) 272
The invention of the metronome (1812) 274
The invention of the miner's safety lamp (1815) 276
The invention of the macadam road surface (1815) 278
The invention of the stethoscope (1816) 280
The invention of the kaleidoscope (1817) 281
The invention of the seagoing iron ship (1821) 282
The invention of the digital calculating machine (1823) 285
The invention of the electromagnet (1825) 287
The invention of the railway (1825) 288
The invention of the friction match (1826) 292
The invention of Braille (1829) 294
The invention of the lawnmower (1830) 296
The invention of the electric motor (1831) 298
The invention of the mechanical reaper (1831) 300
The invention of the sewing machine (1832) 302
The invention of the hansom cab (1834) 304
The invention of the revolver (1836) 306
The invention of the screw propeller (1836) 307
The invention of photography (1838) 311
The invention of the electric telegraph (1838) 314
The invention of the bicycle (1839) 317
The invention of vulcanized rubber (1839) 319
The invention of incandescent electric light (1840) 323
The invention of the submarine telegraph cable (1842) 327
The re-invention of anaesthetics (1846) 330
The invention of the sewing machine (1846) 332
The invention of the safety elevator or lift (1853) 335
The invention of the Bunsen burner (1855) 337
The invention of the Bessemer converter (1856) 339
The invention of the washing machine (1858) 340
The invention of the oil derrick (1859) 341
The invention of the ironclad (1859) 344

The invention of linoleum (1860) 347
The invention of the Gatling gun (1861) 349
The invention of the cylinder lock (1862) 351
The invention of the stapler (1866) 353
The invention of dynamite (1866) 355
The invention of the depth-sounding machine (1866) 357
The invention of the torpedo (1866) 359
The invention of antiseptic surgery (1867) 362
The invention of celluloid (1868) 364
The invention of the typewriter (1868) 365
The invention of margarine (1870) 367
The invention of lawn tennis (1874) 368
The invention of barbed wire (1874) 371
The invention of the telephone (1876) 373
The invention of the phonograph (1877) 375
The invention of the cash register (1879) 379
The invention of lavatory paper (1880) 380
The invention of the skyscraper (1884) 382
The invention of the fountain pen (1884) 386
The invention of the steam turbine (1884) 388
The invention of the fingerprinting identification system (1885) . 390
The invention of the motorcycle (1885) 392
The invention of the electric transformer (1885) 394
The invention of Esperanto (1887) 396
The invention of the electrocardiogram (ECG) (1887) 398
The invention of the pneumatic tyre (1888) 400
The invention of Gestetner typewriter stencil (1888) 402
The invention of the zip fastener (1892) 404
The invention of the diesel engine (1892) 406
The invention of the tractor (1892) 408
The invention of Henry Ford's car (1893) 410
The invention of wireless (1895) 412
The invention of motion pictures (1895) 415
The invention of the x-ray machine (1895) 418
The invention of the escalator (1897) 419
The invention of the submarine (1898) 421

V: THE MODERN WORLD

The invention of the safety razor (1901) 426
The invention of the aeroplane (1903) 428
The invention of animated cartoon film (1906) 431
The invention of the vacuum cleaner (1907) 435
The invention of the helicopter (1907) 437
The invention of cellophane (1908) 439
The invention of Bakelite (1909) 441
The invention of the Geiger counter (1911) 442
The invention of the geological timescale (1913) 443
The invention of the crossword puzzle (1913) 446
The invention of the echo-sounder (1914) 448
The invention of the tank (1915) 450
The invention of camouflage (1917) 453
The invention of autobahn (1921) 457
The invention of the robot (1922) 460
The invention of television (1925) 462
The invention of the liquid-fuel rocket (1926) 464
The invention of the jet engine (1930) 468
The invention of the cyclotron (1931) 469
The invention of the cat's eye (1934) 472
The invention of the Belisha beacon (1934) 473
The invention of nylon (1935) 475
The invention of the ball-point pen (1935) 477
The invention of radar (1935) 480
The invention of the radio telescope (1937) 482
The invention of the photocopier (1938) 484
The invention of the electron microscope (1940) 486
The invention of polyester (1941) 488
The invention of the aerosol spray can (1944) 489
The invention of the atom bomb (1945) 491
The invention of the Identikit picture (1945) 494
The invention of the electronic computer (1946) 496

The invention of Warfarin (1947) 499
The invention of the Polaroid camera (1947) 500
The invention of Velcro (1948) 502
The invention of the transistor (1948) 504
The invention of long-playing records (1948) 505
The invention of the barcode (1949) 508
The invention of the nuclear power reactor (1951) 511
The invention of the hovercraft (1955) 515
The invention of the contraceptive pill (1955) 517
The invention of the smokeless zone (1956) 519
The invention of the videotape recorder (1956) 522
The invention of the man-made satellite (1957) 524
The invention of the laser (1958) 527
The invention of the pacemaker (1958) 531
The invention of the microchip (1959) 533
The invention of heart transplant surgery (1967) 534
The invention of space travel (1969) 536
The invention of e-mail (1971) 542
The invention of the CT-scanner (tomography) (1971) 544
The invention of the cellphone (mobile phone) (1973) 546
The invention of the Post-it note (1974) 549
The invention of the MRI scanner (1977) 551
The invention of the travelator (1978) 552
The invention of the desktop computer (1980) 555
The invention of the World Wide Web (1990) 557
The invention of the clockwork radio (1991) 559
The invention of the global positioning system (1993) 562
The invention of Windows 95 (1995) 564
The invention of the 10,000-year clock (1995) 567
The invention of the Digital Versatile Disc (DVD) (1995) 569

Appendix: Inventions named after their inventors 573

INTRODUCTION

INVENTION IS AN act of creative imagination. In its broadest sense, the word invention includes imagining, designing and creating not only mechanical devices and new modes of transport but also poems, novels, sculptures, pieces of music and theories about the origin of the universe. The French Patent Act of 1791, which in spirit followed the 1789 Declaration of the Rights of Man, explicitly placed economic inventions and social inventions on the same footing as mechanical and scientific inventions. Patents giving legal protection to inventors were therefore duly granted to tariff systems, credit plans and tontine life annuities. It became clear that there was going to be an avalanche of innovative financial schemes requiring patent protection, so in 1792, the French National Assembly decreed that financial inventions were henceforth (and retrospectively) debarred; as far as the laws of France were concerned financial schemes were no longer inventions.

In the USA financial schemes were never regarded as inventions within the law, but in 1930 Congress allowed the patenting of newly discovered or propagated plants. As we shall see later, patent laws create as many problems for inventors as they resolve.

Today we normally understand that an invention is a technological or scientific process, device or object that has been newly developed. There is also a general understanding that invention is an imaginative quality that can be applied in science, technology or any other arena of human activity; it is certainly possible to speak of Stravinsky as an inventive composer or Picasso as an inventive artist, though we would not normally describe their works as 'inventions'.

Invention is a characteristic of certain societies, but not all societies. Societies need to allow individuals a certain level of independence of thought and action as a precondition for invention. Certain archaic societies have such rigid and inflexible structures and mindsets that inventiveness is constrained. In a free-market economy and what we

12

might call a free-thinking society, inventions are likely to happen all the time. Although individual inventors are (rightly) given high-profile credit for their inventions, it is mainly the social, economic and technological preconditions that drive invention. The strange phenomenon of simultaneous invention proves this.

The wheel was invented in China and Europe independently. Carbon paper was invented independently by Pellegrino Turri in Italy and Ralph Wedgwood in England. The screw propeller for ships was invented separately by Ericsson and Francis Smith (both in England) in 1836. The 'Bessemer' steel-making process was invented independently by Henry Bessemer in England and William Kelly in the USA. Charles Hall and Paul Heroult invented the process of making aluminium independently and simultaneously. Tessie du Motay in France and Thaddeus Lowe in America independently devised processes for making street-gas by passing steam over red-hot coal. Photography was invented at the same time by Fox Talbot in England and Daguerre in France. The electric light bulb was invented at the same time by Joseph Swann in England and Thomas Edison in America. The aircraft jet engine was invented more or less simultaneously in Germany by Hans von Chaim and in England by Frank Whittle. In America Alexander Graham Bell and Elisha Gray filed their patent applications for identical telephones on the same day.

Every year there are what are called interferences at the US Patent Office, arbitrations to settle the rival claims of inventors who apply simultaneously for patents on similar devices or processes.

The history of inventions is full of this kind of coincidence. When I was at school, I remember one of my teachers describing an idea for creating a thin cushion of air between a heavy weight and the ground, to enable it to be moved around easily without friction. Another teacher teased him for being 'a mad boffin'. Not long afterwards, the newspapers and television news programmes were full of the news that the hovercraft had been invented, not by my teacher but by somebody called Christopher Cockerell: it was the same device. In the early 1970s I thought of the travelator as a means of travelling within cities, perhaps even between cities. I am glad to see that someone else, thinking along the same lines at the same time in Madrid, invented the travelator and that it is being used to make moving about within airports easier, but disappointed that it has not come into more general use. The travelator could replace tubes, trains, buses and cars.

In the history of inventions there have been many bitter disputes about who was first. Oughtred and his pupil Delamain quarrelled savagely and publicly about which of them invented the slide rule. The dispute between Newton and Leibniz about which of them invented calculus was even more famous.

Often several people are converging on the same solution to a perceived problem at about the same time. From this it follows that it is sometimes difficult to establish exactly where and by whom a particular invention was made. In antiquity this is even truer, because less information is available to us. Often there is an attribution by tradition, and we have to make do with that. With the very earliest tools, such as the lever, pulley, saw and wedge, we have no real way of finding out. Many inventions were the result of a long process of evolution, so it is difficult for us to name one inventor or one particular date. Many inventions lean on earlier inventions too. The microscope was a by-product of the invention of the telescope.

If the idea of simultaneous invention seems uncanny, the phenomenon of repeated invention is even stranger. Sometimes an invention is forgotten and then has to be re-invented later. People chairing discussions often say by way of introduction that the last thing they want the meeting to do is to 're-invent the wheel'; in other words they do not want to waste time devising some way of doing something that has already been tried and tested. Yet this is no mere figure of speech. The wheel very likely was re-invented in prehistoric times, over and over again in many different places. It was a very effective solution to a problem, and the problem was a universal one. On a much lower level, though still relating to the wheel, the pneumatic tyre was re-invented; it was invented and patented early in the nineteenth century and then re-invented again forty years later.

There is often a link between discovery and invention. Once it was discovered that there was a regular rate of expansion and contraction in liquids resulting from variations in temperature, it was a short step from there to the invention of the thermometer.

Inventions are sometimes designed to solve one specific problem, to perform one particular task. Then it turns out that they have all sorts of other applications. Radar was designed to help to detect enemy aircraft; it is now used by police officers to catch motorists in speed traps.

What has been said so far may suggest that inventors are merely the puppets of social, technological and scientific forces, but there is usually much more to them than that. In some cases, an inventor is someone who is conscientiously pursuing a trade or profession and who sees a way – just one way – of doing things far better. There are many inventions that are the result of just a single insight of that kind. Gutenberg's invention of movable alphabet type is a good example; John Kay's flying shuttle is another.

There are other inventors who are creative geniuses functioning on a level with the greatest artists, writers and composers: Galileo, Newton, Leibniz, Pascal, Napier, Maudslay, Faraday, Edison, Kelvin, Tesla. Their minds fizz with creativity to the point where they do not finish working on one project before starting on the next, so that their inventions and ideas overlap chaotically. This, I suspect, is where the 'mad boffin' image comes from, the image that many people have in their minds as the stereotype of the inventor. This second type of inventor goes on inventing almost regardless of the external pressures to innovate, and there the analogy with great novelists, poets, composers and painters is particularly valid. Inventors like Thomas Edison have such an urge to create within them that they fly off at several tangents, often failing to see through a half-invented device. Edison's work on the phonograph is a good example of this. He invented the phonograph, broke off to continue working on the electric light bulb, then returned to perfect the phonograph later. Edison had a head for business, and he was certainly inventing for a living, but he also had what has been called 'the instinct of contrivance' – the urge to invent.

During the course of history, many inventions must have sunk without trace, simply because people had ideas that they were unable to implement simply because they had no capital. In the mid-eighteenth century, James Watt saw relatively easily how to design an improved version of the Newcomen steam engine, but he had very little money of his own with which to build it. Watt got himself into debt, and then allowed the owner of the Carron Ironworks a major share in the proceeds from his steam engine for his financial backing; only then could Watt's engine be built and demonstrated. James Watt came to depend on his financial partner, Matthew Boulton. Alexander Graham Bell was similarly dependent on Gardner Hubbard.

There have always been more ideas around than could be implemented, because of the ever-present problem of finding capital, of making the right business contacts, of the cautiousness of investors. The struggle Trevor Baylis had to get his clockwork radio into production is a classic case. There is commonly a gap of twenty years between a major invention and its commercial exploitation. Edison's concept of a kind of commercial-industrial laboratory was a brilliant concept. It was a great invention in its own right, in many ways ahead of its time. Organized research of that kind has allowed more rapid progress. But there will always be room for – and the need for – the one-off invention generated by freelance inventors. Without the mad and the maverick, the history of inventions would look very different.

Which inventions do we really value? The answer will depend very much on our occupations and which devices make our own individual professional and personal lives easier or more productive. It will also depend on how serious we are as people: on whether we place a higher value on the devices that help us to be more effective at work, or on those that help us to have more fun when we are not at work. If I had it in my gift, I would confer a dukedom on whoever it was who invented gin and tonic.

There have been many attempts to identify the best or the most important inventions. One attempt to define the Top Ten greatest inventions of all time had the wheel as its Number One, then in descending order of importance electric light bulb, printing press, telephone, television, radio, gunpowder, desktop computer, telegraph, internal combustion engine. Another attempt, this time to identify the Top Ten modern inventions, in no particular order, came up with the following list: battery, barcode, ballpoint pen, microwave oven, ring-pull can opener, workmate (the DIY bench rather than the co-worker), sticking plasters, cat's eyes (the reflecting road studs), parking meter and post-it notes. A recent poll organized by a British radio programme revealed that its listeners thought more highly of the bicycle than any other modern invention, on the grounds that it was a simple, ecologically sound means of transport, and universally useful. It was regarded, to use a well-worn phrase, as the best thing since sliced bread – except that pre-sliced bread is a much more recent innovation than the bicycle; it was invented as late as 1927 by Otto Rohwedder.

I
THE ANCIENT
WORLD

THE INVENTION OF STONE TOOLS

(ABOUT 500,000 BC)

DURING THE PLEISTOCENE Ice Age, between half a million and a million years ago, early people (Homo erectus) started making the first well-shaped standard tools. The most obvious and recognizable to us is the stone hand axe. This was a weighty all-purpose tool, the Stone Age equivalent of the Swiss Army knife. To begin with, it was roughly flaked and not very regular but in time, as expertise improved with experience, it became more symmetrical with more regular cutting edges. The hand axe was typically triangular or pear-shaped and rather larger than a man's hand.

It is not known where the hand axe was invented, and it is certainly not known who invented it. We do not know the names of any individual people at all until 3000 BC. By the time the hand axe had been perfected, or at any rate reached this standard form, halfway through the Ice Age, it was being used across a huge area. Hand axes of much the same type have been found from southern Africa to southern England, from southern Europe right across the Middle East to India – a huge continuous region. This suggests that people were actually trading or swapping tools with one another across this whole area, as the distribution of the axes is coherent and finite.

Intriguingly, so far no well-made hand axes have been found in eastern Asia from this early period. There were stone tools in China 500,000 years ago, but they were fairly rough and ready, and used for scraping, polishing and smashing. The better made hand axes of the West were obviously designed for cutting.

By 250,000 years ago, the descendants of *Homo erectus* were still making stone hand axes, out of a variety of hard stones. Swanscombe Man, who lived in the lower Thames valley at about that time, and whose bones were found in river gravels, used hand axes made of flint. In southern England there are lots of places where chalk comes to the surface, such as the North and South Downs, and where there is chalk there is flint. Flint could be picked up on the foreshore under chalk cliffs like the Seven Sisters in Sussex. Flint is an excellent material for making axes and other tools that need to be given cutting edges. It is exceptionally hard and glassy. When it is broken, it gives a naturally sharp edge, like a broken bottle. Usually the toolmakers made lots of small chips along the cutting edge, which made them serrated and even better for cutting.

I found a flint hand axe myself many years ago while I was exploring ancient river gravels in a pit in one of the Thames terraces. It belonged to the Acheulian culture, and it was as sharp when I picked it up as when it had first been made; I found I could still cut meat with it. How many of the things we make today will still be in good working order in 200,000 years' time?

By 35,000 years ago the people of the Advanced Hunting cultures were still making hand axes, but also a range of other tools, including fine chisels and burins or gravers. These were used for working in bone, ivory and antler. This was the moment when the first works of art were produced, such as the beautiful spear-thrower made of reindeer antler found at Mas d'Azil in France, with an expertly carved animal, apparently a young deer, carved at one end.

The craft of stone tool making culminated in the very fine polished stone axes of the New Stone Age. These were not hand axes any longer, but axe heads designed to be hafted into wooden handles and wielded like modern iron axes. The ordinary axes were used for felling trees, but some of the late stone axes were very large and very fragile, especially the ones made out of jadeite. They are so fragile that they cannot have been made for tree-felling or for any useful purpose: they can only have been made for show, to confer status on their owners. They were traded over long distances, which also confirms their very high value. Some of them travelled hundreds of miles halfway across Europe.

These were high-status objects made right at the end of the Stone Age. They were about to be replaced, in the Early Bronze Age, by magnificent gold objects that we today can much more readily see were prestigious. That change marked the beginning of the age of metal and, as metal tools and weapons came in, stone tools were gradually phased out. But for an unimaginably long stretch of prehistory the lives of people everywhere were dominated by stone tools.

THE INVENTION
OF PAINT

(38,000 BC)

PAINT IS USUALLY composed of three things: a pigment, a binder to hold it together and a thinner to make it easy to apply. We use paint is so many situations all around us that it is difficult to imagine a world without it. Our houses are coated in it, sometimes outside as well as inside; many items of furniture are painted; our cars, buses, trains, planes and ships are all covered in paint too. Sometimes the main reason for applying paint is to make an object more durable, more weatherproof, but usually the colour of the paint matters. The colour matters pre-eminently when paint is used in the creation of art works. Virtually every culture in the world has a distinctive tradition of visual art, and paint plays a major role in this. At one end of the spectrum are works such as the *Mona Lisa* and the ceiling of the Sistine Chapel. At the other end are the cave paintings of the Old Stone Age.

Cave or rock paintings have been made since about 40,000 years ago. Among the most famous are the paintings in the Lascaux cave, and these are thought to represent the peak of the cave painting culture in Europe. When modern Europeans first saw the Altamira cave paintings in Cantabria in 1879, they assumed they must be hoaxes, but recent dating tests show that they are authentically ancient. They show a high level of artistry. They depict animals, and were from early on assumed to be connected with some sort of hunting magic. Alternatively, they may have been made by a shaman or witch doctor. The shaman retreated into the darkness of the caves, went into a trance and then painted what he saw in his visions, perhaps to draw power out of the Earth surrounding the caves. This would explain the isolation and inaccessibility of the caves.

It is impossible to be sure, but the paintings give us valuable clues as to the nature of the people and their beliefs and interest. Whatever their original intention the paintings now speak to us eloquently of the people who painted them. And this is one of the most remarkable features of art work – the sheer human immediacy of the communication, person to person, across a seemingly unbridgeable gap of hundreds or even thousands of years.

The oldest of the sets of European cave paintings is in the Chauvet cave, and that dates from 32,000 years ago.

Most of the images are of large wild animals, bison, horses, aurochs (wild cattle) and deer. Some, intriguingly, are tracings of human hands. There are just a few images of people, and these are mostly schematic rather than attempts at portraiture. There are also abstract patterns, which one antiquarian described as 'macaroni'.

The pigment used in the earliest European paintings was made out of haematite, red and yellow ochre, manganese oxide and charcoal. The images were sketched and then painted by lamplight.

Some images were carved before being painted, so that they stand out in low relief, or just lightly etched, like those found at Cresswell Crags in England in 2003. Rock painting was even done on cliff faces, but in the nature of things these have mostly been destroyed by erosion. There are cliff paintings in the Saimaa area of Finland.

In South Africa, the rock art at Ukhahlamba-Drakensberg was being painted about 3,000 years ago. It shows both people and animals. A recently discovered set of cave paintings in Somalia shows the ancient inhabitants worshipping their cattle and performing religious rituals. More ancient cave art, again showing herds of animals, has been found at Tibesti in the Sahara.

There are more paintings in rock shelters and caves in Australia, Thailand, Malaysia and Indonesia. The oldest paintings in Malaysia are at Gua Tambun in Perak, and they are 2,000 years old. Those in the Painted Cave in the Niah Caves National Park date from AD 800.

In the Bronze Age paint was used with great expertise and finesse to decorate the walls of ancient Egyptian tombs and Minoan temples. Again, the subjects painted show us, thousands of years later, what the people of those ancient civilizations thought was beautiful, and what they thought was important.

As time passed, more colours became available. Brown, red, grey, black and red and yellow ochre were available from very early on. The rich elites of the Bronze Age Mediterranean civilizations, the pharaohs of Egypt and the priestesses of Minoan Crete, were able to afford lapis lazuli to grind up for a brilliant piercing blue pigment. Later, the painters working for the kings and princes of the Mycenaean kingdoms were provided with semi-precious malachite to create a bright green. Naples yellow was invented in about 500 BC. In the sixteenth century AD, berries and tree barks from the newly discovered Americas supplied the raw materials for more new colours: Dutch Pink and Crimson Lake. Cochineal was adopted from the Native Americans. The deepest and most intense blue, Prussian Blue, became available in the early eighteenth century.

There was still no strong yellow that would not fade in direct light. That was not to be invented until 1818, as Chrome Yellow. Mixtures of old and new colours made yet more new colours possible. Mixing Prussian Blue and Chrome Yellow created Brunswick Green. A man-made Ultramarine Blue and Alizarin Crimson appeared between 1820 and 1840.

In the 1870s the first washable paint was marketed, as Charlton White. Emulsion paints for painting walls also became available, marketed as oil-bound distemper; these were a marked improvement on the lime washes and colour washes previously available for decorating walls.

THE INVENTION OF
ARROWHEADS

(18,000 BC)

THE BOW AND arrow were invented in about 25,000 BC, probably developing out of the technique of spear-throwing. People soon after that began to use fire to harden the arrowheads, and added feathers to the shaft to improve the accuracy of the flight. The oldest arrowhead so far discovered was found in Africa.

The bow and arrow are a fairly recent invention in North America, coming into use between AD 500 and 900. The American arrowheads were triangular or teardrop-shaped and usually very small, seldom more than 4cm long.

A great many flint arrowheads were made in Middle Stone Age Europe. The people of the Middle Stone Age were mainly gatherers and hunters, so bows and arrows were an indispensable part of their equipment. A refinement developed at that time, around 13,000 BC, was the transverse arrowhead. This was a deadly flying razor blade designed to sever arteries. Earlier arrowheads must often have stuck into a hunted animal, leaving it able to run for miles; the transverse arrowhead meant a swifter death and a shorter chase.

In Italy, a man was buried in 11,000 BC with a flint arrowhead lodged in his hip. Was this a hunting accident? Or do we have here the first evidence of bows and arrows used in aggression? Bows and arrows were obviously a great benefit to people in making it easier to procure meat, but they were also a great blight in that they could be used for murder – on any scale. At the late neolithic fortified enclosure of Crickley Hill in the Cotswolds, swarms of arrowheads were found by archaeologists just inside the two enclosure entrances. They imply a concerted armed attack on the enclosure by people firing at, over and through its gates, and they give us clear evidence of tribal warfare in the late neolithic.

Arrow shafts dating from 9000 BC have been found in Germany and bows from 7000 BC have been found in Denmark. Oetzi, the famous Iceman discovered in the Alps in 1991, died in 3300 BC. He was carrying a quiver containing fourteen arrows, so he was an archer. He also, mysteriously, died from an arrow wound in his shoulder and a lot of discussion has circled round this incident. Was Oetzi the victim of a hunting accident or was he deliberately murdered?

THE DEVELOPMENT OF
SETTLED LIFE

(8000 BC)

THE HUMAN RACE had its barely perceptible beginnings about four million years ago. From then, right through until relatively recently, it survived by hunting and gathering. People led a simple nomadic existence, collecting fruit, berries, meat and firewood as they wandered across the land. They lived for short periods at a time in temporary camps. The surprising thing is that the human race stayed at this Palaeolithic (Old Stone Age) stage for so long. It was only at the end of the last cold stage of the Pleistocene Ice Age, around 10,000 years ago, that things began to change significantly.

As the last cold stage ended, with the onset of large-scale global warming and the consequent large-scale environmental changes that accompanied that warming, people all round the world had to change the way they lived. This new phase, the Mesolithic or Middle Stone Age, represents a response to major shifts in the vegetation boundaries and coastlines. As the temperature rose, so also did the sea. The encroaching sea drew attention to a range of new environments, especially the newly flooded valleys, which made coastlines more intricately indented than they were before, the sort of thing we can still see round the fjord coast of Norway and the ria coast of Cornwall and Brittany. Many of the flooded lowland river valleys, like the lower Thames valley, turned into complex winding estuaries with fast rates of silting. All of this added up to an exciting new range of possibilities for food gathering.

In the Mesolithic, many people gravitated to the coastlines for their sheer ecological wealth. Because of this new-found richness, people did not need to go on such long journeys to find all the resources they needed. They still wandered, but they wandered shorter distances, and on repeating circuits that brought them back repeatedly to the same seasonal camps. Probably each group developed a favourite camp - the one closest to the richest cache of resources – and that was likely to be on the floor of a tidal river valley or at its edge.

We know very little about these early coastal proto-settlements from around 8000 BC, because many of them were later covered by the rising waters of the sea, or buried under many feet of later sediment. We do however know what sort of houses people were building at this time, from the inland sites. In central and eastern Europe, people built tent-like huts. Some camps consisted of houses built on frames of mammoth skulls, arranged so that the long curving tusks pointed upward and inward to make a beehive shape. Skin tents were in regular use in central Asia. From the eastern Mediterranean through the Middle East to India, round huts were built with stone footings. Probably their super-structures were made of branches and daub.

Shortly after this the Neolithic or New Stone Age brought a further radical change. Instead of going about collecting their food, people began to bring the living animals and plants to their base camp and draw on them there as they needed them. This process of plant and animal domestication was one of the greatest inventions of all time, transforming the way people lived. Suddenly it became possible to live in one place. Once seeds were sown, and the germinating plants needed weeding and reaping, people had to stay in the same place. This focus on particular place led directly to the development of villages and the settled lifestyle that went with them. Two of the earliest villages were Haçilar and Catal Hoyuk in what is now Turkey.

This was the neolithic revolution. People still went on hunting expeditions, partly to supplement their ordinary food supply and partly, I suspect, for fun. There is also evidence that people sometimes moved very long distances, perhaps on religious pilgrimages to visit places like Stonehenge, and as a result they died far from home. Something similar happened in the Dark Ages, when it was common for European kings and princes to travel on pilgrimages to Jerusalem or Rome, especially on retirement, and inevitably a number of them died. But for most of the time, people were settled in their villages. This in turn led to the building of more carefully and elaborately designed houses. In around 4000 BC, stout timber longhouses were built at many settlements in central Europe.

From village life, it was but a short step to the evolution of town life. Usually this followed after a few centuries, and the first successful (that is, long-lasting) towns developed in Mesopotamia in around 3000 BC. Town life was itself a major invention, one that was to change the world. It was only possible when food supply out in the rural areas was sufficiently well organized to produce a regular and reliable food surplus. That food surplus also needed to be large enough to support the non-agricultural workers living in the towns. Those non-agricultural workers were to become administrators, craft workers, artists, scribes, architects, engineers, priests and rulers. In other words, the invention of the town made possible the development of specialist occupations and the great leap forward that we know as civilization. The life of the town and city is the epitome of civilized life. This began with the settlements along the banks of the Tigris and Euphrates – Eridu, Uruk, Ur, Akkad, Nippur and Babylon – and the Indus – Harappa and Mohenjo-Daro.

A settlement that looks very much like a town was built long before, as early as 8000 BC. By that incredibly early date, Jericho in Palestine was already a settlement housing 2,000 people who lived by cultivating cereals in the surrounding fields. Jericho was raided, probably by nomads who were envious of their wealth, and the inhabitants had to enclose it with a wall. It was a well-built mud-brick wall with a stone facing, and it stood at least twelve feet high. Inside the defensive town wall the people of Jericho built a sturdy stone watch tower that stood at least twenty-five feet high, with a well-crafted internal stone staircase.

But the town of Jericho really was exceptional, and much earlier than any other attempt at a town – anywhere in the world. It is a great anomaly that still puzzles archaeologists. Jericho was a premature attempt at town life. Perhaps the agricultural basis of the community was insufficiently strong, or insufficiently well organized. The town failed and was abandoned. For some reason, the walls of the town came tumbling down, and it would be centuries before people tried again to build another one.

THE INVENTION OF
TREPANNING

(6500 BC)

TREPANNING, TREPANATION OR trephination is the oldest surgical operation we know about. It was a drastic and dangerous form of surgery, and
yet it seems to have been widespread, and it was practised over a long
period of time. At one burial site in France dated to 6500 BC, there were
120 human skulls, and forty of them had holes in them. These carefully
and deliberately made holes show that 8,500 years ago people were
painstakingly sawing discs out of one another's skulls. The fact that the
sawn edges of the bone healed over shows that many patients survived
this surgery, which in itself is surprising. The surgery must have been
carried out with great skill, precision and a significant amount of anatom-
ical knowledge for the survival rate to have been so high.

Trepanning was carried out throughout the prehistoric period, and it
continued into the medieval and Renaissance periods. Hippocrates and
Galen wrote instructions on how to carry it out. It is impossible to be
certain what the intention of the surgery might have been in 6500 BC,
but we have some evidence for the intentions in the later period. In the
Middle Ages, we know that trepanning was seen as a cure for various
medical problems, including skull fractures and seizures. It is easy to see
why trepanning might have been invented. Some conditions, such as
tumours, haetomas and migraines, give the sufferer a sensation of
unbearable pressure inside the head. Making a hole in the skull, to allow
the brain to expand a little would have been a way of relieving that
pressure. It may even have worked in some cases. In those cases where
it made no physiological difference it may have had some psychological
benefit, as something had been done to address the problem.

Trepanning was carried out in the New World before Europeans arrived there, as we can tell from skeletal remains, art work and reports written in the sixteenth century. It was commonest among the Inca in the Andes, and far less common in Central America. Some of the trepanated skulls found in Central America were as old as 1400 BC.

In the modern period, trepanning seems to have been carried out for philosophical or spiritual reasons, such as the case of Peter Halvorson, who drilled a hole in the front of his own skull in order to gain enlightenment. Most psychiatrists would regard this behaviour as a symptom of a serious mental disorder, but some practitioners of trepanation claim that there are real medical benefits as a treatment for depression and chronic fatigue syndrome.

THE INVENTION OF
THE WHEEL
(6400 BC)

THE FIRST WHEELS were potters' wheels. The earliest pottery was made with either coils or hand-flattened pieces of clay assembled on a wooden plaque, and then the joins smoothed over with a wooden spatula. Turning the pot to work round it more quickly and make it more symmetrical led to the invention of the potters' wheel. This consisted of a small wheel to carry the pot, mounted on a spindle set vertically in a larger and heavier wheel that the potter turned with his feet. Contrary to popular belief, it is this, and not transport, that was the earliest use of the wheel, introduced in about 6400 BC and probably invented independently at several different places at the same time.

The idea of a pair of wheels joined by an axle probably developed out of a simple roller made out of a log. At one time it was generally assumed that the big sarsen stones for Stonehenge were rolled from the Marlborough Downs near Avebury using log rollers of this type. A problem with this technique would have been the variable slopes encountered; the stones would probably have slid sideways off the rollers and jammed. For that particular task a sledge would have been far more likely. Sledges pre-dated the wheel. A light sledge or travois was made out of two poles arranged in a V with a simple frame slung between them. The two splayed ends of the frame rested on the ground, while the point of the V was strapped to a pack animal or a person. This was a simple way of dragging a light though bulky load such as a hay crop.

A wooden sledge was sometimes dragged over rollers to reduce friction and after a time the runners wore grooves in the rollers. The grooves were useful in holding the sledge in position and preventing it

33

from sliding sideways. The wooden rims outside the grooves were, in effect, embryonic wheels. Then came the discovery that the friction could be further reduced by cutting away some of the timber between the grooves, so that the central part of the roller was no longer in contact with the ground; this made a rudimentary axle. The final stage in the metamorphosis from roller to wheel came with the addition of pairs of pegs in the under side of the sledge runners, to hold the sledge in position over the axle or axles. It seems that from the very beginning wheeled carts were made with four wheels, a practice that was to continue for thousands of years.

No one knows where or when the first wheel was made, but it appeared in widely separated places, and it is likely that many communities invented it independently.

The first 'made' wheel was built solid, out of three thick planks held together with braces and cut into a circle. The wheels were mounted usually in two pairs under a wooden cart, and probably used for transporting agricultural produce. The wheel was certainly made as early as 3500 BC in Mesopotamia, where it appears as a pictogram in early writings. Wheeled carts were made even earlier than that in Europe. Clay models of them have survived. The earliest picture of a four-wheeled cart is on the Bronocice pot, made in southern Poland in 4000 BC. In recent years some actual Stone Age wheels have been found, surprisingly well-preserved, in bogs in Germany and Switzerland. The earliest known wheel. The wheel was therefore used for transporting things for almost 2,000 years before the sarsen stones were moved to Stonehenge. But a wheeled wagon would not have been any good for carrying the megaliths; the sheer weight of the stones would have shattered the wooden bearings of the wheels.

The heavy wheeled wagons were drawn by oxen or half-wild asses called onagers.

By 2000 BC, lighter vehicles were being made in Turkey and the Near East. These two-wheeled horse-drawn chariots were fast and could change direction very easily. They were ideal for transporting warriors quickly from one part of a battlefield to another and they transformed the nature of warfare. The kingdoms that had armies with trained battalions of chariots were the nuclear powers of the Bronze Age. The use of bronze made it possible to develop light, almost delicate-looking wheels. They had circular rims and four or six bronze spokes. They were fast, highly manoeuvrable, and they decided the outcome of many a battle. The great early civilizations – the Aryans in India, the Sumerians, the Hittites, the Egyptians, and later the Minoans and the Mycenaeans too – gained much of their power from their chariot forces. They were civilizations based on the militaristic use of the wheel.

The wheel, as used on chariots and supply wagons, lay behind much of the war devastation that spread through Eurasia, from China to Troy and on to Britain. A very fine twelve-spoked wheel dating from 1200 BC has been found in Iran. In the Iron Age the military dominance of the wheel continued. The ancient Britons who tried to stop the Romans from invading were equipped with chariots, but so too were the Romans. Ultimately, it was not a difference in military technology (wheel or no wheel) that led to the defeat of the British but political disunity, an inability to form a coherent and effective defensive confederation.

Iron hubs gave wheels far greater strength than the weaker bronze bearings. Then an iron tyre was invented, to reduce the wear and tear on the wooden rim and also give it greater strength.

Today, we think of the wheel as a benign piece of technology. It is the wheel that allows us to cycle out into the countryside, go on a touring holiday in a car, take a bus or a train to work. But for the first half of its evolution, the wheel was a deadly weapon, the power behind military empires. It was only from the time of Alexander onwards that the chariot began to be outstripped in importance by cavalry.

But the wheel went on to find many more uses. A wheel with buckets or jars strapped to it could be mounted vertically in a river, and the motion of the river turning the wheel could send jars full of water up into a trough and out onto a field. In dry countries like Egypt it became a vital tool for irrigation. Adapted differently, the waterwheel could supply a workshop like a forge with energy. In AD 444, someone had the bright idea of fixing a single wheel to the front of a small cart to make a wheelbarrow. A wheel with large sails could be turned by the wind and used to rotate large grindstones to mill grain. This not only saved a lot of arduous and back-breaking labour and freed people to do other things, it vastly increased the supply of food and made population growth possible. There seemed to be no end to the applications of the wheel – once it was invented. Mechanical clocks contain several wheels, their cogged edges ensuring that they turn one another in a very precisely controlled way, to keep time.

The endlessly versatile wheel can entertain us. In theatres there are revolves to allow quick scene changes. At funfairs there are roundabouts and Ferris wheels to take us round and round or up and down, just for the fun of it. But when the Industrial Revolution arrived, the wheel – for the second time in its history – transformed human society. Endless new and empowering uses were found for the wheel, from the paddle wheels of Brunel's ships the *Great Western* and the *Great Eastern* to the iron wheels of the railway locomotives, and the propellers of the first generation of aeroplanes.

There is no doubt that the wheel is the single most significant invention of all time. The most surprising thing is that we took so long to hit on the idea.

THE INVENTION OF MAPS

(6200 BC)

A MAP IS a two-dimensional representation of part or all of the Earth's surface. A map differs from a picture in being a diagram-like drawing of a landscape as if seen from vertically above. It can vary in scale, from showing a single settlement to a region or a continent, right up to showing the whole world. The oldest map in the world was painted on the wall of a house at Catal Hoyuk in 6200 BC.

Discovered in 1958, Catal Hoyuk was a very large Stone Age village (almost a town) dating from 7500 BC in what is now southern Turkey, near the modern city of Konya. It stands on a raised site, a tell, overlooking the fertile wheat country of the Konya Plain. The excavations of James Mellaert between 1961 and 1965 brought the site worldwide publicity as a focus of advanced culture in the neolithic.

The people of the neolithic village were among the earliest farmers. They had a surprisingly advanced culture, with houses clustered so tightly together that people evidently walked from roof to roof and entered their houses by way of ladders through holes in the roofs. The three-room interiors of their startlingly clean, tidy and smoothly plastered houses had wall paintings and sculpture that are full of religious symbolism. Among the wall paintings is a map. It unmistakably shows a settlement and the distinctive twin peaks of Hasan Dag, a volcano 100 miles from Catal Hoyuk. For a time archaeologists assumed the settlement shown on the map was Catal Hoyuk, but it now looks as if it was another neolithic settlement located closer to the volcano. If so, the wall painting shows a remarkable imaginative leap; it is a map of another place, perhaps the place where the Catal Hoyuk people came from.

There is a tendency today to see maps as scientific and objective depictions of places. In practice they are extremely subjective because they are selective. The cartographer and his editor and publisher make decisions about what to include, what to exclude and what to emphasize. The finished map always reflects the values and interests of the map-maker or his or her patron. In past ages, that subjectivity has been more evident. People's view of the world was strongly affected by their religious belief, just as their view of the shape and working of the solar system was affected by their religious belief; we have only to think of the conflict between Galileo and the Church when he proposed that the Sun and not the Earth was at the centre of the solar system. Medieval 'T-O' world maps, like the Hereford Mappa Mundi drawn in about 1300, showed Jerusalem at the centre of the world, which was thought of (at least by some) as a flat disc and therefore depicted as a circle. The continents were shown as geometric shapes. This map was a statement about faith as much as a map. Today we normally have north at the top of a map. Christian medieval Europe put east, the direction of Jerusalem, at the top.

By contrast, there were other maps drawn at the same time which look quite modern in their respect for the actual form of the coastline. These were the navigation maps of the Mediterranean lands. On these, the coastlines are shown in detail, with headlands accentuated slightly and carefully named, so that navigators could 'bay-hop' round the Mediterranean with ease. So, even in the Middle Ages maps were able to show that different people were thinking about the world in different ways.

Modern maps, as one might expect, are very diverse. They are made for a wide variety of purposes and are often very straightforward in their selectiveness. An atlas often contains a section of specialized world maps, each showing a different characteristic or variable. One is the physical map, showing the heights of the land in different shades of green and brown, the depths of the sea in different shades of blue, and printed over with the names of seas, mountains and rivers. Another is the population density map, showing crowded areas of the world in shades of red and sparsely settled areas in yellow or white.

Specialist interest groups often produce their own maps. Archaeologists interested in megaliths publish maps showing stone circles and chamber tombs, cyclists publish maps of cycle paths, ramblers publish whole books of recommended country walks.

Even the official map series in Britain, the Ordnance Survey, was designed with a specialist need in mind. An accurate one-inch-to-the-mile map of England was needed for artillery to operate effectively in the event of a French invasion. The survey of Sussex was undertaken during the French Revolution, 1789–92, a time of enormous tension between France and Britain. The surveyor, Thomas Budgen, was very careful to record the location of every fort and barracks, and the landmarks that gunners would find helpful, especially the church towers and windmills. Troop movements required that the network of towns, villages and roads was accurately recorded. The result is a map that looks complete and objective – but it is really a map of a land preparing itself for invasion, a war map.

THE INVENTION OF
IRRIGATION

(6000 BC)

THE ARTIFICIAL WATERING of crops to guarantee high crop yields seems to have begun very early in both Mesopotamia and Egypt. The Nile flooded naturally each year, but crops benefit from more frequent watering. Water was lifted from the Nile by means of a simple device called a shaduf, little more than a bag on the end of a lever, and poured into networks of irrigation channels leading out across the fields. A similar situation developed in Mesopotamia. Quite why the Sumerians and Egyptians started irrigating their fields when they did is hard to tell.

One theory is that it was connected with a change in the climate that occurred just as agriculture was getting under way in the north of Iraq. There was a drier and cooler phase known as the Younger Dryas, and it made life very difficult for the hunter-gatherers of the area. This gave them an incentive to find alternative ways of generating a food supply. In other words, necessity was the mother of invention: the cooling episode stimulated the invention of agriculture.

There was another episode of cooling and drying in the upper Tigris and Euphrates catchment areas in 6200 BC. For a time, in this heartland of early agriculture, rainfall decreased to a level where agriculture would not have been sustainable. The farmers of the upper catchment area would have been forced to give up their lands and move out, seeking other opportunities. Some of the displaced farmers saw that the slow-moving waters of the lower Tigris and Euphrates (nearer the Persian Gulf) might be diverted onto neighbouring flat land and used to water crops. They therefore initiated farming under irrigation in their newly adopted homeland.

The irrigation-based civilization of ancient Egypt was in a similar way badly impacted by a natural change of climate in about 2200 BC; a third phase of cooling and drying had such a catastrophic effect on food production that it brought the Old Kingdom to an end. Even with an irrigation system in place, if the rainfall pattern in the Nile headwater regions falls below a critical level the system fails. The change in climate in 2200 BC led to the abandonment of most of the settlements in the Nile Delta. There was a similar drying-out in Syria at the same time; rainfall decreased by as much as thirty per cent, making agriculture there no longer viable. Syrian farmers gave up growing crops and switched to animal herding. This severe economic decline may explain the collapse of the Akkadian empire, which happened at the same time.

The irrigation system in Mesopotamia was certainly among the oldest, if not the oldest, in the world. It was in place as early as 6000 BC. Initially, water was probably simply led away from the lower rivers into the adjacent fields by way of irrigation channels. As time went by the system became more elaborate, until dams and canals in the humid mountain region in the north became the source of much of the irrigation water. A coordinated system of this size required a high level of social organization. It also needed teams of specialist workers capable of constructing and maintaining the various structures. It is possible that irrigation was organized because the social and political structures were in place to make it possible. But it is also possible that the overriding and imperative need to have a fully functioning and reliable irrigation system forced the development of the type of society that could achieve it. In other words, irrigation may have been responsible for civilization, rather than the other way round.

Early European travellers to China were struck by the extensive use the Chinese made of canals. These were used both for transport and for irrigation. China has huge areas under irrigation schemes, and it has been said that the irrigated area in China corresponds to one-third of all the irrigated land in the world. The Chinese canals are difficult to date, but certainly some of them have been in place for more than 2,000 years. The earliest reference in a Chinese document to an irrigation canal dates from the eighth century BC.

Irrigation is often presented as a simple process. Water is introduced into a dry landscape and makes crops grow there. It is the old cliché of 'making the deserts bloom'. The reality is that irrigation is extremely difficult to engineer, and it brings with it a whole set of new problems. For one thing, directing large volumes of water into fields diverts it from wherever it was going, so it may create a water shortage somewhere else. A classic late twentieth-century example of this is the Aral Sea in central Asia. A very ambitious multi-purpose irrigation scheme implemented under the Communist regime involved the diversion of the Amu Darya towards the west and south-west to water fields of cotton. The Amu Darya was diverted close to the point where it emptied into the Aral Sea. The Amu Darya was the main feeder for this once huge lake. The river diversion has had the direct effect of drying the Aral Sea out. In the 1960s the surface level of the Aral Sea fell at a rate of eight inches per year, increasing in the 1970s to two feet per year, and in the 1980s to three feet per year. The Aral Sea is shrinking at an ever-increasing rate. The volume of the Aral Sea fell by eighty per cent after 1960, while the area under cotton doubled.

Another cost that irrigation brings with it is salinization. As 'fresh' river water evaporates on the fields, traces of the minerals dissolved in it are left behind. Over time, these soluble salts build up and become a serious problem to crops. The temptation has often been – right up to and including the twentieth century – to step up irrigation in order to increase commercial crop production, to make more money. But the long-term effect has been the gradual increase in salinization. It was a problem in ancient irrigation systems. It is still a problem in modern irrigation systems. Watering crops in a dry summer as a one-off solution to a specific weather problem is reasonable, but it looks increasingly as if the disadvantages and costs that flow from irrigation are so great that irrigation should never be regarded as a long-term solution.

THE INVENTION OF POTTERY

(6000 BC)

ANY HUMAN COMMUNITY needs containers. The palaeolithic hunters and gatherers needed them for collecting berries. We need them on our tables to hold our food and drink: cups, saucers, plates, jugs, bowls, gravy boats. By the time of the neolithic revolution, when people were sowing seeds, collecting grain, grinding that grain to make flour, the need for a range of containers of different shapes and sizes accelerated. It was possible to use gourds and animal skins, but fairly early on people started to weave natural fibres to make artificial receptacles.

Pottery has several advantages over basketry. One is that it is watertight. Another is that it is possible to control the shape in great detail. One disadvantage is that it does not lend itself to the nomadic way of life. Potters have preferences for particular clays, and develop a feeling for what their local clays will allow them to do. It is also easier to make and store pottery in a settled environment than while on the move. It comes as no surprise that pottery developed as part of the neolithic revolution. As the settled way of life in villages evolved, pottery was one of several crafts invented. No doubt, like a lot of inventions, it was stumbled on accidentally, when people lit fires on the ground and found later that the clay surface underneath the fire had baked into an impermeable crust like a large dish. They might even have seen that accidental dish as a useful object that they could utilize for winnowing. It would be a short step from that observation to initiating and controlling the process by picking up a lump of clay, kneading it, pressing it into the shape of a cup and then burying it in the embers of a fire to bake.

The earliest pots were made in one of two ways. One was by

modelling. Pieces of clay could be pressed flat between the fingers, or flattened with a rolling pin, then pressed together to make a vessel. The alternative was to roll the clay into long thin cylinders or 'worms', as children like to do with plasticine, and then coil the worms to make a circular pot.

From very early on, pottery was decorated. Neolithic pottery was sometimes given rings of impressed decoration by using a thumbnail or the end of a bird bone. Potters regularly used simple tools like spatulas for smoothing the outside surface of a pot, and these too could be used to make patterns. The surface of Grooved Ware, high-status pottery made in Britain at the end of the neolithic, was often elaborately decorated with triangles, chevrons and zigzags in relief, apparently to make the surface look like basketwork. It was a very rich effect.

Paint was available, and many pots were given painted decoration. Often the painted designs were geometric shapes: straight lines, stripes and zig-zags were common, and later zones or belts of different textures were tried out. The Minoan painted pottery that was made in Crete in around 1700–1400 BC included elaborate images of birds, fish, seaweed and octopus, and shows the tastes and preoccupations of a nature-loving civilization. The classical Greek pottery made from about 500 BC onwards was also often pictorial, showing the Greeks' preoccupation with detailed and very realistic scenes from Greek myth and legend.

Right from the start, potters did rather more than just supply containers. Potters certainly supplied cups, beakers, jars, jugs, ewers, bowls and amphorae (wine containers) for everyday domestic use. But they also knew that their products acted as a medium for cultural reinforcement, a means by which a community could show itself, and its neighbours, what sort of a community it was. Pots were a statement about identity and self-perception. In the late neolithic stone houses at Skara Brae in Orkney, a big stone dresser was regularly positioned facing the doorway, so that visitors entering the house would at once be impressed by the fine Grooved Ware and other pots on display on its stone shelves.

The more distinctive and beautiful pots made in the Bronze Age and later were often coveted by the people of other communities, and they became important trade items in their own right. Pots tell us many important things about the people who made them. Most neolithic pots are round-bottomed, which shows that they were routinely rested on the ground; that tells us something important about the lifestyle of the people who used them. By the time we reach the bronze age we find that pots are mostly flat-based, which shows that they were routinely set down on flat surfaces, like table tops. Beakers, with flat bases, therefore imply that their owners had furniture.

As the skill of the potters increased, with knowledge being passed down through many generations, incredible feats of craftsmanship became possible. These reached a pinnacle of refinement in the eighteenth and nineteenth centuries AD. The eighteenth century potter Josiah Wedgwood experimented with different types of stoneware and he is probably best known for his most successful invention, jasperware, which has an unglazed green or blue background ornamented with white relief portraits or classical scenes. The idea was inspired by the magnificent Portland Vase, a glass vase made in the reign of the Emperor Augustus. Really, Wedgwood's jasperware was made as a kind of homage to this single remarkable vessel, the pinnacle of the glass-maker's craft. The poet John Keats paid his own, equally memorable, tribute to the Portland Vase in his 1819 *Ode on a Grecian Urn*, addressing it as the 'foster-child of Silence and slow Time.'

The very finest pottery can end up symbolizing an entire culture. The Grooved Ware pottery with its bold and exotic basketry patterns has come to symbolize the people who completed Stonehenge. Wedgwood's jasperware sums up the English eighteenth-century upper middle class. Clarice Cliff's tea sets with their bold orange, green, blue and cream designs on equally jazzy pottery shapes have come to symbolize the popular art deco movement of the 1930s. And, just like the most beautiful and most culturally distinctive pottery of antiquity, Clarice Cliff is much sought after by collectors, and has become very expensive.

THE INVENTION OF
COPPER SMELTING

(5500 BC)

COPPER WAS SMELTED for the first time in Iran in about 5500 BC. The metal-workers used a naturally occurring copper carbonate, malachite, as an ore. The smelting process simply involves roasting a metal ore, a rock that contains a percentage of metal, until the metal melts and runs out. Smelting still is the commonest method for separating a metal from the waste rock. In 5500 BC, smelting was a newly invented technique, which eventually became the standard technique for extracting not only copper but other metals too.

Copper was found to be an accommodating metal: one that could be drawn into wire, cast in moulds to make ornaments, or hammered into shape to make blades and points for weapons. In later centuries, experimental axes were made out of copper (instead of stone), but the metal was too soft to hold a sharp and usable edge. Oetzi, the Iceman, owned a copper axe, but it may have been more of a status symbol than a useful tool. In an age when most tools were made out of stone, a cast metal tool must have seemed a truly wonderful thing, even if it was not of much practical use, rather like a prestigious open-topped sports car that makes a lot of noise and uses a lot of petrol.

In time, bronze would prove to be a far more useful metal. Bronze is an alloy, a mixture of copper and tin. It is a stronger metal than copper, and it holds an edge far better. With the invention of bronze towards 3000 BC, it became possible to make serviceable swords, knives, axes and a range of other weapons and implements. The technique for making bronze was developed in the same region of south-west Asia. The invention of bronze was a major breakthrough; bronze was the first

46

manufactured metal that was hard enough to take a sharp edge. Gold, also exploited in the Bronze Age, was even softer than copper, and only useful for decorative work. Tin also was soft, but the experiment of combining copper and tin to make bronze, in effect an artificial metal, proved to be a great success.

It was not until 2500 BC that Middle Eastern metal workers developed a smelting technique that was suitable for iron. The furnaces needed to burn hotter. Temperatures of 1,500 degrees Celsius were needed to smelt iron – much higher than those needed for smelting copper. This was still very new technology, though, and it would not be until around 700 BC that iron smelting would become common. The swords that were made of iron were much more reliable than those made of bronze, and far less likely to snap in combat, with probably fatal consequences. The gradual development of metallurgy made way for the development of weapons technology – and for the gradual scaling up of warfare.

THE INVENTION OF MONUMENTS

(4700 BC)

THE FIRST MEGALITHIC monuments were built just as agriculture was being adopted as a lifestyle in Europe. The first true megaliths (literally 'gigantic stones') seem to have been raised in Brittany. The Dissignac passage graves and the Kercado chamber tomb were among the earliest of these monuments. At some sites, like Barnenez, small chamber tombs were built and then, after a lapse of time, much larger mounds were raised over them, sometimes burying two or more of the earlier chambers and making a more impressive monument.

There were also isolated standing stones, and the largest of these was the Grand Menhir Brisé (literally 'the great broken standing stone'). The Grand Menhir was a huge monolith weighing about 300 tons. It is estimated that it must have taken 3,000 people to pull it upright in 3880 BC, when it would have stood sixty feet high. It was an artificially dressed and smoothed hexagonal pillar of non-local granite that must have been transported over two miles to the place where it was erected. This was a major feat of engineering and communal organization in itself. The Grand Menhir now lies broken into four pieces and it looks as if it was deliberately pulled down in the megalith-building period. Other megalithic monuments in the area were also dismantled and in some cases the component stones were re-used in the building of new monuments. Some say the pillar slipped, fell and broke accidentally while an attempt was being made to raise it, which is possible. Others say it fell as a result of an earthquake, which is also possible. But the fact that the fragments of the pillar point in different directions strongly implies deliberate destruction rather than an accidental fall. The Grand

Menhir originally stood at one end of a long barrow, a classic neolithic burial monument, and was therefore part of a tomb design; the associated tomb itself has been destroyed.

What the history of these monuments tells us is that the ancient past was turbulent and volatile, that there was colossal commitment to ideologies, that there were crises of faith and large-scale changes of mind.

At Carnac, also in Brittany, there are multiple stone rows – rows and rows of standing stones, over a thousand of them in all. No one has successfully explained their purpose, but they were probably connected with funeral ceremonies. The Breton name of the place, Kermario, means 'The House of the Dead', which may contain an ancient folk-memory of the area's purpose.

In Brittany an entire, and to us rather mysterious and alien, culture grew up around these monuments, which became larger and more ambitious with time. They consumed an enormous amount of the Stone Age community's time and effort. The last great megalithic monument to be raised in Brittany was the Gavrinis passage grave, in about 3250 BC. At about the same time a huge burial mound, La Hougue Bie passage grave, was built in Jersey in the Channel Islands.

During the flurry of megalith building that focused on Brittany, megaliths were going up in other areas too. The Cueva del Romeral passage grave was built in Spain and some very fine stone temples were built on the island of Malta, at Mgarr and Ggantija in 3500 BC, at Mnajdra and Hagar Qim in 3400 BC and the most spectacular of all, at Tarxien in 3300 BC.

Just as this Maltese sequence came to an end, around 3250 BC, the first megalithic monuments were raised in Britain, the earliest stone circles. The North and South Circles at Avebury were built shortly after this. The famous stone circles and horseshoes at Stonehenge were a later creation still, although the earth circle there was laid out at the same time as the stone circles at Avebury, in about 3100 BC. It is possible that people in Britain were responding to an environmental disaster, that they were making a desperate appeal to the gods to give them better weather. Cores from the Greenland ice cap tell us that there was a major volcanic eruption in 3200 BC, and narrow tree rings in ancient Irish bog oaks show that the climate cooled suddenly and significantly in the British Isles around 3150 BC. Here we can see a connection between a big volcanic eruption and a major climate change; and a possible human response to environmental disaster.

That brings us to the fundamental question: what were the megalithic monuments for? They involved a colossal amount of work, took a long time to build, and preoccupied a large proportion of the community. One obvious point is that megaliths are fixed in the landscape, and that implies that the people who built them had a focus on the places where the megaliths stood. It is not a coincidence that megaliths – and other neolithic monuments like the long barrows that were made of earth – started to appear at the same time as agriculture. People were settling down to live in fixed and permanent farmsteads, and that must have given them a new view of landscape, especially particular tracts of landscape, as 'home'.

Some of the monuments were family tombs that were deliberately built so that they would be silhouetted on the skyline when seen from the farmsteads where those families lived. They may have been intended as an ancestral claim to the territory they overlooked. If people were sowing seeds, they had a vested interest in claiming the fields and gardens where a harvest could be expected later. Claiming territory and defining boundaries became much more important once agriculture started. People were investing in land and therefore needed to announce their claim to it.

Ploughing, sowing and reaping were key stages in the farming year, and some monuments were clearly intended to mark certain calendar dates. The orientation of Stonehenge on the midsummer sunrise is the most famous of these. But the monuments were doing more than marking the calendar: they were willing the succession of the seasons, and no doubt people thought that by carrying out the right ceremonies they were helping the gods to shift the sun and moon across the sky, and bring the next harvest round at the appointed time.

The megaliths are important in marking the beginning of public architecture. Already at this very early stage people were experimenting with designs that would impress others. They even used optical tricks to make their monuments look bigger than they really were. Some long barrows and chambered long cairns were built higher and wider at the end where the burials were placed, and where visitors arrived to take part in ceremonies; an observer looking from the forecourt in front of the 'business' end of a barrow would have got the impression, looking along the tapering mound, that it was twice as long as it really was. That same taper would later be used by the Greeks to make the columns of the Parthenon look taller than they really were. It is a trick we scarcely notice, though it is very noticeable when a clumsy modern architect forgets to use it, as in the 330-feet high chimney of the new Shoreham power station (built in 2000).

THE INVENTION OF THE ARD

(4500 BC)

THE PLOUGH IS one of the simplest yet most important inventions of all time. It had humble beginnings as an ard. This consisted of two wooden poles strapped together to form a cross. The lower end of one pole had a pointed tip which made the furrow in the ground; its upper end was steered by the ardman. The upper end of the other pole was pulled along by a draft animal, usually an ox, but probably at first by another man.

The ard or scratch pole was in use as early as 4500 BC in Mesopotamia. There the soil was light and sandy and the wooden tip of the pole was adequate to make a furrow. In higher latitudes, in more humid areas like Europe and China, the soils were heavier and stickier and the point needed to be reinforced. The next big development was made by the Chinese in 3000 BC. They strapped sharpened stones onto the toe of the ard, to make a primitive ploughshare. This made it possible to plough a furrow in moist and heavy soil. It seems that this invention was made independently and simultaneously in north-west Europe. Deliberately sharpened stone ploughshares have been found in Britain, on the small neolithic farmsteads of Shetland.

The ard was only capable of making a fairly shallow furrow to receive seeds for grain crops. To sow seed across a whole field meant making many parallel furrows. But the furrows could not be ploughed too close together, or the ard slipped back into the previous furrow. To cope with the berms between the furrows a second set of furrows had to be ploughed at right angles to the first. The characteristic criss-cross pattern of ard-marks has been found, preserved for 5,000 years, under neolithic long barrows in Britain.

The ploughed land yielded far more food and it seems the process of ploughing acquired a mystical status. Magic stone talismans were made bearing the criss-cross pattern.

The ard could nevertheless do little more than hoe the soil. To dig deeper and turn the soil a curved ploughshare was needed, and that could only be made in metal. That development had to wait until the Iron Age. In 600 BC the Chinese made the first iron ploughshares. They would not be made in Europe for another 400 or 500 years. The iron ploughshare had a sharp point that dug deep. It also had a curved mould-board designed to turn the earth over to one side. The Iron Age ploughshare was so well-designed that it has remained virtually unchanged to the present day. Only the power source has changed.

THE INVENTION OF
PAPYRUS

(4000 BC)

THE ENGLISH WORD paper derives from the Egyptian papyrus. The original meaning of the word papyrus was 'what belongs to the house', the bureaucracy of ancient Egypt. Papyrus was invented primarily for record-keeping. Just as ancient Egyptians invented a written language the necessity arose for an appropriate medium on which to write. They found what they were looking for in the papyrus plant, a reed that grew plentifully all along the banks of the Nile. The material they created out of the reed was an excellent writing medium. Light, strong, hard-wearing, easy to carry, it was a far better solution than the clumsy and not very portable clay tablets used in Mesopotamia and Minoan Crete.

Because of its excellent qualities, its fitness for purpose, papyrus continued in use in Egypt for 5,000 years from 4000 BC until AD 1000. Papyrus had other uses besides writing; it was also used for making furniture, mats, baskets, sandals, boxes, boats and rope. The root of the plant could be made into perfume or medicine; you could even eat it.

The papyrus grows to about ten feet high. After it is harvested, the skin is peeled away and the core sliced into very thin strips. The strips are soaked in water to remove the sugar, then placed side by side, slightly overlapping. A second set is positioned at right angles to the first. The mat is then pounded and left under a heavy weight for a week. After drying, the surface of the sheet is polished with a shell to give a smooth finish.

The ends of rolls were reinforced by gluing them to an extra strip of papyrus; sometimes the ends were wound round sticks which had cords to stop the roll from unravelling.

The way papyrus was made varied but little over the 5,000-year-long period when it was produced on a large scale – and is still made, for fine art purposes. The quality of papyrus varied, just as the quality of modern paper varies. The highest quality papyrus, which came from the Delta, was reserved for literary or religious works. The ancient Egyptians normally wrote on one side of the papyrus only.

In the Greek period, a standard paper size was in use, twelve inches by nine, which is within half an inch of the A4 that is today's standard paper size in the EU. For a long document, though, these standard sheets were joined together to make a long roll, of up to twenty sheets. Later, papyrus books were made and this bound form, called a codex, was found more convenient to read and store than the roll.

The pharaohs recognized the importance of papyrus and made the manufacture of it a state monopoly. Exactly how it was made was maintained as a state secret. It was such a successful invention that a demand for it quickly developed outside Egypt. It was exported. Unfortunately very little of the exported papyrus has survived. The dry conditions in Egypt favour its survival there, but the seasonal humidity of places further north, such as Crete, means that little has survived there. Only a few fragments of papyrus from the classical period have been found in Greece, although many drawings of papyrus rolls can be found on vases painted at the same time.

The remains of papyrus documents are found at nearly every ancient Egyptian archaeological site. The ancient rubbish heaps are full of them, showing that papyrus was fundamental to the Egyptian civilization. It seems everything was written down, everything was inventoried.

The Egyptians even recycled their papyrus. When mummies were being made, the corpses were first prepared and wrapped in linen. Then the linen was encased in a kind of papier mâché made of several layers of papyrus usually thrown away by administrators. This papier mâché layer was then coated in plaster and painted in bright colours.

The heyday of the use of papyrus was from the fourth century BC to the seventh century AD. After Egypt's conquest by Alexander the Great, most of the administration was conducted in the Greek language, and the papyri reflect this. Curiously, Greek continued to be the language of administration in Egypt after the Romans took control, and even after the Arab conquest of 642.

Some of the most exciting papyrus finds that have so far been made are some very old Bible texts. At Wadi Murbaat, near the Dead Sea, a papyrus dating from about 650 BC has been found. Another dating from the fourth century BC was found above Jericho. These finds confirm that the so-called 'books' mentioned in the Bible were really written on papyrus, and were probably stored as rolls rather than what we would call books. The texts found at Qumran, the famous Dead Sea Scrolls, consisted of more than 800 scrolls.

The thousands of scraps of surviving papyri have been studied by papyrologists, who during the course of the twentieth century in particular have been able to glean a great deal of information about the ancient world. But the work is frustrating. Since 1788, about 50,000 papyri have been published (and are therefore accessible for research), out of an estimated 400,000 that exist in private and museum collections all round the world.

An environmental crisis in Egypt at the end of the seventh century AD had some unforeseen effects. So much papyrus had been harvested from the wetlands of Egypt that the papyrus plant virtually disappeared from the landscape. The Egyptian authorities banned further exports of papyrus. This created a paper famine, and as in many similar situations that we will come across in this book, the crisis was met by a new invention. European administrators and scholars switched to parchment, and then to modern paper made out of wood pulp.

So-called 'ground paper', which is the ancestor of modern paper, was actually invented in China in AD 105, by Ts'ai Lun, who was a Chinese Imperial Court official. The paper was made by grinding plant material into a pulp, and spreading the pulp out into a thin sheet. This was then dried in a form. Ts'ai Lun originally used the waste from silk production, but a great variety of plant material might be used. Some early Chinese paper was made out of the bark of the mulberry tree but, as was discovered centuries later, there was no difficulty in using pulp from the conifers that covered thousands of square miles of Eurasia.

The Chinese had been making paper by this method since 105, but it only reached western Asia when the Arabs conquered Turkestan in 751. The Arabs were then able to introduce a paper-making process which they had extorted from their Chinese prisoners. The paper was not as good as papyrus – for one thing it was not nearly as durable – but it was much cheaper. Gradually, as parchment and paper took the place of papyrus, the Egyptians gave up making it; the papyrus plants had all gone.

The process of making papyrus has only been revived in recent years. In 1969 the papyrus plant was reintroduced from the Sudan into Egypt, where a papyrus plantation has been established, on the Delta not far from Cairo. The papyrus manufacturing process has been reconstructed from the description in the *Natural History* written by Pliny the Elder.

THE INVENTION OF SUN-DRIED BRICKS

(4000 BC)

BRICKS ARE SO old that it is hard to be sure where or when they were invented, but they are certainly as old as civilization itself. The earliest bricks that we know about were made along the banks of the Tigris and Euphrates by the Sumerian peoples. When flood waters receded, deposits of wet river silt were periodically left coating the fields. As the sun dried this silt out, the mud deposits turned into a cracked and hardened cake that could be broken up and the irregular fragments used to build the walls of huts. Wet mud could be used as mortar; if a damp-proof course was required, a layer of bitumen was used.

From this it was a short step to making simple four-sided wooden moulds, filling them with wet mud and leaving them to dry in the sun to make a concrete-hard rectangular brick. It produced a shape that was easy to build with. It also produced a smooth flat wall surface that ended in a sharp, neat, right-angled corner. This feature led on to shaping buildings in the orderly rectilinear shapes that we now take for granted. Until this point, the plans of buildings could be round, roughly oblong with rounded corners, or of any irregular shape. The rectangular sun-dried brick nudged builders towards building straight walls with well-defined ninety-degree corners. It also encouraged builders to build walls vertical instead of leaning. The brick, in other words, made building design tidier and straighter than it had been before.

But the brick was also versatile, as for special purposes it could be made in other shapes. An early written account describes how sun-dried bricks were used in the Sumerian city of Ur to make the first true arch. The arch was as important an invention as the brick. It was made of wedge-shaped bricks. By placing a series of these wedge-shaped bricks together in an arc, the builder was able to distribute the weight of the structure evenly on each of the bricks. This invention made possible the arched doorway, the semicircular vault, which is really a lot of arches placed one behind the other, and bridges and aqueducts. Eventually, in the European Middle Ages, the arch almost completely replaced the horizontal lintel for spanning spaces.

That Mesopotamian brick arch has long since turned to dust in Iraq. But the sun-dried bricks were used to make one structure after another, including the monumental ziggurats, as well as the humble dwellings of the poor. As the bricks disintegrated and houses crumbled, old buildings were levelled and new ones built on their remains. In this way, towns and cities rose up on mounds of their own debris. Nineteenth- and twentieth-century archaeologists became expert at identifying these tells, which can be found all over the Middle East, including Egypt, where the sun-dried brick was of major importance for building.

It was at Ur that a further development took place, the discovery that oven-baked bricks were much harder than sun-dried bricks. The potters and brick-makers developed ovens that were capable of producing the very high temperatures needed to make fired or ceramic bricks and tiles. This use of kilns brings us very close to the modern brick and tile, and the technology was already invented and in use as early as 1500 BC in ancient Ur. The Sumerians often encased the more vulnerable sun-dried brick walls in a facing of fired bricks, to make them more durable.

The brick- and tile-makers of ancient Ur even invented glazing for their bricks and tiles, producing some really spectacular effects. Probably the most notable of the buildings to receive this treatment was the Ishtar Gate in the city of Babylon. This gate marked the main entrance through the inner city wall of Babylon, approached by a processional way, and was only one of hundreds of structures made of brick. Many other great buildings were given the glazed brick treatment. There was an elaborate multi-coloured facade to Nebuchadnezzar's throne room, showing a procession of lions, lines of rosettes and panels of stylized palm trees.

The Ishtar Gate is the only well-preserved monument that survives on the site itself, and what can be seen there is only the first stage in the development of the elaborate gate built on the orders of King

Nebuchadnezzar II (604–562 BC). This is a well made structure of rectangular bricks. The flat vertical walls are punctuated by lions and dragons moulded in relief on the faces of the bricks. The lion and dragon designs extended across thirteen courses of bricks, so the designs were carefully planned beforehand. The earliest stage of the Ishtar Gate was made of unglazed brick. Later, this was encased in glazed bricks. These have been removed from Iraq and reconstructed in a Berlin museum to make a structure forty-five feet high, though the original gate may have been as much as seventy-five feet high. The gate is a lofty arch flanked by towers and the walls are again punctuated with bulls, the symbol of the god Adad, and dragons, the symbol of the god Marduk. It would be an impressive structure, even in unglazed brick. The brilliant blue glaze of the walls, with detailing in white, yellow and red, is still as startling to look at as when it was newly built. The Ishtar Gate is a dazzling display of the potential of brick. When newly built in the sixth century BC, brick Babylon must have been one of the wonders of the world.

Brick was extensively used in the Mediterranean world too. It found its zenith in the form of terracotta, meaning baked earth. The Romans used brick a great deal to make some fine structures such as aqueducts. Often the facing walls were made of brick while the fill was made of concrete. They also used brick in conjunction with concrete to make the Pantheon in AD 123; this was fitted with a dome made out of this previously untried combination of materials. The Pantheon is still standing and intact.

The houses of ordinary people were still made mainly of wood and wattle and daub, though as the centuries passed the shortage of timber made this harder to sustain. The growth of towns and cities during the Middle Ages and later also brought an ever-greater risk of catastrophic fire. The cities of Bergen and London suffered serious fires. After the Great Fire of London, much of the rebuilding was done in stone and brick to prevent a repetition of the destruction. One of brick's outstanding qualities is its flame-resistance, and this made it the obvious choice of material for house-building in Europe in the eighteenth and nineteenth centuries.

Brick was always a favoured building material in China too. The first emperor, the Qin emperor Shih Huang Ti, who was born in 259 BC, decided to connect a series of defensive walls together to make one massive wall more than 4,000 miles long. Behind this wall, the various provinces were unified into one large empire. The Great Wall, which dates to around 220 BC, is of mixed materials, but sun-dried and fired bricks constitute a large part of it.

In England, as deforestation forced a change to brick in the sixteenth century, the properties of brick were rediscovered. It could be carved, it could be moulded to create relief decoration. The brick chimneys of the Tudor Hampton Court Palace were treated to elaborate decorative mouldings to make them look like the pillars of a Norman cathedral. Even ordinary undecorated rectangular bricks could be laid in different ways to make patterns: running bond, English bond and Flemish bond.

The raw materials for making bricks, whether freshly deposited river mud, consolidated floodplain alluvium or outcrops of clay, are to be found in many regions. Bricks are very easy to make, and exceptionally versatile. It is not at all surprising that brick continues to be a very widely used building material all round the world. It was one of the truly great inventions of mankind.

THE INVENTION OF WRITING

(3000 BC)

ONE OF THE characteristics of the New Stone Age way of life was its lack of writing. Thought had developed to a level of complexity where handing-on became a major priority, yet writing had not been developed. Maybe we could coin the word geoglyphics to describe the use of monuments for writing large on the landscape.

Ritual is an aid to learning and memorizing. In ritual, past reaches through present to future, often by way of an unusual medium, such as song, dance, archaic language or drama. The physical performance is part of the way both the ritual and its content are memorized. The ritual cannot be changed; this protects the belief system from discussion and dissent, and protects the content from the possibility of change. Monuments have forms inextricably bound up with rituals, and therefore are also mnemonic devices – aids to memory.

Stonehenge was fitted out with a variety of architectural features, many of them long gone, some still faintly visible, which held meanings for people in the past. There were the Heel Stones telling of the monument's key relationship with the midsummer sunrise; the A posts summarizing the most northerly positions of the moonrise; the bluestones and sarsen stones and the stories they told about their journeys to Stonehenge; the trilithons representing entrances to tomb chambers; the almost chanced-upon summative image of the roundhouse in decay.

Just as a cathedral contains within its architecture a mass of symbolic statements, with every arch, door and pillar carrying a meaning, as well as the rituals that take place among them, so it was with the monuments of the New Stone Age. Stonehenge was a stone, timber and earth mnemonic encapsulating the community's belief-system, its collective identity, its communal pride, its technology, its faith, its past, its tryst with the past, its status, its selfhood. In an age before books Stonehenge was an entire library.

Writing as such had not developed in Britain in 3000 BC, but that brink was being crossed in eastern Europe, and there were marks and signs that recur often enough in the British Isles to show that writing was about to happen. Most famously, there are the decorative stone carvings found in passage graves in the Boyne valley in Ireland. The complete vocabulary of symbols used in the Boyne tombs can be seen, though locally fragmented, all over the British Isles. There was a repertoire of recurring motifs: rayed circles, flowers, serpents, shields, multiple arcs, chevrons, lozenges, triangles, spirals. Some of these are found on Grooved Ware pottery, which shows how they found their way from place to place, and over long distances at that.

In the south of England, chalk tablets were made, very often covered with sets of parallel lines, or even two sets of lines at right angles. This is reminiscent of the patterns made in the ground by cross-ploughing and strongly suggests that some form of sympathetic magic was involved. At Maxey, on the edge of the Fens, two sceptres or batons made of red deer antler were carefully inscribed with zigzags, and the cuts were highlighted with red paint. They evidently held some magical significance, and were deposited in the ditches of ceremonial enclosures.

No one knows what the neolithic symbols meant, but they were evidently pictograms, signs with some verbal or conceptual meaning, rather than just decoration. With these simple signs from 3000 BC, people were on the brink of writing.

By 2500 BC, when Britain was at the end of the New Stone Age and entering the Early Bronze Age, and people were still living in isolated farmsteads or villages, in Egypt and Mesopotamia, town life was well under way along with the craft specialisms that went with town life. A major new development in Mesopotamia was cuneiform script. There was an earlier written script that consisted of thousands of picture-signs or pictograms, but it was clumsy and unwieldy. The new script was based on the older pictograms, and it was quicker to write.

With writing, other things became possible. It became possible to keep records of commodities that were being bought and sold, and one of the earliest uses of writing was for simple record-keeping of this kind. It was also possible to keep a record of agreements, for instance agreements between individual people regarding land holdings or political treaties between kings, write prayers, poetry, stories and histories. People would no longer have to commit information to memory and carry it in their heads. With writing, vast quantities of information could be stored and passed on to the next generation.

Writing made possible a great leap forward, because an ever-increasing volume of information and ideas could accumulate, year by year, generation by generation. Cultures were no longer limited by the amount of information that could be transmitted by oral tradition. Ever since it was invented, and right up to the present day, writing has been one of the principal ways in which civilization has developed.

THE INVENTION
OF SOAP

(2800 BC)

SOAP WAS INVENTED in about 2800 BC. The earliest evidence for the use
of soap is found in clay cylinders from Babylon. A clay tablet written in
Babylon in about 2200 BC contains a formula for soap: it included alkali,
water and cassia oil.

Ancient Egyptians washed themselves regularly. The Ebers papyrus,
written in 1550 BC, tells us that they used soap made of a combination
of animal and vegetable oils with alkaline salts. There are also documentary references from ancient Egypt to a kind of soap used to prepare
wool for weaving; washing the fibre was evidently even then a
preliminary to spinning and weaving.

The early reports that a soap factory complete with bars of Roman
soap were found at Pompeii were simply untrue. It seems that the
Romans were unaware of soap or its possibilities. They used oil instead
for cleaning themselves. Soap as such is not mentioned until Pliny talks
about it in his *Natural History*; he discusses the manufacture of soap
from ashes and tallow, and it seems as if this was used as a hair dressing
rather than for cleaning. Pliny said rather scornfully that among the
Germans and Gauls it was the men rather than the women who used it.
He was probably right, and it would explain how the warriors of the
north were able to make their hair bristle so alarmingly when they went
into battle.

Soap was made in Arab countries from olive oil or aromatic oils like
thyme oil. Sodium Lye came into use in the Muslim world and the
formula for soap has not really changed significantly since then. From
AD 600 onwards soap was manufactured at Nablus in Palestine, and at
Kufa and Basra in Iraq.

Modern soaps are the direct descendants of those historic Arab soaps, which were already perfumed and coloured, just as modern soaps are. Some countries in the West were reluctant to adopt the use of soap. Personal hygiene has not always been a priority. Many people have needed to be educated by health authorities in the importance of hygiene, both in self-washing and washing clothes, in order to reduce the populations of pathogenic micro-organisms. It is sobering to realize that bars of soap only became available in the USA in the late nineteenth century. In the first half of the twentieth century there were advertising campaigns to try to make people wash with soap. It was only by the 1950s that there was a broad public acceptance across North America and Europe that soap was necessary for personal hygiene. Given that soap was invented in 2800 BC, it had taken a long time.

THE INVENTION
OF SILK

(2640 BC)

SILK WAS FIRST created in China in 2640 BC. The legendary inventor of silk was the Empress Xi Ling-Shi, and at first this exotic and glamorous fabric was used exclusively by the Chinese emperors. Gradually, more people were allowed to wear silk, and it spread through Chinese society. At the same time its use spread geographically, so that it could be seen in every town. From there it was a short step to its spread outside China.

Silk became a highly sought-after luxury product in every country that had trading contacts with China. The smoothness, lightness, fineness and lustre of the cloth had no equal. Everybody who could afford it wanted silk. The rich and powerful were prepared to pay to have it transported long distances.

The earliest evidence of long-distance trade comes from ancient Egypt in 1070 BC, when silk was included in a mummy burial. Chinese silk was exported to India, the Middle East, North Africa and Europe. It became so important as a high-status good that well-defined trade routes across Eurasia were established. They became known as the Silk Road. There was no question about it: silk was startling. When Julius Caesar first appeared at the games in Rome wearing purple silk, he was the centre of attention rather than the games.

In order to maintain their monopoly, the Chinese emperors tried to keep the secret of silk manufacture (sericulture) from other countries. This was not entirely successful, as Chinese settlers in Korea in 200 BC took the secret with them, and by AD 300 the art of silk-making reached India.

Silk has a long history in India and it is still manufactured there today. It played an important part in the finely graded caste and social system.

Only the rich and important people could wear silk; the rest wore cotton. The Kanchi silk is traditionally hand-woven and hand-dyed and usually has some silver threads woven into it. A lot of this silk is now used to make saris. Silk garments are a very important element in ceremonial occasions in Indian culture.

In the Roman Empire, silk was traded. Tiberius tried to enforce sumptuary laws to prevent men from wearing silk, but this was a failure. In spite of its popularity with the Roman elite – of both sexes – the craft of silk-making did not arrive until the Roman Empire had imploded. It was not until 550 that sericulture arrived in Europe, by way of the Byzantine Empire. According to legend, monks working on orders from the Emperor Justinian smuggled silkworm eggs into Constantinople inside hollow canes.

The Byzantines recognized the value of secrecy just as much as the Chinese, and they too kept the process of silk-making to themselves. The looms and the weavers were all confined within the palace complex in Constantinople. The silk fabric produced there was, just as in the early days in China, reserved mainly for the use of the emperor, either for wearing or for high-status diplomatic gifts for foreign dignitaries. The rest was for profitable export.

In the Middle Ages, the silk secret spread to Italy. By the thirteenth century, silk had become a vital part of the Italian economy. The wealth of Florence, and therefore all the art work generated there, depended largely on the textile industry. It was therefore silk that made a good deal of Renaissance art possible. In a similar way, it was silk that funded some of Richard Wagner's music. One of Wagner's most loyal and steadfast patrons was Otto Wesendonck, who made his fortune as a silk merchant; without silk, there might have been no *Tristan and Isolde*. Silk was indispensable to Wagner in another way too: like Julius Caesar and the Chinese emperors, he just had to wear it.

THE INVENTION OF CURRENCY

(2200 BC)

CURRENCY IS AN agreed unit of exchange to make the transfer of goods and services easier. It may take the form of money in coins or notes, or it may take the form of some common resource such as grain. If the latter, currency is a halfway house between barter and money.

Traditionally, each country has had its own individual currency, but increasingly these are being harmonized to make trading internationally easier. Many European countries have resisted the idea of a common currency, but one after another has gone along with the idea of the Euro, the common unit. Panama and El Salvador have declared US dollars to be legal tender within their borders. In centuries gone by the common currency was gold.

Currency emerged partly from the use of counters to ensure that the cargo unloaded at the destination was the same as the cargo put on board at the start of the voyage. It was a short step from there to using the counters themselves as the equivalent in value to the commodity they were measuring. This was have made the transactions of buying, selling and exchanging commodities on the quayside far easier. This practice was introduced in about 2200 BC. It was also common practice to equate silver ingots with the value of a certain amount of stored grain. Silver ingots therefore became a form of money that could be used as a receipt for grain stored in temple granaries in Mesopotamia and Egypt.

This use of metal to represent the value of stored commodities was the basis of trading in the Middle East for 1,500 years. The collapse of that trading system exposed a problem with this approach to currency. If there was nowhere safe enough to store value, the value of both the ingots and the grain would fluctuate, and that led to an unstable trade system. By the late bronze age, around 1400 BC, international treaties had been negotiated allowing merchant ships to navigate safely round the eastern Mediterranean and the Red Sea, from Mycenae to Bahrain. From finds at various locations, and a couple of shipwrecks in particular, it looks as if the common currency, the Euro of its day, was the copper ingot made in Cyprus. This was made in what is called an ox-hide shape, a rectangular slab with projections at the corners. It is likely that the shape was designed to make the edges concave, allowing people to pick the ingots up and put them down without risking pinching or crushing their fingers in the process. These bronze ingots were used to buy and sell all kinds of produce, from food to luxury goods.

This well-developed early trading system was brought to an end shortly after 1200 BC. The Mycenaean civilization seems to have been critically weakened by the Trojan War. Other cultures in the region appear to have been raided or invaded by the mysterious Sea Peoples, pirates from the North who have still not been satisfactorily identified, but who were described in Egyptian archives.

There seems to have been some kind of systems collapse after that. There was a recovery of the trading network from about 1000 BC, led by the Phoenicians, and this marked the start of a continuous development of commerce in Europe and the Middle East right through the iron age and into the classical period. It led on in stages to the development of real coinage.

THE INVENTION OF
THE AQUEDUCT

(1700 BC)

THE EARLIEST AQUEDUCTS were made to supply water to the Bronze Age city of Knossos in Crete. The town stood on a low hill at the confluence of two streams along with its spectacular temple complex, today misleadingly called 'the Palace of Minos'. To get water from the hill country south of the city across the Vlychia ravine and into the city a bridge-like structure must have been built. No traces of this have survived, though the clay pipework on each side of the ravine has been found. Certainly the Minoan masons were able to build a stone aqueduct. A little further downstream from the place where the aqueduct must have crossed are the substantial remains of a contemporary viaduct, a very similar structure for taking a roadway across the valley, but at a slightly lower level. The aqueduct and viaduct were built between 1800 and 1400 BC.

There was another early aqueduct, which few people know about, at the city of Troy. Below the western walls of the Bronze Age city, there was a place where the women, and probably slaves of both sexes, went to wash clothes. The stone washing basins there were fed by three springs in a cave under the city. Perhaps there was originally a natural fissure leading from the springs to the stone basins, but it was certainly artificially enlarged in the Roman period – and probably in the Bronze Age long before.

An aqueduct is a man-made channel for conveying water from place to place. Many of the well-known aqueducts from antiquity are raised above the landscape, designed to carry water at a gentle gradient right across valleys and resembling stone bridges. Some aqueducts built by the Romans were built as a series of arched bridges one on top of the other to reach the required height. An aqueduct may be built large enough to accommodate boats or ships.

The Assyrians built an ambitious raised aqueduct out of limestone thirty feet high and 900 feet long to carry water across a valley to their principal city of Nineveh. The rest of the aqueduct was at ground level and therefore less striking as a piece of engineering, but it ran for fifty miles in all. In the Aztec capital of Tenochtitlan two aqueducts were created to ensure water supply.

The Romans are most famously associated with aqueducts, largely because of the multi-tiered water bridges, which look spectacular. Almost miraculously, about 200 of them have survived to the present day. There is a very fine two-tiered aqueduct at Segovia in Spain, built of granite without any lime or cement. It is completely intact, in amazingly good condition and still functioning today, bringing water to the city of Segovia from twelve miles away. Perhaps even more striking to look at is the Pont du Gard in France, a three-tiered aqueduct built to supply the city of Nimes with water.

The water moved from the water source to the city by gravity, so the channel bed had to be carefully engineered with a very gentle gradient to keep a steady flow. The aqueduct channels were often three feet wide and six feet deep, made of brick or stone and then lined with watertight cement.

Not all of the aqueducts the Romans created were high-level water bridges. They also made an aqueduct from the Pools of Solomon to the city of Jerusalem. This was an exactly circular tunnel hewn out of the rock. When aqueducts ran through tunnels, shafts were usually added at intervals to prevent airlocks and ensure a smooth flow. The Romans made lots of aqueducts, all over the empire. In the city of Rome itself there were 260 miles of aqueducts and there were at least forty other cities in the empire fitted with similar aqueduct systems. The Aqua Marcia, built in 144 BC, carried drinking water fifty-seven miles from springs in the Anio valley twenty-three miles away. This was mostly a covered conduit, but for the last seven miles into Rome it was hoisted up on stone arches.

Water was usually carried across valleys on stone aqueducts, but in some parts of the empire lead or earthenware pipes were used, just as at bronze age Knossos. Pipes were taken down one side of a valley, then up the other side to a lower level. The pressure from the fall on the higher side forced the water across the valley floor and up the other side. The Roman water supply to the city of Lyons was organized in this way, across the river valleys of the Garonne, Beaunant and Brevenne; the pipes for this system took 12,000 tons of lead.

In recent years, Roman aqueducts have been discovered in unexpected places. A severely weathered and degraded aqueduct has been discovered leading from the west into the Roman town of Dorchester in Dorset.

THE INVENTION
OF COINS

(640 BC)

COINAGE WAS A development of currency. As we saw earlier, in the late bronze age traders were using big slabs of bronze as currency. They were heavy, unwieldy, and their projecting corners must occasionally have damaged the hulls of the merchant ships that carried them to exchange for goods. Coins were far more flexible. They have been so successful as an invention that they have been in continuous use ever since they were invented in 640 BC.

There was an attempt in China in around 1000 BC to use cast copper coins. There has been a similar isolated find in ancient Thrace in Turkey of 400 coins dating from perhaps as early as 3000 BC. At the moment it is not clear whether these early appearances of coins were a false dawn. Coinage in the West is usually taken as beginning with an issue of coins by King Croesus of Lydia in 640 BC. Herodotus tells us that the first coins were made of pure gold and pure silver, but the archaeology tells us something slightly different; the first coins were made of electrum, which is a natural alloy of gold and silver. The first coins were made of gold and silver together.

Some coins were cast, like the Chinese coinage, but mostly coins have been struck with a hammer. The ancient Greek coins carried sophisticated and often beautiful low-relief art work. Some carried a portrait profile of a god or goddess, Pegasus the flying horse, the owl of Athena, or some other mythological icon. Others carried portraits of rulers in profile, with their name helpfully embossed. The coins have an extra value to us now, in telling us something about the culture of the time. Roman coins, imitating the Greeks, invariably carried the portrait of the emperor. At the time this was part of the image-branding that went with the celebrity cults surrounding emperors. Today the emperors' portraits are very helpful to archaeologists wanting to date sites.

Peoples living beyond the edge of the Roman Empire saw and understood coins, and made their own. In Britain in the Late Iron Age, some very crude images were applied to coins including what looks very much like the exploded image of the Uffington White Horse.

By the time of the Roman Empire, coins were the basis of all commerce. From very early on coins were made of different metals, gold, silver and bronze, reflecting their different values. This threefold, three-tier monetary system was an integral part of the Roman economy. The gold libra, the silver solidus and the bronze denarius survived right through to the middle of the twentieth century in Britain as the initials LSD, standing for pounds, shillings and pence. The (Roman) tradition of putting profile heads of rulers onto coins has been maintained in England for well over 1,000 years.

THE INVENTION OF
THE STIRRUP

(500 BC)

STIRRUPS ARE A pair of rings hung from each side of a saddle to make footrests for the rider. The advantage of stirrups is that they increase the rider's ability to control the horse. This in turn greatly increases the usefulness of the horse and rider in communication, transport and warfare. The invention of the stirrup was one of the great spurs to civilization, and it was as influential in its own way as the invention of the wheel or the printing press.

The horse was domesticated as early as 4500 BC. It is not known how early people experimented with riding horses, but it is surprising that the stirrup was not invented until as late as 500 BC. That innovation was made in India. After that a nomadic group called the Sarmatians used a single stirrup as an aid to mounting. True stirrups, in pairs, were adopted in Central Asia in around 50 BC.

We do not see a horse rider using paired stirrups in a work of art until as late as AD 322, in an image found in a tomb in China. The use of the stirrup was spread right across Eurasia by the horsemen who had devised it.

Curiously, the Greeks and Romans did not think of using stirrups, which one might have expected them to have invented independently. In the Graeco-Roman world the usual method for mounting a horse was by vaulting onto it or climbing on from a mounting block. Whether the Huns used stirrups or not is uncertain. If they did, it may account for their advantage over the Romans.

The use of the stirrup reached Scandinavia in the sixth century. The Norse sagas indicate that horses were very important in the Northern culture of the time. The sixth-century Swedish King Adils was said to be a great horse lover; in battle he fought on horseback. The later Vikings were not horsemen, but preferred to fight on foot. Invaders from Central Asia brought the stirrup with them into Europe in the seventh century, and by the eighth century Europeans seem to have adopted it.

This set the scene for the development of cavalry units in armies, and undoubtedly changed the outcome of many a battle thereafter, in the middle ages and later. The last cavalry charges in history took place as recently as the beginning of the Second World War, when in desperation Polish and Russian cavalry regiments charged the German invaders.

THE INVENTION OF GEOMETRY

(300 BC)

GEOMETRY IS AN investigative discipline, enquiring into the properties and interrelationships of lines, surfaces and solids. Before geometry could evolve, there had to be a concept of measurement and therefore a concept of number. It seems just from the layout of ancient monuments that the people who built them had a counting system. It is not a coincidence that the 5,000-year-old stone circle of the Stones of Stenness in Orkney consists of a ring of twelve stones and that each stone is separated from its neighbour by a walk of just twelve steps. For thousands of years before that, people had been recording numbers as tallies by carving on bones.

By 4000 BC calendars were in use in Babylon and Egypt, and by 3400 BC the Egyptians were using tally symbols for numbers, developing within a few hundred years into a hieroglyphic system of numerals. By 3000 BC, both a decimal system of counting and the abacus were in use in the Middle East and the Eastern Mediterranean, though for some reason the Babylonians used a counting system with a base of sixty for their financial transactions. This is perhaps an ancient equivalent to the anomalous situation in Britain in the first half of the twentieth century, when a base of ten (the decimal system) was used for counting most things, but money was an exception; there were twelve pence to a shilling and twenty shillings to a pound. By 2000 BC the people of Harappa in the Indus Valley were adopting a decimal system of weights and measures.

Shortly after this, mathematics was under way in the Near and Middle East. By 1800 BC the Babylonians were solving quadratic equations, using multiplication tables, and knew about Pythagoras's Theorem. They were

78

compiling tables of square roots and cube roots. They were also starting to apply their new mathematical knowledge to the science of astronomy. The Egyptians too were exploring mathematics. They devised schemes of multiplication based on repeated doubling and division based on repeated halving.

In 575 BC, Thales of Miletus took the mathematical knowledge of the Babylonians from Egypt to Greece, which initiated a new phase in the development of geometry. Thales was accorded high status in antiquity; he was known as one of the Seven pre-Socratic Sages. Thales used geometry to solve everyday problems like calculating the height of the pyramids and the distance of ships from the shore. It is thought that Thales discovered experimentally that any angle formed inside a semicircle is a right angle, though some people ascribe this discovery to Pythagoras.

In 529 BC, one of the great figures of ancient Greek mathematics, Pythagoras of Samos, moved to Croton in Italy, where he taught mathematics, geometry and music. He also set up a religious community and seems to have suffered some kind of persecution; he was driven out under the tyranny of Polycrates. Pythagoras acquired a reputation as a learned, inquisitive and universally informed man.

Pythagoras wrote nothing and his life is shrouded in legend and mystery. He seems to have established a distinctive way of life for himself and his followers, rather than a system of philosophy. But it is for his mathematical discoveries that Pythagoras is remembered now. He arrived at a mathematical analysis of musical intervals and, with astonishing daring, proposed that those musical intervals lay behind the structure of the universe. It seems that Pythagoras thought of the universe as a series of concentric spheres. His model had the Earth at the centre of the universe, but it was the first important step towards the sun-centred model Copernicus put forward hundreds of years later, and we know that Copernicus consciously built on the system proposed by Pythagoras. Today we take it for granted that the solar system consists of planets travelling round the Sun along almost exactly circular orbits, those orbits tracing paths round a set of notional, invisible concentric spheres with the Sun at their centre. It seems perfectly natural that the solar system should have turned out to have this shape. Pythagoras somehow knew that the solar system was this shape without any real scientific evidence. We could almost say that the structure of the solar system is an invention of Pythagoras; he did not in any modern sense discover it.

Pythagoras said that numbers lay behind space itself; 'one' was a point, 'two' was a line, 'three' was a surface, 'four' was a solid. He discovered that the angles of a triangle always add up to 180 degrees. He also famously stated the properties of the right-angled triangle, already known in Egypt, but now formally encapsulated in the Pythagoras Theorem. Oddly enough it was not to be the right-angled triangle that became an object of veneration among the Pythagoreans, though it should have been, but the equilateral triangle made up of ten points, arranged $4 + 3 + 2 + 1$.

Pythagoras started an analytical approach to numbers and trigonometry which inspired, fed and stimulated scholars, laying the foundations for all later mathematics. His school is considered to be the first known source of logical, deductive thought: the birthplace of reason itself.

Hippocrates of Chios (470–410 BC) was the author of the first Elements of Geometry. It is likely that Euclid used this work as the starting-point for his own *Elements Books I* and *II* over a hundred years later. Hippocrates gave geometric solutions to quadratic equations. He studied the classic problem of squaring the circle. He worked on duplicating the cube, which he showed was equivalent to constructing two mean proportionals between a number and its double. Hippocrates was the first mathematician to show that the ratio of the areas of two circles was equal to the ratio of the squares of their radii.

Plato (427–347 BC) was a much higher-profile figure. He founded his Academy in 387 BC, a kind of early university, which flourished for over 700 years. In his book *Phaedo*, Plato set out a theory of forms, in which mathematical forms are considered as having perfect attributes; a line, for instance, is considered as having the property of length but no width. Plato emphasized the crucial importance of proof in mathematics. He insisted on clear hypotheses and accurate definitions. All of this paved the way for Euclid and the formal foundation of geometry though, perhaps surprisingly, Plato never came up with any important mathematical discoveries himself.

Plato's students were profoundly influential in the development of geometry. One of them, Theaetetus of Athens (417–369 BC), was the creator of solid geometry. He was the first to study the icosahedron and octahedron and construct all five regular solids. His work was incorporated in *Book XIII* of Euclid's *Elements*. Eudoxus of Cnidus (408–355 BC) found ways of calculating the volumes of pyramids and cones. Menaechmus (380–320 BC) was a pupil of Eudoxus. He discovered the conic sections. He was the first to show that ellipses, parabolas and hyperbolas are obtained by cutting across a cone in a plane that is not parallel to the cone's base.

Then came Euclid. Euclid of Alexandria (325–265 BC) gathered together the various theorems of Pythagoras, Hippocrates, Theaetetus, Eudoxus and other mathematicians, forging them into a logically connected systematic whole – Euclid's *Elements*. Euclid was probably trained as a mathematician in Athens by pupils of Plato, though little is known of his life. It was in Alexandria that he founded his own school of mathematicians at the time of Ptolemy I.

Euclid is remembered for just one work, called *Elements*, though often referred to as *Elements of Geometry*. The work, divided into thirteen books, makes up the most substantial treatise on mathematics that has survived from ancient Greece or indeed from any ancient civilization. Euclid wrote other books about geometry, astronomy, optics and music, but unfortunately most of these have been lost. *Elements of Geometry* is the best-known mathematics textbook of all time. It has been printed in countless editions, with modifications and simplifications, and was still in use as a school textbook in the early part of the twentieth century. Only then did rival textbooks start to appear.

King Ptolemy asked Euclid whether there was any shorter way in geometry than that of his book. Euclid answered, 'There is no royal road to geometry.' Euclid's book was a great landmark in the development of Western civilization. It was the first book on mathematics to be printed, and so acquired a major importance in the post-medieval world. It also stood as a model of rigorous mathematical argument for hundreds of years.

THE INVENTION OF THE ARCHIMEDEAN SCREW

(265 BC)

ARCHIMEDES, BORN IN about 287 BC, was on very friendly terms with the king of Syracuse, Hieron. After travelling to Alexandria, the great seat of learning of the ancient world, Archimedes returned to Syracuse to devote himself to mathematics.

Archimedes has a central place in the study of mathematics, but remains more famous for his inventions, on which he himself set no value at all. He regarded applied mathematics as a trivial pursuit. It was probably while Archimedes was in Egypt that he invented the Archimedean screw which is still in use there for raising water to irrigate fields. It consists of a hollow cylinder, inside which a spiral surface (originally made of pitch-coated willow) is rotated by a handle at the top. The lower end of the cylinder is placed in the water, and when the handle is turned the water is drawn up, pouring continuously out of the top.

The Archimedean screw was a most original and useful invention, one that has been in constant use since Archimedes' lifetime for raising water from rivers and ditches up onto fields, and out of drainage ditches into embanked rivers. By the middle ages it was in regular use as a bilge pump, enabling sailors to empty the seawater leaking into a wooden ship. When, whether and how the wood screw evolved out of the water screw is not known. Like all good inventions, Archimedes' screw is very easy to operate, efficient and adaptable. In the Netherlands, large steel Archimedean screws, electrically powered, are used today to raise water from drainage ditches. It is a versatile invention with staying power, a truly great invention.

THE INVENTION OF
THE ASTROLABE

(150 BC)

THE ORIGINS OF the astrolabe are by no means certain, but it appears to be far more ancient than the earliest surviving examples, which date from the ninth century AD. The astrolabe is an ancient instrument designed to take the altitude of heavenly bodies, in other words to measure their distance up the sky as an angle above the horizon. By making these observations with an astrolable, it was possible for navigators in antiquity to compute both latitude and time. The astrolabe is believed to have been invented by Hipparchus in 150 BC; some believe it was invented even earlier, by Apollonius of Perga in 250 BC. It is the oldest scientific instrument in the world.

The underlying theory for its construction, the stereographic projection, may well have been known to Hipparchus and the instrument seems to have been well known in the first century AD. There is an old Muslim tradition that it was not Hipparchus who invented the astrolabe but Ptolemy, and that he invented it in the second century. Ptolemy, this legend goes, was riding on a donkey while pondering on his celestial globe. He dropped the globe, the donkey trampled on it and the result was the flattened disc, the celestial sphere reduced to two dimensions.

Several different types of astrolabe have been manufactured. The commonest type is the planispheric astrolabe. This is a flat disc, on which the celestial sphere is projected onto the plane of the equator. A typical astrolabe of this type was a brass disc about six inches in diameter, though they were made in a variety of sizes, some much bigger than that.

The principle of the astrolabe is to show what the night sky looks like at a given place at a given time. This was done by drawing a schematized representation of the sky on the astrolabe's face and marking it so that positions in the sky are easy to find. The movable components were adjusted to match the situation on a particular date at a particular time. Once it is set, the visible sky is represented on the instrument's face. Various problems could then be solved visually. A typical use was finding out the time of day or night, finding the time of sunrise or sunset on a particular day, by extension also the length of the day and night. The astrolabe could be used to simulate the movements of heavenly bodies. It could be used for surveying. It could even be used for astrology. It was an incredibly versatile multi-functional instrument.

The mariner's astrolabe was a simpler instrument, often little more than a metal ring with degrees marked off on it for measuring celestial altitudes.

True astrolabes were made before AD 400, and its use was highly developed in the Islamic world by 800. It was introduced into Europe by Muslims colonizing Spain in the twelfth century. The name astrolabe seems to be an Arabic form deriving ultimately from a Greek phrase meaning 'star holder'. The astrolabe was extremely popular in the middle ages, and remained the most popular scientific instrument until 1650; after that it was gradually supplanted by more accurate instruments, falling out of use altogether in about 1800.

Astrolabes are still greatly appreciated for their unusual capabilities, for their usefulness in astronomical education – and for their sheer beauty as objects.

The earliest surviving texts explaining the use of the astrolabe date from the seventh and eighth centuries. Eighth-century texts from Damascus and Baghdad show that by then astrolabes were in use throughout the Arab world. Islamic influence stretched from Spain and Morocco in the west to India in the east. This meant that a wide range of astronomical observations could be incorporated in the instrument. The earliest surviving astrolabes date from the ninth century, and their high quality implies that they were part of a long tradition; the astrolabe was by no means newly invented.

By the twelfth century there are many surviving instruments and texts describing their use. The style of the craftsmanship varies, but the function and basic design remained the same. Astrolabes made at this time had a range of useful extra features. One was a shadow square, for measuring the heights of things that were out of reach, such as the summits of hills, the tops of trees or towers. There were also scales for calculating the calendar, and for calculating the correct direction for Muslim prayers. One text describes over forty uses for the astrolabe. It was an instrument, used together with tables of the sun's declination, that was absolutely essential for successful navigation at the time of the great voyages of discovery. From the time of the great map-maker Martin Behaim (about 1480) right through to the middle of the eighteenth century, navigators relied very heavily on the astrolabe; this was the key instrument behind those great voyages, behind the first crossing of the Atlantic, the first rounding of Africa, the first circumnavigation of the globe. The reign of the astrolabe only ended when superior instruments were invented, such as John Hadley's quadrant.

THE INVENTION OF
THE ENCYCLOPEDIA

(35 BC)

TODAY WE UNDERSTAND an encyclopedia to be a compendium of all knowledge, or at least a summary of all knowledge. Historically, the range of knowledge has varied. For instance, before 1492 European scholars had no knowledge of North and South America, which gave them a very different world view from scholars after that date. When looking at works written hundreds of years ago, we therefore have to allow for this and ask ourselves whether an author was comprehensively covering the state of knowledge at the time when he was writing.

The word encyclopedia comes from a Greek phrase meaning 'rounded education', which is rather softer and more generous in approach; an encyclopedia does not, after all, have to include absolutely everything. This is just as well, as the sum of human knowledge now, at the beginning of the twenty-first century, is far too great to fit inside a book or even a series of books like the *Encyclopedia Britannica*, which is still one of the most ambitious printed encyclopedias in existence. In the eighteenth and early nineteenth centuries, human knowledge was still manageable in scope, and it was possible to encompass it in a single multi-volume publication, though it was necessary even then to draw on a range of minds and have it written by a committee.

The latest date was probably 1600, when a single mind could encompass all that was known by man, and it is no coincidence that some people in the sixteenth century did attempt to accomplish great things in all areas. During his imprisonment in the Tower of London, Sir Walter Raleigh began, though he did not finish, a *History of the World*. He was what has become a cliché, a Renaissance man, a polymath, someone who was accomplished at a high level in many areas of human endeavour. Raleigh was well fitted to attempt to write a comprehensive, all-encompassing history of the world. It was, with hindsight, the last moment in history when one man could try to accomplish such an ambitious project – and legitimately hope to succeed.

The title of the Mel Brooks comedy *History of the World, Part 1* is a reference to the fact that Raleigh had only completed the first part of his history at the time of his execution in 1618. The first volume, out of the five volumes planned, only took the history of the world from the beginning to 130 BC. Raleigh's encyclopedic coverage of history was in any case going to be incomplete, even if he had lived another twenty years and finished *Volume Five*. He had a political agenda and that made him see the world through a distorting prism of bias. He wanted to illustrate that only good kings have the right to rule. Queen Elizabeth was going to be portrayed as a good and just monarch, and the implication that would hang in the air, probably unspoken and unwritten, as it would be overt treason to put it into as many words, was that the current monarch, King James I of England and VI of Scotland, was a bad one.

Raleigh saw the world through the eyes of Tudor propaganda. We can see this in his short aside on Richard III: 'To Edward the Fourth succeeded Richard the Third, the greatest master in mischief of all that forwent him.' As far as the Tudor dynasty was concerned, Richard III had to be portrayed as a bad king with no legitimate entitlement to the English throne – simply in order to justify the usurpation of the throne from Richard by Henry VII, who was the founder of the Tudor dynasty. Richard III had to be vilified in order to validate the reign of the Tudors through the sixteenth century. It has subsequently emerged that Edward IV, Richard III's older brother, was illegitimate and therefore in reality only his half-brother; Richard III in fact had a better claim to the English throne than either Edward IV or Edward IV's son, one of the princes in the Tower who was legitimately deposed by Richard before he could be crowned king. Richard also had a better claim to the throne than Henry Tudor. Raleigh's view of the past was not objective, but it was very much the orthodox view of his time, and that is what has to be borne in mind in a reading of any of the encyclopedic works of antiquity.

The idea or even ideal of an encyclopedia has existed for hundreds of years, though the word itself was first used as the title of a book in 1541 by Joachimus Fortius Ringelbergius (*Lucubrations, or rather the Most Complete Cyclopedia*) and Pavao Skalic in 1559 (*Encyclopedia*, or *Knowledge of the World of Disciplines*).

In the eighteenth century, the idea of the dictionary came to the fore. In a dictionary the emphasis is on words and their definitions, but it is a short step from there to a discussion of the background and history of the word and the way it has been used. That fleshing-out of background takes the dictionary halfway towards being an encyclopedia. The alphabetical layout of the dictionary has also been adopted for most modern encyclopedias. The *Shorter Oxford Dictionary* is arranged in alphabetical order, starting with A, Aal and Aam, and my 1957 edition of the *Encyclopedia Britannica* is organized similarly, starting with A, A1 at Lloyd's and Aa. The difference between the two is in the amount of background information given.

The idea of a compendium of all the world's knowledge was in the air in antiquity. From the way Aristotle wrote, on a great range of subjects, it looks as if he was attempting to create an encyclopedia informally, simply by the aggregate of his collected works. Other writers were more overtly writing encyclopedias. One of the most noteworthy of these was Pliny the Elder. In the first century AD, Pliny wrote his *Naturalis Historia* (*Natural History*), which was an account of the natural world running to thirty-seven volumes; it was very popular in Western Europe in the middle ages. Pliny's *Natural History* and the way in which it was used by readers give it a claim to being the first encyclopedia.

Diodorus Siculus has a prior claim, with his *Library of History*, which he is known to have been working on at least as early as 56 BC and may have completed twenty years later in about 35 BC. He mentions Tauromenium, a town in Sicily, being made a Roman colony, and that event took place in 36 BC; no later event is mentioned in the book. It was therefore completed several decades before Pliny's work. Diodorus's work is incredibly ambitious. It is nothing less than an attempt to write a comprehensive universal history of the world from the creation down to the writer's own day. Along the way he includes a lot of material other than history. He is keen to tell us about places, mythology and anthropology too. Because of the huge scope of his work it deserves to be regarded as the earliest encyclopedia – or at least the earliest encyclopedia that we know about. Diodorus's great work demands attention because of his seriousness as a writer, and encyclopedias have to be serious in intention and content.

Travel was difficult and dangerous in the first century BC, yet Diodorus seems to have travelled widely in order to gather information first-hand. He visited Egypt in 56 BC, and noted in his book that while he was there he saw with his own eyes a mob of Egyptians successfully demanding the death penalty for a man working at the Roman embassy who had killed a cat. He rightly thought this was interesting as he observed that the Egyptians greatly feared the Romans, yet they wanted justice for the dead cat even more. There is no positive evidence in the book that Diodorus went anywhere other than Egypt, but he claimed that he visited all the most important parts of Europe and Asia. He also leant heavily on a great reservoir of existing literature, which he used very selectively and judiciously.

Diodorus's *Library of History* consists of forty books, opening with the myths, kings and customs of Egypt and the history of Assyria. The later books are chronological, each dealing with two or three decades. It is a powerful achievement, enormously impressive, and full of exciting detail described in a lively style – the very best encyclopedia.

The first encyclopedia to present a world view from the Christian standpoint was the *Institutiones* written by Cassiodorus in AD 560. This in turn was the inspiration for the *Etymologiae* of Isidore of Seville in 636, the encyclopedia that was to become the most influential in Europe in the Early Middle Ages. In the Late Middle Ages, the most widely read encyclopedia and the most widely quoted was *De proprietatibus rerum*, written by Bartholomeus de Glanvilla in 1240. The most ambitious of the medieval encyclopedias was the *Speculum Majus*, written in 1260 by Vincent of Beauvais: it was three million words long, about ten times longer than this book.

The medieval Islamic world was equally interested in the acquisition of knowledge, equally interested in compiling encyclopedias. Abu Bakr al-Razi wrote an encyclopedia of science and Ibn Sina wrote a medical encyclopedia which became a major reference work. There were many other Muslim scholars writing similar large-scale works, and they had a major influence on standards of research and scholarship generally. This was in large measure due to the Islamic emphasis on a spirit of sceptical enquiry and faithfulness to written record, so that earlier sources were correctly quoted.

In 1408 Yongle, a Chinese emperor of the Ming dynasty, oversaw the writing of the *Yongle Encyclopedia*. This was one of the biggest encyclopedias ever written, and when complete it consisted of more than 11,000 volumes; only 400 of them have survived. A later Chinese emperor, Qianlong, contributed 40,000 of his own poems to another large-scale encyclopedia entitled *Watching the Waves in a Sacred Sea*. These early Chinese works, and the similar works written in Japan, were intended for use by the privileged few: only the rich and the scholars in their monasteries had access to them.

By the eighteenth century in the West, encyclopedias were seen as ways of equipping a much wider readership with the sum of human knowledge. From the time of Sir Walter Raleigh onwards, right up to the present, the process has been one of democratization of knowledge. In 1646, the English physician and philosopher Sir Thomas Browne wrote a book called *Pseudodoxia Epidemica* (Vulgar Errors), in which he tried to refute some of the commonest errors of his time. In the preface he rightly referred to his ambitious work as an encyclopedia. Going through several editions, and translated into several European languages, Browne's book was to find its way onto the bookshelves of many educated readers in Europe through the seventeenth and eighteenth centuries.

The now-familiar alphabetical layout of the encyclopedia was first used by John Harris in his 1704 *Lexicon technicum*. Interestingly, Harris himself thought of his book as a technical dictionary, and it really lies halfway between dictionary and encyclopedia. When the second edition came out in 1710, Harris got Sir Isaac Newton to contribute an essay on chemistry; it was the only piece on chemistry Newton ever wrote.

A real landmark in the history of the encyclopedia was the publication in 1728 of Ephraim Chambers' two-volume *Cyclopaedia*. This used the alphabetical layout, used many contributors and incorporated cross-referencing within each article to other entries – a major innovation in its own right. This was a profoundly influential book. Its translation into French inspired the writing of the great *Encyclopédie*, which became the most famous of all the early encyclopedias. This astonishing work, which was undoubtedly the greatest publication of the eighteenth century, was edited by Jean le Rond d'Alembert and Denis Diderot and published in seventeen volumes of text between 1751 and 1765, followed by eleven volumes of illustrations published between 1762 and 1780. The final part, the index, came out in 1780.

It was no accident that this great thrust in the democratization of knowledge came immediately before the French Revolution: the *Encyclopédie* was really part of the French Revolution. The scope of Diderot's encyclopedia was enormous, and the quality of many of the contributions was very high. It was not surprising that its social and political impact was enormous. Diderot himself was dissatisfied with it. He wanted the encyclopedia to be far more than the sum of its parts. He believed more than anything else in the interconnectedness of knowledge; an ideal encyclopedia was really a massive index of connections. Diderot understood that human knowledge had reached a stage where it could not all be satisfactorily amassed in one publication, but he hoped that at least the relationships among the subjects could be.

Chambers inspired Diderot. Diderot in turn inspired the *Encyclopedia Britannica*, the third of the quartet of giant modern encyclopedias. This was produced in Scotland in 1768–71. The first edition consisted of just three volumes, but it quickly expanded. By the third edition, in 1797, it had grown to eighteen volumes. Then came the six-volume *Conversations-Lexicon*, completed and published in Leipzig in 1808. This endeavoured to be comprehensive, with shorter summaries aimed at general readers; it was widely followed as a model for later encyclopedias and is really the closest in form and style to the encyclopedias of the twentieth and twenty-first centuries, such as the well-known Funk and Wagnalls.

Today, encyclopedias are as popular and useful as ever, with experiments in internet encyclopedias. The Microsoft *Encarta* encyclopedia was a notable landmark, in existing as a CD-ROM but having no print version. There is also Wikipedia, which goes one significant step further in democratization, with articles of very variable length and completeness by a multitude of different and anonymous authors, allowing anyone to add or modify content.

THE INVENTION OF
PAPER FROM PULP

(AD 105)

THE EARLIEST WRITING was inscribed with a stylus onto clay tablets. These were not very portable, and used mainly for archives and inventories; the tablets might be stacked on shelves and rearranged, but they could not easily be carried around. As early as the bronze age, around 1500 BC, small notebooks were being made. These consisted of hinged pairs of wooden plaques with their inner faces coated with wax. They could be used to record a business deal, a truce or a message, but the communication had to be kept fairly brief.

Papyrus, the first paper-like writing material, came into use in Egypt as early as 4000 BC. Papyrus was made by pounding a bundle of reeds until it turned into a thin, flexible, fibrous sheet. The great virtue of papyrus was that it was very light, could hold an enormous quantity of writing, and could be rolled up into scrolls for storage or for portage.

Paper as we use it today was invented in China in AD 105 by a man called Ts'ai Lun, an official at the Imperial court. Before the invention of paper, there had been two alternative writing materials in China. One was silk, which was very expensive. The other was bamboo tables, which were too heavy. Ts'ai Lun's prototype paper was made of bark, fishnet and bamboo, hammered flat to make a very thin film. In fact, like many another later inventor, Ts'ai Lun was improving on an earlier method. It looks as if there was an earlier version of paper in use in China from around 50 BC, but that had been made from hemp.

The Chinese used paper for writing on, but also for other things too. It was used as a wrapper, just as it is today, and as a medium for decorative artwork. It was also used, perhaps less appropriately, to make clothing. But the superiority of paper as a medium for writing was very plain to see, and it was not long before it replaced all the other media.

In about AD 600, Chinese Buddhist monks took the art of paper-making with them to Japan. There, too, paper was adopted for writing and for making decorative objects – including fans and dolls. The Japanese even adopted paper for making partitions or screens to divide the rooms in their homes. In 750, the Chinese passed the art of paper-making on to the Arabs. It was the Arabs (Moors) who built the first paper-mill in Europe, though the Europeans were slow to adopt the new medium, perhaps because of its association with the Moors. In England and France in the early middle ages, vellum made from sheepskin was the preferred material, but by the late middle ages the cheaper new material, paper, was in use. French monks started using paper for making copies of the Bible.

The Germans greatly improved the craft of paper-making, and together with the invention of the movable type printing press the higher-quality paper made possible the large-scale production of books.

For a long time paper was made by recycling old clothes and other fabrics, but the increasing demand for paper outstripped the supply of these raw materials. It looked as if it should be possible to make paper out of wood, but it took a long time to develop the technique. It was an Englishman called Hugh Burgess who invented a method of making high-quality wood pulp for paper-making – in 1852 – and now nearly all paper is made of wood pulp.

THE CREATION OF THE FIRST WORLD MAP

(150)

TODAY WE TAKE it for granted that there are world maps. We grow up surrounded by them and the layout of the continents and oceans is as familiar to us as the arrangement of the rooms in our homes. But there was a time before maps, and living in that world must have been very different. The first true world map was the Ptolemy world map. We see this now in the form in which it was drawn in the middle ages, but that is really a reconstruction based on the description of the world given by Claudius Ptolemaeus, who was born at Ptolemais Hermii, a Greek colony town in Egypt, in about AD 90.

Ptolemy was very dependent on other scholars for information. He did great work as an astronomer, and in that area he was dependent on Hipparchus. In geography he was similarly dependent on Marinus of Tyre. He did a lot of checking, correcting and updating of Marinus's work. This led to his great work, the *Guide to Geography*. This work, along with a lot of other classical learning, was forgotten or lost in Europe and only kept alive by Arabic scholars. It was the rediscovery by Europeans of those Arabic copies that fuelled the Renaissance. The astronomer Hipparchus pointed out that the only way to construct a reliable and trustworthy map of the world was to use astronomical observations to fix the latitude and longitude of all the key points on its surface. This was admirably scientific in principle, but the means of acquiring this sort of information were not in existence.

Then, just before Ptolemy's time, Marinus of Tyre started to collect determinations of latitude and longitude from itineraries. It is not clear how far Marinus got with this project, but Ptolemy evidently used Marinus's work and continued where Marinus left off. Ptolemy borrowed the system Hipparchus had invented, and which we still use, for dividing the equator into 360 parts, and these became our modern degrees of longitude. Ptolemy drew lines through these points, connecting them to the unvisited North and South Poles, to make lines of longitude. He drew another set of lines parallel to the equator to mark the latitude. Now that he had a grid, he could begin to locate his known points on it.

At this point, Ptolemy made an odd mistake. Eratosthenes had correctly calculated the circumference of the Earth as 25,000 miles. Posidonius had wrongly reduced this to 18,000 miles, and both Marinus and Ptolemy unfortunately followed Posidonius instead of Eratosthenes. As a result, significant errors were built into the ground-breaking world map from the start, errors that had epoch-making consequences in the fifteenth century.

Ptolemy was also working on relatively little data on longitude. Given the flaws in his approach, Ptolemy made an honest attempt at a world map. Europe, Asia and Africa are recognizable. He also produced regional maps, and in effect invented the atlas at the same time as the world map. He established regional and world maps, at two different scales, as ways of looking at the world. This was a major step in changing people's perceptions of the world they lived in. In remote antiquity, long before Ptolemy, people never thought in cartographic terms. After Ptolemy, most educated people did think in cartographic terms.

It is all too easy to find fault with the result – its lack of geographical information such as climate, inhabitants, resources, vegetation, relief, drainage – but Ptolemy was primarily an astronomer and he was constructing a map of the world in much the same spirit as he might have made a map of the Moon. The world was still almost as alien a place. Even so it is odd that he left so much out. Strabo had, after all, shown that he was well aware of the need to include the rivers and mountains that, to use Strabo's own word, 'geographize' a country. In Gaul, Ptolemy included a stream that happened to be the boundary between two Roman provinces, yet left out the major tributaries of the River Rhine.

The *Guide to Geography* included tables of places and their locations, showing an admirable rigour of approach, one that would be a model to the Renaissance world. Ptolemy's maps (or at least the instructions for drawing them) were for all their faults better than most of the maps that were constructed in medieval Europe. The rediscovery in about 1300 of a copy of Ptolemy's *Guide to Geography* together with the statistical tables enabled the maps to be reconstructed from his thousands of reference points. The reconstructed world map in particular was a turning-point in the European Renaissance, giving the early explorers and their sponsors a clear perception of the world they were investigating.

That perception was not entirely accurate, and the smaller planet envisaged by Posidonius meant that some seriously wrong estimates were made about the distance westwards from Europe to China. The Columbus voyage of 1492 was predicated on the Ptolemy map, and that implied that only the width of the Atlantic separated Spain and Portugal from China. As luck would have it, there was another land mass in between, so Columbus was able to make a landfall. Ptolemy's reconstructed world map was a sensation in itself, but when the text it was based on was translated from Greek into Latin shortly after 1400, it reignited enthusiasm for a global system of coordinates using latitude and longitude as a scientific means of fixing and finding locations. It revolutionized geographical thought in Europe, put it on a scientific basis and prepared the way for the great voyages of discovery that followed within a few decades.

THE INVENTION OF WOODBLOCK PRINTING

(200)

WOODBLOCK PRINTING WAS invented in China in the ninth century AD. This was a simple method for printing text, patterns or images onto cloth and later paper, and it was widely used in Eastern Asia. The Japanese also adopted woodblock printing: ukiyo-e is a well-known type of Japanese woodblock art print. By around AD 600 the woodblock technique was being used in Egypt. It is not known whether the Egyptians developed woodblock printing for themselves or imported the idea from China, perhaps via India. Given that wood is scarce in Egypt, it is likelier that the technique was imported rather than home-grown. A parallel tradition was probably available in Mesopotamia. As early as 3000 BC there were cylinder seals in Mesopotamia, which were used to create impressed images on clay tablets. It may be that an idea for woodblock printing later developed out of that use of seals. Curious-ly, the Egyptians seem never to have used woodblocks to print onto papyrus, although it would have been technically possible.

By the middle ages, the woodblock was in use in Europe, for text and illustrations, and the art form became known as the woodcut.

A woodblock has one face carved to make the image. The parts of the image that are to show black after printing are left as untouched flat surfaces. The parts that are to show white or blank are cut away, so that they do not get inked during the printing. The block image was made in reverse, as a mirror image. This might not be much of a consideration in an illustration, where a 'back to front' result might not matter, but when text was concerned, as it often was in the later middle ages, the image had to be carved back to front in order to come out the right way round.

Black and white or single-colour printing only required one block. Multi-colour printing required a block for each colour.

Printing could be done in more than one way. Stamping was one possibility. The woodblock was inked, placed ink side down on the paper, and pressed down by hand or tapped down with a hammer. Rubbing was another method, especially popular in the Far East. The block was inked and placed ink side up on a table. The paper or fabric was placed on top and the upper surface rubbed with a leather pad. A third method was to use a printing press, and this was only used relatively late in the evolution of the technique, from about 1480 in Europe.

The earliest printed fragments to have been made by the woodblock technique are from China. They consist of silk printed with a flower design in three colours, and date from perhaps AD 200. The Chinese were also the first to use the woodblock technique for printing text. It was a technique that lent itself to the Chinese language. Although the Chinese invented movable type using baked clay in the eleventh century and metal type was available from Korea 200 years later, the Chinese printers preferred to carry on using woodblocks. Typesetting from a font of 40,000 Chinese characters was a daunting prospect.

The use of the printing, the target of the task, was different in China. Mainly the purpose was to produce standardized ritual texts, and in China woodblock printing is especially associated with Buddhism. Placing this revered ritual text on woodblocks would have the effect of preventing the text from being corrupted. It would also be possible to return to the blocks time and again without any need to revise.

The whole philosophy of printing in medieval Europe was different. Woodblock for text was used in Europe but only for small amounts of text at a time. Alphabets were made out of woodblocks, and these could be 'typeset'. The use of woodblocks to print substantial amounts of text together with images in block-books only took place after the invention of movable type in the 1450s.

The oldest printed book in the world was a woodblock printed in the ninth century. It was a Chinese scroll containing the text of the Diamond Sutra, measuring sixteen feet in length. It was found by Sir Marc Aurel Stein in a cave and is now in the British Museum. At the end it reads, 'Reverently made for universal free distribution by Wang Jie on behalf of his two parents on the thirteenth of the fourth moon of the ninth year of Xiantong.' That is equivalent to 11 May, AD 868.

In the eleventh century, the complete Buddhist text *Tripitaka* was woodblock printed by Chinese printers. The text is 130,000 pages long and the task took twenty-two years, from 1080 until 1102.

In Europe the most famous block book to be produced was the *Ars Moriendi*, with some pages text and some pages illustrations. The next most popular title was a biblical picture book called the *Biblia pauperum*. Obviously a lot of work was involved in making these block books, so they tended to be fairly short, at around fifty pages.

THE INVENTION OF
THE HORSESHOE

(450)

THE HORSE WAS domesticated very early on, more than 6,000 years ago. It is clear that by the time of the great chariot civilizations – ancient Egypt, Mari, the Hittite empire, Minoan Crete and Mycenaean Greece – the horse was of vital military importance. Horses were regularly used in pairs to pull chariots along. Warrior princes in the second millennium BC were regularly carried into (and out of) battle by their charioteers, as vividly described in Homer. By the time of the Roman Empire, chariot warfare was becoming old-fashioned. Roman historians commented on the fact that the barbarian tribes of northern Europe, people like the Iceni, with their queen Boudicca, were still fighting in this old-fashioned way.

Drawings and sculptures from the Bronze and Iron Ages show that both men and women at least occasionally rode on horseback. This led to a new mode of fighting, and the chariot ranks quickly gave way to cavalry, battalions of fully mounted warriors.

Saddles probably originated in the steppes of Central Asia, along with the horse collar and stirrups. The horseshoe came in later. It is not known who invented the horseshoe, or exactly when, or exactly where. From fairly early on it was seen that there was a need to protect horses' hoofs to some extent. Horsemen in early Asia made sandals out of reeds for their horses and even little leather boots. During the first century AD, and as so often evidently copying an earlier Greek practice, the Romans manufactured shoes for their horses out of soft leather with metal blakeys.

The horseshoe as we recognize it was invented in Europe about AD 450, and by the sixth century, horsemen were regularly nailing curved metal shoes onto the base of their horses' hoofs. In 1000, bronze horseshoes complete with nail-holes were being cast all over Europe; at that time, the horseshoe made its appearance in art, literature and the archaeological record. In the thirteenth century, iron horseshoes were being made by professional horseshoe-makers, or farriers, in every town and village; part-blacksmith, part-vet, the farriers became key figures in the life of the community. Medieval warhorses were trained to paw at the enemy and punch with their forelegs; they could also trample men once they were down on the ground. Adding iron shoes to their hoofs meant that the horses could inflict far worse injuries. They were in effect fitted with knuckle-dusters.

The process of heating the horseshoe before shoeing the horse, hot-shoeing, became common in the sixteenth century.

Over the centuries the horseshoe has acquired a symbolic significance. The nursery rhyme *For Want of a Nail* portrays the simple horseshoe and its nails as key to success; without them, kingdoms could be lost.

> *For want of a nail the shoe was lost.*
> *For want of a shoe the horse was lost.*
> *For want of a horse the rider was lost.*
> *For want of a rider the battle was lost.*
> *For want of a battle the kingdom was lost.*
> *And all for want of a horseshoe nail.*

The horseshoe has also been cast as a symbol of good luck. In Britain the superstitious mount a cast horseshoe open end up, so that it collects good fortune; in mainland Europe it should be mounted open end down, so that good fortune pours out.

There is no trace of the name of the original inventor of the horseshoe. In 1892, a US patent was granted for a horseshoe to Oscar E. Brown of Buffalo, but his invention was a double shoe. This consisted of a normal, upper, shoe with a lower shoe attached to it. The idea was that the lower shoe would take all the wear, and when it needed replacing it could simply be unlocked and removed so that a new lower shoe could be locked onto the permanent upper shoe. It would make re-shoeing easier and quicker. Oscar E. Brown is frequently listed as the inventor of the horseshoe, which he clearly was not. There were in any case thirty-nine earlier patents for double horseshoes, the first of them issued to J. B. Kendall of Boston in 1861.

THE CREATION OF THE CHRISTIAN CALENDAR

(525)

IN ROME IN the sixth century AD there lived a monk called Dionysius Exiguus, an impressive-sounding name that literally means 'Dennis the Small' or 'Little Dennis'. Dionysius Exiguus was a Roman scholar, theologian and mathematician. He originally came from Scythia, in the east, but he lived and worked in Rome and died there some time before AD 550. He is sometimes described as an abbot, but more usually as a monk. He had a great reputation as a theologian, was well versed in canon law and was an accomplished mathematician and astronomer. Among other scholarly activities, he translated many Greek works that have subsequently been lost, including the Lives of several saints.

If those were all the things that Dionysius did, it is doubtful whether we would remember him now at all, but he did one thing that changed the world in a very particular and all-pervading way. It was Dionysius who created the modern calendar, in use today all round the world. The calendar hinged on his decision that Jesus was born on 25 December in Year 743 of the Roman calendar, in other words 743 years after the legendary date of the founding of Rome.

The first year of the life of Jesus in the calendar of Dionysius was 'the first year of Our Lord', Anno Domini 1. Christian communities have been content to use this numbering system, and also the corresponding negative calendar, in numbered years Before Christ (BC). There was no Year Nought: AD 1 followed 1 BC. Non-Christian communities, Jewish communities in particular, have been less than comfortable with the religious labels, preferring the more non-committal 'Christian Era' (CE) and 'Before Christian Era' (BCE) instead. But regardless of religious beliefs, the world seems to have accepted the numbering scheme, counting forwards and backwards from the time when Jesus was born. We all take it for granted, whatever we believe about Jesus.

Dionysius was almost certainly wrong about the year when Jesus was born. The canonical Gospel accounts are not consistent with one another in terms of the historical detail they give. The Roman census that is used in the Gospel accounts to explain why Mary and Joseph were in Bethlehem rather than Nazareth happened a few years after AD 1. King Herod the Great who is mentioned as being alive at the time of Jesus's birth actually died in 4 BC, a few years before AD 1. So those two pseudo-historical elements in the story of the Nativity of Christ are incompatible.

The arrival of the magi, the wise men from the East, guided by a star shortly after the birth of Jesus, has led people to look for an astronomical event that could connect up with this element in the story. It is possible to reconstruct the positions in the sky of the planets and the stars at any time in the past, and with the use of computers it is relatively easy to find those moments when there were perhaps two or three bright heavenly bodies close together in the same part of the night sky, giving the illusion of a very bright new star.

In the year 7 BC the planets Saturn and Jupiter were very close together in the constellation of Pisces (The Fish). The conjunction gave the appearance of a bright new star. Interestingly, Jewish astrologers at Sippar in Babylon had prophesied the arrival of a long-awaited Messiah at a time when Saturn and Jupiter met, and this in turn would explain why magi, or astrologers from Babylon, were roaming about looking for the Messiah. And it would have been in 7 BC that they were searching.

Some astronomers have found a different solution to the puzzle. They looked at Korean and Chinese annals which describe an exploding star or supernova that blazed in the skies for seventy days in the spring of 5 BC. This too could have been interpreted as a sign by the astrologers in Babylon that something momentous was happening in the world, and sent them out to find it. Either way, the astronomical events would tie in with Jesus being born in 5, 6 or 7 BC.

If Jesus was really born in 7 BC, that is the year that should have been Year 1 of the Christian Era. It means that the entire modern calendar is seven years out. The Millennium was celebrated seven years too late, and any superstitious numerology attached to particular dates has no meaning whatever. The year when I am writing this is not really year 2007 of the Christian Era but perhaps 2012, 2013 or 2014, or maybe 2750 in the old Roman calendar. The date turns out to be no more than a man-made number – and a number resulting from Little Dennis's big mistake.

THE INVENTION OF
THE WINDMILL

(600)

THE EARLIEST KNOWN use of wind power was the sailing boat. From the earliest times, people understood that wind caught in a sail could be made to push a boat along. The application of this principle to windmills came much later. The first known windmills were built in Persia (Iran) in about 600. It is believed that they were built to pump water. It is not known what they looked like, because there are no surviving plans or drawings. But they were described. They had vertically mounted sails that were made not out of cloth but out of bundles of reeds or wood. These were attached to a central, vertical shaft by horizontal struts.

The first documented windmill was built to grind grain into flour. In this design, the millstone was attached to the vertical shaft. The milling equipment was housed inside the building, so that it would be out of the wind. The same arrangement was maintained in windmills in Europe right through to the early twentieth century.

Windmills are recorded in China in 1219, and some people believe that windmills were being built by the Chinese long before 600. Windmills first made their appearance in Europe in 1300, and it was not long before the Dutch started to adapt the design, to improve it. One change they made was to mount the post mill on top of a tower that was several storeys high. This was to exploit the higher wind speeds that are usually experienced well above ground level, where frictional drag slows the wind down. The different floors inside the tower mill were then assigned different functions, such as storing the grain, separating the chaff, milling the grain, storing the flour. Usually the ground floor was the dwelling for the windsmith and his family. The tower and the post mill mounted on top of it had to be moved manually when the wind changed, so that the sails were always facing directly into the wind.

When the Dutch installed dykes and drains to reclaim land, windmills were used to pump water from the drains out into canals and rivers. As the reclamation projects became more ambitious, the number of windmills multiplied.

The windmill was extremely versatile as a converter of energy. It could be used to pump water or mill grain. It could also be used to saw timber, drive hammers for a forge and do all kinds of commodity processing.

Windmills were still in widespread use in America and Europe in 1900, but there was sharp competition from steam power, and they quickly went out of use in the early decades of the twentieth century. With hindsight, given current concerns about burning fossil fuels, it may be that we gave up windmills just when we ought to have been building more of them. Perhaps they will return.

II
THE MEDIEVAL AND RENAISSANCE WORLD

THE INVENTION OF
GUNPOWDER

(900)

IT IS THOUGHT that it was in about AD 900 (though possibly earlier) that Chinese alchemists accidentally invented gunpowder. They mixed together potassium nitrate, charcoal and sulphur and when they put a flame to this concoction they created a bang, an expansion of gases and a cloud of white smoke. The first use the Chinese made of this accidental invention was for signalling and to make fireworks. They found that by igniting the gunpowder inside a container the rapid expansion of gases could be made to throw an object a considerable distance, though the Chinese saw no application for this property.

Things changed when gunpowder found its way to Europe in the thirteenth century. The European mindset was very different from the Chinese, and gunpowder was quickly harnessed to war technology in the manufacture of siege cannons and bombs.

The Mongol forces used gunpowder-fuelled firearms to conquer much of eastern Europe in 1237. The Mongols, led by Khan Ogadai, Batu Khan and General Subutai laid waste to Poland. In 1240 they conquered Kiev, sweeping through Ukraine and southern Russia. In 1241 the Mongols of the Golden Horde won the Battle of Liegnitz in Silesia. They were unstoppable. Such was the military superiority of the Mongols, with their gunpowder-fuelled firearms, that they could have gone on to conquer western Europe too. The accident of Ogadai's death caused the invasion to implode and the Mongol horde retreated back into central Asia.

Until then gunpowder had not been known in Europe. Military weaponry was based on swords, halberds, axes and arrows – virtually a prehistoric armoury. There were no European firearms. The brush with oriental culture changed all that. The thirty-five-year-old Franciscan Roger Bacon made the earliest known European reference to gunpowder in a letter written in 1249. Bacon had discovered how to make gunpowder. He discovered some of its properties, but did not realize that it could be used as a propellant.

The transfer of knowledge of gunpowder from Asia to Europe started a slow revolution in the nature of European warfare. From the mid-thirteenth century on, gunpowder came to dominate warfare. From 1250, gunpowder was used to shoot projectiles. By the sixteenth century, gunpowder had made castles obsolete, because it could fairly easily undermine walls, no matter how thick they were.

The original hand-ground dry mixture, called serpentine, was dangerous stuff. It could fail to ignite properly, or explode without warning. There was also a problem with the ingredients separating when shaken during transit, sulphur settling at the bottom and charcoal rising to the top. That meant that the powder had to be remixed when the battlefield was reached, creating clouds of poisonous and potentially explosive dust.

By the fifteenth century, a range of different mixtures had been tried out and techniques for making the material more stable too. When the ingredients were ground into a watery mud, there was less dust in the air and less likelihood of random explosion. The mud could then be dried in a sheet and broken up into granules or pellets that retained the right proportions of ingredients.

Gunpowder was always unsatisfactory as a weapons explosive. In spite of refinements, gunpowder always produced a sixty per cent solid discharge which tended to wear out the barrels of guns and cannons, and made them impossible to keep clean and safe. By the time gunpowder had been refined as near to perfection as it would ever be, it was being replaced by other propellants. Nitrocellulose propellants or 'gun cotton' were preferred because, apart from other properties, the explosion did not create a plume of white smoke that gave the shooter's position away to the enemy.

In the seventeenth century, gunpowder took on a major new use, in the mining industry. In 1696 it was used to blast away rock in a road-widening project in Switzerland. More and more such applications were found, until gunpowder was overtaken by dynamite and nitroglycerine, which became available at a competitive price in around 1870. Probably the biggest single project in which gunpowder was used was the destruction of Pot Rock, a navigation hazard in New York Harbour, in 1851–53.

Significantly, gunpowder is now used almost exclusively for signals and fireworks. The Chinese had, after all, long ago found the most appropriate use for gunpowder.

THE INVENTION OF
PAPER MONEY

(960)

PAPER MONEY WAS used locally and informally in China as early as the
seventh century and adopted on a national scale in the tenth century.
Some block-printed banknotes have survived from this early period, the
Song dynasty. Coins had holes in the middle so that they could be
strung together, but Chinese merchants were often so wealthy that
carrying their cash around became a serious problem. The coins were
left with a trustworthy agent in exchange for a slip of paper stating how
much money he had left with his agent. The promissory note was much
lighter and easier to carry. The formal issue of paper money, jiaozi,
developed from these notes.

Genghis Khan was intrigued by the paper money when he conquered
China, thought it was a good idea and adopted it for use throughout his
empire. The use of paper money spread south and westwards through
the Arab world, arriving in Europe very late. The Europeans did not
make or use paper, preferring parchment. The first paper mill in Europe
was not founded until 1151, by Moors in Spain. In spite of the obvious
advantages of paper, it was not widely accepted for a long time, mainly
because of religious prejudice: paper had been introduced by Muslims.
Only gradually did paper become accepted as merchants found it an easy
medium for recording contracts and promises of payment.

Paper money did not really get under way in Europe until the Spanish
siege of Leyden in 1574. During this severe siege, the Protestant residents
of Leyden took sheets of paper out of their hymn books and struck them
with the same dies that they had previously used to strike metal coins.
They became in effect paper coins. The first formal European banknotes
were issued in Sweden in 1660.

THE INVENTION OF THE MAGNETIC COMPASS

(1086)

VARIOUS VERSIONS ARE given of the invention of the compass. At one
time the Chinese were credited with inventing it as far back as 2634 BC.
The Emperor Hwang-ti was trying to attack an enemy enveloped in
thick fog and he built a 'chariot' for finding south, and this in turn
enabled him to find his enemy. But this account is purely legendary, and
in any case a south-finding chariot is not at all the same thing as a
magnetic compass.

It was much later, in AD 1086, that the magnetic compass was
invented in China. The waterworks engineer Shen Kua was the man
who was responsible for devising the compass. He wrote about his
invention, explaining that it was possible for magicians and geomancers
to find directions by rubbing an iron needle on a lodestone and allowing
the magnetized needle to hang freely on a thread. The needle, he said,
pointed south, which is a point of view; curiously, and no more
objectively, in Europe compasses have always been regarded as pointing
north. In reality they simply align with the Earth's magnetic field and
therefore point both north and south.

The peculiar characteristics of lodestones and their ability to
magnetize needles had been known about for a long time. The lodestone was actually mentioned in a Chinese dictionary dating from AD
121: 'a stone with which an attraction can be given to a needle.'

The invention soon found practical applications. By 1150 Chinese
caravan masters were using Shen Kua's compass to find their way across

the steppes and deserts of Central Asia, and sea captains were using it to navigate in open water. The first mention of a magnetic compass on a Chinese ship was in 1297. The simple compass of Shen Kua meant that sailors knew in which direction they were sailing, even when they were out of sight of land. By a method called 'dead reckoning', they could calculate where they were in relation to the home port; this involved using the compass direction combined with the ship's speed and the number of hours' sailing. The direction could be drawn as a line on a map. The distance along that line could be marked to give the ship's exact position.

The Chinese navigators regularly sailed south to trade with the islanders of the East Indies, and occasionally sailed west from there into the Indian Ocean. At some point in the middle ages, the compass was passed on to the Arabs, and they in turn passed it on to the Europeans. In a treatise on lodestones by Albertus Magnus, the north and south poles are referred to as Zoron and Aphron, which are names of Arabic origin, so it is probable that the earliest commentaries on the properties of lodestones and compasses were also of Arabic origin. It would be likely, given the Arabs' great reputation as mariners, that they acquired the compass at an early date.

The new technology seems to have spread across Eurasia very fast, as one would expect from its practical usefulness. As early as 1218, Cardinal Jacques de Vitry, a bishop in Palestine, spoke of the magnetic needle as 'most necessary for such as sail the sea.'

In 1498, Vasco da Gama was shown a map of the entire coastline of India. The bearings were 'laid down after the manner of the Moors, with meridians and parallels [lines of longitude and latitude] very close together.' Da Gama was left in no doubt that this fine cartography was the work of Arabs and that a compass had been used to fix the cardinal directions: north, south, east and west.

A different form of compass was in use in the East Indies in the sixteenth century. Instead of a needle suspended from a thread, it was described as 'a sort of fish made out of hollow iron which, when thrown into the water, swims upon the surface and points out the north and south with its head and tail.' A similar model using a needle mounted on a wooden float, so that it could rotate freely on the surface of water, was in use in Europe in the thirteenth century. The first detailed description of a medieval compass was written in 1269 by Peregrinus, who describes an improved floating compass with a circle marked out with ninety degrees in each quadrant and fitted with movable sights for taking bearings.

In 1558, Queen Elizabeth I's magician, mathematician and geographer, Dr John Dee, succeeded Robert Recorde as technical adviser to the Muscovy Company. The ever-ingenious Dr Dee devised an improved magnetic compass to assist with navigation along the trade route between England and Russia. In the great age of discovery and exploration, no ship could afford to be without the most accurate and up-to-date compass possible.

The great importance of the magnetic compass as an invention was immediately recognized by everyone who used it. It is clear that it made the voyage of exploration more purposeful, because it became possible to locate islands or shorelines on a map with a fair degree of accuracy. Filling in the map of the world could begin. People travelled far and wide without the compass, but once they were equipped with the new technology a progressive, cumulative world map could be created fairly quickly and efficiently.

THE INVENTION OF ANAESTHETICS

(1236)

IN ANTIQUITY, PEOPLE probably learned to tolerate the levels of pain caused by conditions such as arthritis. They also had to undergo surgery intermittently to repair broken bones; 4,000 or 5,000 years ago people were, it seems, injured while felling trees, fighting and building megalithic monuments and pyramids. From very early times trepanning was carried out. This entailed sawing out a disc of skull, possibly to release evil spirits. The pain involved must have been severe and it is not known how patients endured it. Many recovered, though some died shortly after the traumatic surgery.

It is possible that surgery was carried out using a primitive anaesthetic technique that was in use in eighteenth-century Italy. It was described as a technique used for castrating boy singers but it could have been used for other operations equally well. The boy's jugular vein was pressed until he fell into a semi-coma. In that state the operation could be performed 'with scarce any pain to the patient'.

There were also herbs that could have been used as painkillers. Seeds of a known hallucinogen, black henbane, were found in late neolithic pots in Scotland. The tattoos on the Iceman coincide with acupuncture points, in particular with master points against pain and rheumatoid arthritis. This suggests that acupuncture was used to relieve the pain caused by this widespread neolithic ailment.

Throughout antiquity, the search for a cure for pain continued. During surgery in particular there was a need to find ways of reducing pain.

Theodoric of Lucca, a Dominican friar, experimented with anaesthetics in 1236. Theodoric was a teacher at Bologna and also the son of a surgeon to the Crusaders. Theodoric recommended using sponges soaked in narcotics applied to the patient's nose. This induced a deep sleep, during which surgery could be carried out without the patient experiencing any pain.

Theodoric favoured using mandragora and opium, both known as opiates, for his pioneering anaesthetic. The idea of using anaesthetics came and went repeatedly through history. Some technologies and techniques were progressively built on, while others were devised, forgotten, and then later reinvented. It is perhaps surprising that something as basic as pain relief should be forgotten and require reinvention.

For hundreds of years after Theodoric's invention, surgeons went on operating on conscious patients, who sometimes had to endure terrible pain. After Joseph Priestley discovered oxygen and nitrous oxide in 1772, there was continuing research into these and other chemical agents – and this research eventually led to the rediscovery of anaesthetics over the course of the next eighty years. This process of rediscovery was, as we shall see later, surprisingly slow, and it was the result of work by several independent researchers: Humphry Davy and Michael Faraday in England and Crawford Long and William Morton in America.

When, as late as 1846, William Morton gave a public demonstration of the use of ether to induce unconsciousness, he administered it in exactly the way that Theodoric had recommended, more than 600 years earlier, by the application of a drug-saturated sponge or cloth to the patient's nose. It was a case of knowledge found, lost and then found again.

THE INVENTION OF
SPECTACLES

(1265)

IN ANTIQUITY, POOR eyesight or eyesight that deteriorated with age was something people had to live with. When wealthy and literate Romans 2,000 years ago reached a point where their eyesight was so poor that they could no longer read, they had slaves who read aloud to them. Magnifying glasses existed and it seems that craftsmen used them to enable them to execute very finely detailed work. Indeed, it looks as if crystal magnifiers were used even by Minoan craftsmen in the bronze age, in 1600 BC.

Making pairs of relatively low-powered magnifying glasses in frames that people could wear did not happen until much later. The invention came in about 1265. The Oxford scientist Roger Bacon wrote in 1268 that it was possible to examine 'letters or minute objects through the medium of crystal or glass' in order to magnify them. This reference suggests that he had yet to come across spectacles as such – in England. But an Italian called Sandro di Popozo wrote in the following year, 'I am so debilitated by age that, without the glasses known as spectacles, I would no longer be able to read or write'. He added usefully, 'These have recently been invented for the benefit of poor old people whose sight has become weak'.

The name of the man who invented spectacles is not known, though there is a reference to him in a sermon delivered in Pisa in 1306. 'It is scarcely twenty years since the art of making spectacles, one of the most useful arts on earth, was discovered.' The preacher continued, 'I myself have seen and conversed with the man who first made them.'

The problem with spectacles was to find a way of keeping them on. From early on the two lenses were mounted in a wire frame that rested

117

on the bridge of the nose, but there was a difficulty in fastening the frame to the face at each end. There were loops like shoelaces that hooked onto the ears. It was not until 1730 that a London optician called Edward Scarlett designed rigid wire side-arms that rested on and hooked over the ears.

Refinements in lenses also came in. It was possible to make glasses for reading and glasses for distance. Benjamin Franklin tired of having to use two different pairs of spectacles, so he invented bifocal glasses. 'I formerly had two pairs of spectacles, which I shifted occasionally, as in travelling I sometimes read and often wanted to regard the prospect. Finding this change troublesome, I had the glasses cut in half and a half of each kind associated in the same circle. By this means I have only to move my eyes up or down, as I want to see distantly far or near, the proper glasses being always ready.'

We take spectacles for granted, but in past centuries a person's working life, especially a craftsman's, might be ended prematurely if his eyes failed. The great blossoming of artistic achievement in the Renaissance was made possible by extending artists' creative lives by giving them spectacles. Today, the ordinary pleasures of our everyday lives are greatly enhanced by being able to see everything more clearly.

THE INVENTION OF
THE CLOCK

(1280)

FROM THE PREHISTORIC period on, people have devised ways of telling the time. The megalithic monuments raised in neolithic Europe sometimes had built into them alignments on sunrise and sunset positions on key days in the year, notably midsummer and midwinter. On a shorter timescale, people have known since antiquity that when the sun reaches its highest point in the sky it is noon, the middle of the day.

Later, candles were used to mark off hours. In still air, a candle burns down at a regular rate, so it was a simple matter to mark the sides of candles with hour marks. An hourglass could in a similar way be used to note the passing of an hour. The sundial too was widely used in ancient times.

By the ninth century a mechanical clock had been invented that was only short of an escapement mechanism. Once that refinement was added, in 1280, the modern clockwork clock was in existence. Unfortunately no actual clocks have survived from the middle ages, but there are enough references to them in church records to confirm that they existed in churches and to give us some idea of their mechanisms.

Necessity was the mother of this invention. All the religious communities in Europe, whether great abbeys or minor priories and nunneries, required their members to pray, work and eat at strictly regulated times. This could only be achieved by using a variety of means of time-telling and recording methods. Water clocks, sundials, hourglasses and marked candles were all in use, and they may well have been used in combination, perhaps with someone checking by using a second time-keeping device. When an important time was reached, such as the hour of a service, a bell was rung, at first by hand, later by a mechanism attached

119

to a clock. This is where the idea of a chiming clock originated.

In many records a general term, the Greek word *horologia*, was used for time-keeping devices, so it is often hard to tell what was going on. The early mechanical clocks may not have had faces with hands, or even a single hand, but were fitted with sound effects like bells to indicate the time at intervals. When the word clock, which comes from the French *cloche*, was used in the records, it is certain that the mechanical clock with bells fitted was intended. These clocks appeared in the thirteenth century.

There was an increase in the number of mentions of clocks in the records from 1285 onwards, so it is likely that in 1280 a new type of clock mechanism came into existence. The early clocks were unreliable and it was only when a regularly oscillating regulator called the escapement was invented that the fully reliable mechanical clock came into being.

When clock faces were introduced the symbolic value of the clock was seen by the clergy. The round disc of the clock face could be the universe, the world, or the wheel of fortune. Decoration accompanied these symbolic interpretations, and some of the church clocks became major set pieces. Sometimes, on the hour, an elaborate procession of mechanical figures, automata, would emerge from the clock face on one side and parade briefly before disappearing on the other side of the clock face. These too had their symbolic and educational value, reminding people of the shortness of their lives.

In 1283, Dunstable Priory had a big new clock installed. It was set up above the rood screen inside the church, and therefore cannot have been a water clock; it must have been one of the new mechanical clocks with an escapement mechanism. In 1292, 'a great horloge' was installed at Canterbury Cathedral. In 1322, a brand new clock was set up at Norwich Cathedral, which seems a great extravagance given that a new clock had been installed in only 1273; that too suggests that a major technical improvement had come about in the meantime. A new generation of clocks had come into being.

So far, no one has been able to find out where the first of the new mechanical clocks was built, or who invented it.

THE INVENTION OF THE WATERMARK

(1282)

A WATERMARK IS a faint mark that becomes visible in a sheet of paper when it is held up to the light. It shows up lighter than the rest of the paper, because the paper is thinner there.

The earliest watermark was created by accident at the Fabriano paper mill in Italy, where paper was manufactured from 1260 onwards. A mould was used to press the water out of the wet paper, and the mould had a small piece of wire sticking out of it. The finished paper was thinner where it had crossed the wire, and the distinctive mark of the wire was visible when the paper was held up to the light. It meant that anyone scrutinizing that sheet of paper would know that it had come from that mould, and therefore could only have come from the Fabriano mill.

From this it was a short step to making a mark deliberately, to a chosen design, by inserting a pattern made out of a length of wire and attaching it to the base of the mould. In 1282 the first deliberate watermark was made by Fabriano, in the form of a cross. It was a subtle and discreet way of marking merchandise, and a producer of high-quality paper would want customers to know where it came from. It was the discreetest form of advertising.

Today, much the same process is used for making watermarks. The wet paper is squeezed by a roller called a dandy roll. The raised design that creates the watermark is either soldered or sewn onto the dandy roll.

Watermarks have been used for 700 years to identify the manufacturers of fine stationery. Foolscap paper got its name in the early eighteenth century from the watermark of a fool's cap, used on paper thirteen and a half inches wide and seventeen inches long. Watermarks have proved invaluable in recent decades in identifying forgeries. Even the most expert forgers have overlooked detail such as watermarks. Authentic pages from the Gutenberg Bible would have to display watermarks from Caselle in northern Italy, which was one of the most important centres for paper manufacturing in the fifteenth century. About seventy per cent of the paper in the Gutenberg Bible has the watermark of an ox's head with a star over it, twenty per cent has a bunch of grapes, ten per cent has a walking ox. If a page alleged to be from the Gutenberg Bible comes onto the antiques market with some other watermark, it cannot be genuine.

THE INVENTION OF THE MOVABLE TYPE PRINTING PRESS

(1440)

THE PRINTING PRESS allowed the multiple production of printed pages, and therefore the easy production of many copies of a book. For the first time in history it became possible for ideas and information to be communicated to the masses, or at any rate to those of the masses who could read. The increasing availability of multiple copies of books made it more worthwhile for more people to learn to read, so the invention of the printing press actually stimulated the growth of literacy. The Gutenberg Bible was the great landmark, but many people decided to learn to read in order to gain access to an increasing range of books, from romantic fiction to scientific and philosophical works.

The great breakthrough that Gutenberg made was the use of movable type. It was an invention that had in a sense already been made, by the Chinese. In 1041, the Chinese Bi Sheng devised movable type printing, but for Chinese characters. Gutenberg can only be credited with inventing the modern Western system of movable alphabet type. It is doubtful whether Gutenberg knew about the Chinese movable type; this looks like a case of the same invention being made twice over.

The movable type technique involved the carving of cubes of wood with raised letters on one side. The cubes were then arranged in order inside a frame, inked, and then a sheet of paper was pressed against the blocks. When the paper was peeled away, the imprint of the letters was left on it. Instead of a scribe taking a year to write out a copy of a book, a printer could run off a copy of a book in an afternoon.

There was a problem with the woodblocks, though. With many inkings and much use, they began to disintegrate and had to be replaced. What Gutenberg did was to devise a simple method for casting the blocks in metal; this metal type had a much longer life and there was no falling away in the quality of the printed copies. Gutenberg's invention was so good that it remained the standard method for printing right into the twentieth century.

Johannes Gutenberg was born in Mainz in 1400, probably beginning work as a printer in Strasbourg by 1439. By 1448 he was back in Mainz again and in 1450 he was working in partnership there with Johannes Fust, who financed Gutenberg's press with 800 guilders. It was not a happy partnership. It came to an end after five years with Fust bringing a suit against Gutenberg for the 800 guilders and receiving the printing equipment in lieu of payment. During this time, Gutenberg must have prepared the plates for his great project, the first printed Latin Bible.

Johannes Fust then continued the printing business with Peter Schoffer, his son-in-law, and the two of them completed the famous Forty-two Line or Gutenberg Bible. This historic feat was accomplished in August 1456. It involved the printing of 1,282 pages in two columns with gaps left for hand-painted illuminated initials. Gutenberg himself meanwhile set another press in Mainz in partnership with Konrad Humery.

Gutenberg is often credited with inventing printing, but the printing process already existed in a rudimentary form. What Gutenberg did was to refine the technique and use it for a highly significant landmark project. The Bible is a very long book, one that tested the new technique to its limit. Gutenberg remains the best claimant to the title 'inventor of printing', even though his contribution was really one step in the development of the process.

Until Gutenberg's time, bibles were hand-made, scarce and expensive. Producing whole pages at a time on a press made bibles cheaper and therefore more accessible. At the same time, the age of mass production was some way off. Printers usually worked on copies of only five books in a typical year. Gutenberg's activity led to an explosion of literacy, the democratization of Christianity, the reduction of the power of the priests who recited and interpreted the Bible for the rest of the Christian community, and the consequent reduction of the power of the Church itself. The publication of the Gutenberg Bible led by a short route directly to the Reformation and the Enlightenment.

Another significant development was the creation of many different styles of type. The most important of these in the sixteenth century was the one called 'Roman', which lent itself to the particular qualities of steel type, and it replaced the older 'Gothic' styles throughout Europe. The availability of other typefaces was something printers revelled in, and they often used several different types on their title pages, which developed into extravagant displays of their craft. This availability of many styles was re-explored in the late twentieth century, with the advent of word processors with a choice of scores of different fonts.

THE INVENTION OF
THE RAIN GAUGE

(1441)

THE RAIN GAUGE was invented in 1441 by a Korean scientist called Jang Yeong Sil.

There had been ad hoc measurements of rainfall by the ancient Greeks in 500 BC and in India in 400 BC, but it appears that Jang Yeong Sil's rain gauge was the first attempt at a scientific instrument that would gather data as part of a systematic study. Jang Yeong Sil devised his gauge as part of an initiative by King Sejong to measure the rainfall pattern across the whole of his country. It was a far-sighted initiative, several centuries ahead of comparable undertakings in the Europe of the Enlightenment. In Britain, Christopher Wren would not invent his tipping-bucket rain gauge until 1662. King Sejong's intention was to build a picture of weather patterns across his country with a view to planning agriculture better and making adjustments to tax demands by his government officials.

A rain gauge is an instrument for collecting and measuring the amount of rain that falls at a particular place, usually measured every twenty-fours. As some precipitation falls as snow, sleet or hail, that too is collected, melted and added to the rainfall. Technically the device should be called a precipitation gauge, but it would be pedantic to do so. The modern gauges have graduated sides, marked in either inches (and tenths) or millimetres.

Rain gauges are simple, low-technology devices which function very well. Problems sometimes arise because the geographical distribution of rainfall is often uneven. Rain may fall in one part of a town, and not in another. The official weather station for Seattle is at the airport, but that is the driest area in the city and the gauge was by chance positioned at the driest site at the airport. This means that official figures are incorrect, giving Seattle 250 mm per year less rain than actually falls on the city. The problem is that it would not be practicable to put rain gauges everywhere.

Measuring rain systematically in many locations has enabled geographers and climatologists to identify significant spatial patterns, such as rain shadow effects. Taking the measurements over a long period enables them to identify significant pattern through time. By chance finding a continuous run of rainfall measurements in the English Midlands going back over 120 years enabled me to analyse patterns during that period. It turned out that there was an eleven-year cycle that closely followed the sunspot cycle. When there was more sunspot activity, there was more rainfall. Could small variations in solar weather have a perceptible effect on our terrestrial weather pattern? This is the sort of enquiry that becomes possible when we gather data systematically over a wide area and over a long period; we begin to see the wood instead of the trees.

THE INVENTION OF THE PADDLE WHEEL

(1450)

WE TEND TO associate the paddle wheel with the steamers of the Industrial Revolution, with ships like Brunel's *Great Western*, but the paddle wheel has been around for rather longer than that. The medieval Harleian Manuscripts contain a book of Italian sketches drawn in the fifteenth century. One of the sketches includes a man-powered paddle boat. Horse-powered paddle boats were also tried. At Barcelona in 1543, Blasco Garay fitted paddle wheels to a vessel, one on each side. The paddles were powered by forty men. The experiment was not followed up.

The experiment with paddle wheels was repeated from time to time without being taken any further forward. The Comte d'Auxiron and M. Perrier experimented with a vessel powered by paddle wheels in 1774, and the Comte de Jouffroy made another attempt in 1783. In Britain Thomas Savery tried too. Then in 1787 Patrick Miller of Dalswinton invented a double-hulled boat with paddles on each side. His paddles were worked by a group of men turning a capstan, which powered the paddles on each side. The men providing the power quickly became exhausted with this work. Patrick Miller mentioned his experiment to William Symington, who was exhibiting his road locomotive in Edinburgh at the time. Symington's answer was inevitable. 'Why not use steam power?' And the resulting combination of steam power and paddle wheels was to prove a major success through the nineteenth century. In the end, the screw won over the paddle wheel in terms of efficiency, but the paddle made a major contribution to marine transport through ships like the *Great Western* and *Great Eastern*.

Even after the contest with the screw propeller had been lost, the paddle steamer went on being used for its charm and picturesqueness. In 1835 Casimi Friedrich Knorr set up a steamboat company to provide pleasure steamer trips on Lake Lucerne in Switzerland. His first steamer, the *Stadt Luzern*, went into service on the lake in 1837. The present generation of five paddle steamers (called *Uri, Unterwalden, Schiller, Gallia* and *Stadt Luzern [III]*) is now 100 years old, but still operating perfectly well. The *Schiller* (built in 1906) was completely refitted and restored to its original state as a historic monument in 2000. These incomparably elegant white paddle steamers that still ply daily on the lake are as much a part of the Swiss summer tourist scene as Mount Pilatus, Tellskapelle, Triebschen and Lake Lucerne itself.

THE INVENTION OF
THE ARQUEBUS

(1475)

THE RIFLE IS one of those devices with a gradual evolution, so that it is hard to put a specific date to its invention, and impossible to say who invented it. Semi-portable long-barrelled guns were invented in the fifteenth century in Europe. This early firearm was called an arquebus or bow-gun. It was fired by a trigger, which let loose a spring, which in turn threw a match forwards into a pan holding the priming. A small explosion sent a bullet along the barrel. This new weapon, which was really the ancestor of the musket and the rifle, was invented in about 1475.

The arquebus was used at the Battle of Morat, fought in Switzerland between the army of the Duke of Burgundy and the Swiss Confederates in June 1476; a contemporary account of the battle mentions it specifically. It was in use in England in 1480.

The problem with the arquebus was that it was hard to aim. The butt or handle was dead straight, and supported against the soldier's chest to hold it steady as the gun was fired. From that position, it was impossible for the soldier to squint along the barrel at his target. It was also common for the powerful recoil when the gun was fired to knock the soldier over, while the chances of hitting the enemy were small. Overall, the initial design for the arquebus seems to have been as a weapon of self-harm.

German engineers saw that the solution was to bend the butt downwards and slope it so that it could be supported on the shoulder. This meant that the barrel could be raised to eye level; accurate aiming became possible. This was a major invention in itself, one that was adopted for a whole succession of weapons, including the modern rifle. The arquebus fitted with this bent butt was in use in England in the reign of Henry VIII. At the same time a smaller version of the arquebus, and known as a demi-haque, was manufactured; this seems to have been the forerunner of the later pistol.

The arquebus was developed into the musket. This was invented in Spain, was quickly adopted in France and then in England. Like the arquebus, the musket was semi-portable, but with its five-feet long barrel it was so heavy that each musketeer needed to have a boy with him to help him carry it. Further work refined the weapon so that it became light enough for the musketeer to manage it single-handed; but it was still sufficiently heavy to need to be supported on a rest while firing. This rest consisted of a shoulder-height stick with an iron fork at the top to support the musket barrel. In addition to his musket and rest, a musketeer also had to carry a flask of powder, a bag of bullets and a burning match. He also had to be ready to drop everything and defend himself with his sword if his shot missed.

The musket was a clumsy piece of weaponry. One of the many problems was that the light of the burning match gave the musketeer's position away to the enemy, and the imminence of firing; there was little possibility of surprising an enemy in these circumstances. In 1615, some advice for musketeers (*L'Art Militaire pour l'Infantrie*) was published by Captain Walhuysen of Danzig. 'It is necessary that every musketeer should know how to carry his match dry in moist or rainy weather, that is, in his pocket or in his hat, by putting the lighted match between his head and his hat, or by some other means guard it from the weather. The musketeer should also have a little tin tube, about a foot long, big enough to admit a match, and pierced full of little holes, that he may not be discovered by his match when he stands sentinel.'

Despite all its drawbacks, the musket saw a great deal of use in battle, including the English Civil War and the Thirty Years War. Europeans also took muskets with them and used them to overcome any resistance in their colonization of the New World.

Lighting the slow-match was an annoyance to musketeers, especially in damp weather. What was needed was a means of starting a fire inside the gun itself. In 1515, the first wheel lock gun was invented. The wheel lock was a small wheel which, when spun with a finger, made sparks by rubbing on a flint. The principle was the same as in a modern cigarette lighter. The sparks fell on the touch hole and ignited the powder. It was an excellent solution to the problem, but unfortunately relatively expensive, so the infantrymen in most armies continued for a long time to be equipped with the cheaper matchlocks. Only the wealthy could afford wheel locks.

The flintlock was a further ignition invention. Developed in the early seventeenth century, it combined the good features of the matchlock and wheel lock. As the trigger was pulled, the pan cover automatically drew back to expose the powder. A flint mounted on the tip of the cock struck against steel to make a spark. The flintlock gun was a step forward, but it was still fairly inaccurate, and individually it could not be relied on to fire. Flintlock guns and other muskets depended on sheer numbers for their effect in battle.

The Americans developed their own version of the flintlock gun, which became known as the Kentucky rifle. This was used for hunting. It was also used by the white colonists against the Native American communities.

Technically a rifle in the strict sense has a rifled bore, that is to say its barrel has fine grooves cut into it in spiral form in order to give a spinning motion to the bullet as it is fired. This rifling makes the bullet fly straighter and therefore it also makes the weapon more accurate. The principle of rifling was also applied to big artillery guns for the same reason.

Rifles first went into widespread practical use in America. The main drawback of the rifle was that it was expensive to manufacture and it had a slow rate of fire; for those reasons, infantrymen in European armies were regularly equipped with muskets until about 1850. By the beginning of the nineteenth century, firearms were greatly improved by having a percussion-cap method for igniting the powder. The percussion-cap was a little metal capsule that exploded when it was struck and so fired the gun instantly; this very quickly replaced the flintlock. There was another major improvement in the 1840s, which was the development of gas-expanding bullets.

In 1855, the Americans adopted a new hybrid weapon, the rifled musket. This looked like a musket, used gas-expanding bullets, had a rifled barrel, was fired by percussion-caps and was muzzle-loaded. One of the key differences between the musket and the rifle in the nineteenth century was that the musket was muzzle-loaded, that is, loaded at the front end of the barrel, whereas the rifle was breech-loaded, loaded at the rear end. The new rifled musket was in use on both sides in the American Civil War. From then on most weapons had rifled barrels, with the exception of the shotgun, a smooth-bore weapon intended for short-range firing of slugs or small shot. Shotguns, whether single-barrelled or double-barrelled, were and are used principally for hunting.

THE INVENTION OF
THE GLOBE

(BEFORE 1490–92)

THE OLDEST SURVIVING globe representing the Earth was made by the map-maker Martin Behaim in 1490–92.

Martin Behaim was born in Nuremberg in 1459. He moved in 1484 to Portugal, which was at that time the repository of all the geographical knowledge there was to be had. Prince Henry the Navigator had made it his business to gather all the information he could, to assist in the voyages of exploration he was sponsoring. Behaim was given a warm welcome in Lisbon. King John appointed him to his Mathematical Council and he was able to join one or more expeditions along the West African coast, probably to Gambia. He was knighted by the king and lived for some years in the Azores.

Behaim returned to Nuremberg in 1490, where he was welcomed home as a distinguished scholar and traveller. It might have ended there, with an honourable retirement, but for the enthusiasm of one of Behaim's fans. George Holzschuher was a member of Nuremberg's City Council, and himself a famous traveller. He proposed to the Council that they should formally commission the construction of a globe from Martin Behaim, to incorporate all the latest discoveries. On the globe itself, inside the *terra incognita* of the Antarctic Circle, there is an inscription confirming that the work was undertaken on the authority of three distinguished citizens, Gabriel Nutzel, Paul Volckamer and Nikolaus Groland. But for this commission, the globe would probably not have been made, and we would not remember Martin Behaim's name at all.

Behaim collated all his knowledge and set about expressing it in one summative map, his Nuremberg Terrestrial Globe, or as he nicknamed it his Earth-apple. Specifically, he prepared a flat world map from which a hired painter would be able to fill in the vellum surface of the globe. The major responsibility of painting the globe was given to an artist called Glockenthon, and the painting took Glockenthon fifteen weeks. When newly finished, twenty inches in diameter, the globe must have looked spectacular in its bright colours and crammed with fine detail. It is in a poor and darkened state now, and quite a lot of the names on it have been corrupted by well-meaning restorers in the nineteenth century. But it has at least survived.

There are 1,100 place names, but also lots of blank spaces, which the artist has ingeniously filled in with heraldic devices, flags, pictures of elephants, ostriches, mermaids, saints, ships, kings seated on thrones, and missionaries converting natives.

What we most notice about the Behaim globe today is the absence of the New World. When Behaim and the city fathers of Nuremberg set about the project, they did not know about the existence of the New World; that revelation was about to burst upon Europe before the paint on the globe was dry. They did not know that the globe was short of three continents, two of which were to come to light shortly. We, with the benefit of hindsight, can see the irony. The fact that it was a globe that he was constructing also forced Behaim to face the crucial issue of the width of the ocean separating Europe and Asia. Much of the information incorporated on the Behaim globe was probably taken from a pre-existing globe, so Behaim cannot be credited with any great originality. But his globe acquired instant high status, perhaps because of his Portuguese credentials and his trip along the African coast. In spite of the absence of the Americas, later map-makers followed his outlines fairly slavishly, trying to force new discoveries into the geographical framework that he had created.

Behaim's flat *Map of the World*, his preliminary drawing, was later put on display in the clerk's office in the town hall. The globe summed up not only everything he had been able to ascertain, but everything that was known – by anybody in Europe – as the middle ages came to an end. In a sense that era ended with the Columbus voyage, and that came just as Martin Behaim completed his globe.

The truth is that there were earlier globes. Giovanni Campano, who was alive in 1260, was a mathematician at Novara. He wrote a *Treatise on Solid Spheres*, in which he described manufacturing globes out of wood or metal. In 1474, Toscanelli wrote a letter in which he said that a globe was the best way to demonstrate the distance between western Europe and eastern Asia. We also know that Columbus had a globe on board the *Santa Maria*. It may have been made for him by his brother Bartholomew, who was said at the time to be a maker of charts and globes. The Columbus globe had Cipangu (Japan) marked on it, and it seems it was Cipangu Columbus thought he was heading for as he crossed the Atlantic.

Maps had been in existence for a long time and some had even attempted to draw maps of the whole world, but there were almost insuperable problems in representing a sphere on a flat piece of paper. In the middle ages there were probably many uneducated people who thought of the Earth as flat, though that does not seem to have been a tenet of the Christian Church. Certainly some educated people in the middle ages thought of the Earth as a sphere. It is often forgotten, amongst all the grotesqueness and picturesqueness of *The Divine Comedy*, written in about 1300, that Dante Alighieri not only thought of the Earth as a sphere but assumed his readers shared that belief.

Thinking of the Earth as a sphere was no heresy. The Columbus voyage of 1492 was an explicit avowal of a belief that the Earth was spherical. If Columbus was sailing west in order to arrive at the Far East by a new route, he can only have thought of the Earth as round. The sponsors who let him go on that voyage must have shared his belief; they cannot have thought of the Earth as a flat disc, or they would have been sending Columbus to certain death by sailing off the edge of the world.

Martin Behaim's globe was a solid affirmation that the Earth was indeed round. The great advantage of the globe was that it did not distort the shapes of continents or oceans in the way that flat maps always did, whatever projection the map-maker used.

Globes were made to embellish and decorate the libraries of aristocrats all over Europe in the eighteenth and nineteenth centuries. Modern globes are often made to a scale of 1:40 million. Usually, globes are mounted at an angle. The axis leans to reflect the way in which the real Earth leans in space. The Earth's axis of rotation does not stand at ninety degrees to the plane of its orbit but at sixty-six and a half degrees. The inclination of the manufactured globes is therefore not decorative, but true, and it helps us to understand more easily phenomena such as the seasons. Most globes have lines of latitude and longitude printed on them. These are not real features of the Earth at all but man-made constructs. But having them drawn on the globe helps us to find places. Often locations are pin-pointed by giving their co-ordinates, an angle for latitude and another for longitude.

All of these globes are terrestrial globes, globes representing the Earth. As knowledge of the topography of other heavenly bodies has become better known, other globes have become possible. Globes have been manufactured of the Sun, the Moon and Mars.

Some very large terrestrial globes have been made, largely as curiosities or showpieces, as their size serves no particular purpose. The largest rotating globe in the world, at forty-one feet in diameter, is Eartha at the Delorme headquarters in Yarmouth, Maine. The Unisphere in Queen's, New York is 120 feet in diameter. This is the largest globe in the world, apart from the world itself. They shrink to insignificance beside Martin Behaim's globe, which defines an epoch; it shows us exactly how the world appeared to Europeans at the moment before the New World was discovered, the world we left behind as we left the middle ages.

THE INVENTION OF
THE GREGORIAN
CALENDAR

(1582)

FOR MANY CENTURIES, the Western world ran to the Julian calendar. In 1582, Pope Gregory XIII proclaimed a new calendar, which was from then on known as the Gregorian calendar. The Julian calendar reckoned a solar year, the length of time it took the earth to travel once on its orbit round the Sun, as 365.25 days. The quarter of a day was a minor problem, easily solved by having three years of 365 days followed by one year of 366 days, the leap year. The idea was to maintain an exact match between the calendar, in terms of days, months and years, and the solar year.

But there was a flaw in the Julian calendar, which was that the catching up by adding one day every fourth year did not quite bring the two into line. The result was that the man-made Julian calendar was falling gradually out of phase with the solar year. The result, if the Julian calendar was allowed to continue indefinitely, would be that the months would be out of phase with the seasons. The calendar dates for the seasons were retreating by one day every century. By the middle of the sixteenth century, the Julian calendar was ten days out of line with the seasons.

The solution, adopted in 1582, was to drop ten days. This brought the spring equinox back to 21 March and Easter also back to its proper position in the calendar. In 1582, the day after 4 October was not 5 October but 15 October. There was no 5–14 October that year, at least in the countries where the Pope's commands were obeyed. And here some new problems arose.

Most of the Catholic countries in Europe went along with the change as instructed. Hungary followed suit in 1587. But there were major difficulties as a result of the Reformation. Protestant régimes in northern Europe could not and would not acknowledge the overlordship of the Pope in any matter. The Protestant states of what is now Germany held out for 100 years, but decided in the end to adopt the calendar reform in 1700.

England under Henry VIII had similarly made a great point of having an autonomous Church. The English monarch was the head of the English Church, and so it was politically difficult for the English establishment to comply. The result was that a different calendar was in operation in England compared with that in mainland Europe. Eventually, the practical advantages of working to the same calendar prevailed, but it was not until 1752 that England, now Great Britain, changed to the Gregorian calendar. By then the calendar was eleven days out. Tsarist Russia carried on using the old calendar, mainly to assert religious independence of Rome, for much longer. It was not until the Russian Revolution of 1918 that Russia switched to the Gregorian calendar.

Apart from the ten-day displacement, the Gregorian calendar only differs from the Julian calendar in one detail. In the Gregorian calendar, no century year counts as a leap year unless it is exactly divisible by 400. The years 1600 and 2000 were therefore leap years, 1900 was not a leap year.

Years that are divisible by 4,000 are not counted as leap years. These minor exceptions are designed to keep the calendar accurately in line with the solar year for 20,000 years, at least to within one day.

It was a necessary step, but the great mass of people, uneducated, suspicious and superstitious as they were, opposed it bitterly. Some believed it was a swindle, an attempt to cheat them into paying a week and a half's rent for nothing. Some believed their lives had been shortened by the ten (or eleven) missing days. In fact, in calendar terms, their lives were being lengthened. In England, when Wednesday 2 September 1752 was followed by Thursday 14 September, there were widespread street riots. People chanted, 'Give us back our eleven days!'

THE INVENTION OF
THE KNITTING
MACHINE

(1589)

ACCORDING TO POPULAR folklore in Nottinghamshire, it was a nineteenth-century curate from Calverton who invented the knitting machine. He did it, it is said, in order to win the love of a young lady who was too preoccupied with her knitting to pay him any attention. The story is given credence by the coat of arms of the Worshipful Company of Framework-knitters, which shows the knitting frame in all its complicated glory in the middle, William Lee to the left and the object of his affections to the right.

Attractive though the story is, it seems to have no basis in reality. There is no evidence that William Lee was a cleric, or at Calverton in the nineteenth century.

The knitting frame was invented long before, in 1589, and its inventor was indeed called William Lee. Lee was first named as the inventor of the machine in a document formally setting up a partnership between him and George Brooke on 6 June 1600. The agreement was that Brooke was to supply the capital, £500, and Lee the technology. The first £200 of profits were to be Lee's, and thereafter the profits were to be divided equally between them for a period of twenty-two years. The perennial problem that inventors have had over the centuries is that capital is nearly always needed to turn an idea into concrete reality, and that organizing the pairing of invention and investment is often difficult to get right.

So it turned out for William Lee. In 1603 Lee lost his partner in the most spectacular way imaginable: Brooke was arrested for treason and executed. Suddenly Lee was without a backer. For the next ten years he went on trying to find a financial partner, with no success. Maybe Lee's knitting machine was perceived as a threat to vested interests, and it would certainly have shaken up London's weaving industry. Maybe Lee himself was lacking in the ability to persuade others of the worth of his invention. Maybe the machine was in some way mechanically suspect, perhaps seen as unlikely to stand up to the rigours of mass production. Advisers to the Crown were for some reason unimpressed, challenged the quality of the product and refused Lee a patent. Behind this negative response from the London establishment may have been a simple fear of the consequences. The introduction of mechanization would probably have created a significant level of unrest amongst the hand-knitters, especially at a time when enclosure in the rural areas was turning a lot of people off the land and large numbers of people were drifting into the towns without work.

William Lee was convinced of the usefulness of his machine, in spite of the indifference and opposition to it he was encountering in London. He eventually moved to Rouen. In France he and his invention were made welcome. The French king granted him a patent for the knitting frame. In February 1612, Lee signed a contract with someone called Pierre de Caux to supply knitting frames to make stockings out of silk and wool. He undertook to supply English workers to operate the machines and train French apprentices. A factory for the production of knitting machines was to be founded in Rouen.

The last we hear of William Lee is a French document dated 1615. He was still in Rouen, and described as an English gentleman; his occupation was the knitting of stockings. It is said that Lee's brother worked with him in France and that after Lee died the brother returned to London, there setting up a factory to make stockings for the wealthy using William Lee's knitting frames.

The knitting frame was a major step forward in the industrialization of textiles. When it was fully developed, the knitting frame could sew at a rate of 600 stitches a minute. Interestingly, it used a needle design that has remained a key feature of knitting machine design to the present time. The knitting frame was a remarkable invention for the sixteenth century. In some ways it was 150 years ahead of its time, belonging more to the first decades of the Industrial Revolution than to the heyday of Gloriana. It took two working days to knit a stocking by hand. With William Lee's machine it took only one-twelfth of the time. At the time, William Lee and his knitting frame must have seemed curious anomalies. Now we can see that they were a glimpse of the future. Lee's machine was the shape of things to come.

THE INVENTION OF THE FLUSHING WATER CLOSET

(1589)

THE IDEA OF a flushing lavatory has come and gone several times in history. The earliest known flushing lavatory was made by the Indus Valley civilization as early as 2500 BC. There were well-made stone-seated lavatories in the houses of Bronze Age Akrotiri, the Aegean city that was destroyed in the eruption of Thera in either 1520 or 1620 BC. These lavatories were connected by vertical ducts to drains under the streets. Whether or how they were flushed it is not possible to tell.

Not far away to the south, in Crete, well-made lavatories were installed in the temple at Knossos in about 1500 BC – and therefore probably at the other great Minoan temples too. The best preserved one, in the East Wing, has both its floor plan and the stone 'plumbing' beneath the floor still intact. It consisted of a box-shaped seat, probably originally made of wood or gypsum, with an aperture above a void that opened onto a well-designed and well-built square drain. This drain led away to the east, was joined by other drains and opened onto the valley side below the outer walls of the temple complex. It is thought that rainwater guided down from the roof was used to flush the entire system of drains. The drainage system of the temple was evidently designed as a coherent whole. It is very impressive for a building complex of such an early period.

Another feature of the Knossos lavatory is a channel under the floor, leading up from the lavatory to a small circular hole just outside the lavatory door. There was a kind of ante-room to the lavatory, with a bench, and it is possible to imagine an attendant sitting there with filled ewers, pouring water into the round hole to flush the short vertical drain leading down below the lavatory seat. There was also a kind of hood between the seat and the bottom of the drain, and this may have been put in as a baffle, to stop sewer gas percolating up into the temple.

The technology was fairly sophisticated, thoughtful and ingenious, but it was clearly intended for the use of the elite group who ran the temple – the high-status priestesses who are shown in the frescoes in all their finery. No doubt the ordinary people had to make do with far less sophisticated latrines. The advanced technology of these early civilizations was later lost, and lost for a very long time.

The water closet was invented again by John Harrington, one of Elizabeth I's elegantly shallow young courtiers. He installed his useful appliance at his house at Kelston near Bath, and described it in a lively and facetious publication which got him into serious trouble. It was entitled *A New Discourse upon a Stale Subject*, or *The Metamorphosis of Ajax*. It was divided into three sections, but its main purpose was to describe in detail, with diagrams, the mechanism of his water closet. His invention was seriously useful but unfortunately Harrington presented it very badly. He undermined it by including badly judged unsavoury digressions about other courtiers, who did not share his earthy sense of humour. The name he gave his invention, Ajax, was a pun on the phrase 'a jakes', the Elizabethan expression for a privy. His insulting asides about the queen's favourite, the Earl of Leicester, caused particular offence and he was for a time in danger of facing a charge in the Star Chamber. Elizabeth I decided to exile him from court to get him out of Leicester's way.

Instead of promoting and selling his brilliant idea, John Harrington in effect undermined it by being silly, inviting people not to take it or him seriously. The queen eventually forgave him, and in 1592 visited him at Kelston. There she graciously tried out her godson's invention and was deeply impressed with it. She ordered one of Harrington's closets for her own use at Richmond Palace. This royal seal of approval did not lead to the widespread adoption of the water closet, and for a long time everybody in England went on using chamber pots and earth privies.

Harrington's personal reputation suffered a further eclipse when he set off with the Earl of Essex on his ill-fated expedition to Ireland the very next year. Among other offences, the Earl of Essex conferred knighthoods on his associates, Harrington among them, and the Queen was extremely angry at this; giving knighthoods was her prerogative. 'Sir John' Harrington was unfortunate enough to be present at Essex's interview with the incandescent Queen on his return from Ireland, and had to share the fury directed mainly at the treacherous Essex. It was not safe to be a bystander at Elizabeth's court. She ordered Harrington to be taken to Kelston; he was in effect under house arrest. Harrington commented, 'I did not stay to be bidden twice.'

John Harrington's Ajax consisted of a bowl with an exit at the bottom which was sealed with a leather-clad valve. There was a system of handles, levers and weights that poured water into the bowl from a cistern. The water closet was not adopted on a large scale for another 200 years. The design was improved in stages. One major development was the retention of water in the bowl when it was not in use, to create an odour trap; this was the invention of Alexander Cummings in 1775. In 1778 Joseph Bramah introduced the pour flush, which used water and gravity to flush the bowl out, and these two refinements took the flushing water closet close to the peak of modern design.

THE INVENTION OF
THE THERMOMETER

(1593)

THE EARLIEST TYPE of thermometer was called a thermoscope. The earliest person that we know of to put a calibrated scale on a thermoscope was Santorio Santorio, who did so with a view to using it in medicine.

Santorio came from an aristocratic Italian family and was educated in Venice and Padua, where he graduated as a doctor at the age of twenty-one in 1582. Santorio worked for a time as physician to a Croatian nobleman, and in 1599 set up in practice in Venice. There he joined a circle of highly educated people, including Galileo. He became Professor of Theoretical Medicine at the University of Padua, where he taught until he retired in 1624. Santorio's approach to medicine was innovative. He took the original view that the fundamental properties of the human body were mathematical properties such as number, shape and position. At a time in history when medics liked to talk about qualities, essences and humours, Santorio was thinking radically differently; he thought of the body as mechanical, more like a clock than anything else. Santorio had a passion for describing phenomena in terms of numbers and this naturally led him to invent instruments for measuring phenomena. He invented a wind gauge, a current meter for measuring the flows of water in a river, a device called a pulsilogium – and the thermoscope.

The thermoscope Santorio Santorio invented was mentioned by Galileo and, as so often, there is some dispute among historians as to whether the great Galileo himself invented the device, or saw it and described it, or saw it and copied and improved it, as he did with the telescope a few years later. But even if Galileo had a hand in devising the thermoscope, it is certain that it was Santorio who applied a numerical scale to it.

146

In 1593, Galileo invented his own version of the thermoscope, a rudimentary water thermometer, which used the contraction of air to draw water up a glass tube. Galileo discovered that liquids with lower densities than water could be suspended in glass balls within it and that they would float at different levels within it depending on the temperature. This led to the invention of Galileo's thermoscope; alcohol enclosed in glass spheres floating in a column of water. The changes in level of the alcohol spheres reflected changes in temperature, though still it seems without actually measuring it. This still was not quite a thermometer in the truest sense. It was more a scientific toy than an instrument, both looking and functioning rather like a 1960s lava lamp (which was invented by Edward Craven Walker in 1964 as the Astro Lamp); it is not a coincidence that people today buy modern versions of Galileo's thermoscope as interior decoration.

Something nearer the sealed thermometer was invented in 1641, by the Grand Duke Ferdinand II. His thermometer was a glass tube containing alcohol instead of water. Alcohol does not freeze until a temperature of –115 degrees Celsius is reached, so it remains liquid under all conceivable weather conditions.

The earliest thermometers were seriously hampered by there being no numerical scale for measuring temperature. The liquid going up and down inside the glass tube showed that temperature was rising or falling, but there was no objective measure of temperature. Temperature was recognized by scientists as a significant property of gases, liquids and solids, but without a numerical scale of measurement they could only treat the fluctuation of temperature as a curiosity. There are curiosity areas like this in all branches of study. When geography was in its infancy, topographers, as they called themselves, were uncertain what the significant properties of places were. In one entirely serious topography, Robert Plot's *Natural History of Oxfordshire*, published in 1677, the author discussed the variability of echoes, listing places where a single echo might be heard, places where two, three or four echoes might be heard. Plot wrote excitedly of 'a new echo' that he had just discovered at Headington. Today we would not regard the echo as a serious subject for study at all.

Temperature became the object of legitimate scientific study once it could be objectively measured, and that only came with Daniel Fahrenheit. When Daniel Gabriel Fahrenheit (1686–1736) invented his thermometer in 1714, he used mercury in combination with a chemical solution to prevent the mercury from sticking to the tube of the thermometer – and he devised a temperature scale.

THE INVENTION OF
THE MICROSCOPE

(1595)

A COMPOUND MICROSCOPE is one that has more than one lens. The first compound microscope was made by a Dutch lens- and spectacle-maker called Zacharias Janssen of Middleburg in Holland in 1595. Janssen may have been helped in the construction of his microscope by his father Hans. The Dutch were well placed to invent the microscope because of their familiarity with the properties of both single lenses and double lenses; they therefore understood the principles on which the microscope is based.

It was common practice in the sixteenth century for inventors to make several copies of their inventions and give some of them to royalty. This was partly as a sign of gratitude and patriotism, but also a way of giving the invention some publicity and eliciting an official seal of approval. One of Janssen's microscopes did survive until the first decades of the seventeenth century, but unfortunately none of them has survived to the present day.

Janssen's microscope was made out of two tubes, one sliding inside the other, and with a lens fitted at each end. The tubes were slid in or out by a screw until the image came into focus. The eyepiece lens was bi-convex, that is, bulging outwards on both sides. The lens at the lower end of the microscope, called the objective lens, was plano-convex, that is, bulging outwards on one side and flat on the other. This compound microscope gave a magnification of up to nine times.

One of Janssen's friends, Cornelius Drebbel, described the microscope as being composed of three tubes that slid inside one another, with diaphragms between them to cut down the glare from the lenses. Inside there were two lenses. Fully extended, it was eighteen inches long

149

and two inches in diameter. It gave three times magnification when it was closed, and nine times magnification when it was extended. The general design was remarkably close to that of the telescope.

Shortly after Janssen, the microscope was improved by adding a third lens. The news about the microscope spread quickly and, as with the telescope, scientists began to make their own versions of it – and use it. Galileo was one of them. Galileo announced that he had invented the microscope in 1610, and this may be one of the many examples of parallel coincidental invention. Galileo made major improvements to the telescope and may, while working on that invention, have adapted its principles and invented the microscope as a kind of by-product. Some people have suggested that this is what happened. Certainly Galileo was interested in exploiting what the microscope could do, as were many other scientists, such as G. B. Amici, Nehemiah Grew and Robert Hooke.

Over the succeeding years, improvements were made. The main problem was the poor contrast in the image, and various ways were tried to enhance the contrast. The stand was the main feature to evolve. In the end a flat stand was preferred and a light source was placed under it to help with contrast.

Just as the telescope opened up the universe outside the earth to study and scrutiny, the microscope opened up a new inner world. The compound microscope is still in use today. It has been used a great deal in bacteriology, biology and medicine, and has greatly aided advances in medical science It was to lead to a better understanding in several branches of science, such as biology, medicine, geology and metallurgy. It opened up a world that otherwise we would never have been able to see, let alone understand.

III
THE
ENLIGHTENED
WORLD

THE INVENTION OF
THE TELESCOPE

(1608)

THE INVENTION FOLLOWED on from the invention of spectacles in the thirteenth century, which in turn depended on the production of high-quality glass at the major glass manufacturing centres of Florence and Venice. By around 1450 the technology was available to make a telescope – both glass lenses and glass mirrors of a high enough quality were being made – but it seems that no one thought of building one. The means existed, but as yet the idea did not.

It is thought that in the 1570s Leonard and Thomas Digges built a telescope using a mirror and a convex lens, but the device was not developed beyond an initial experiment.

The telescope's history really begins in 1608, when there is documentary evidence that it had been invented in the Netherlands. In fact there were two applications for the patent, one from Hans Lippershey of Middleburg and one from Jacob Metius of Alkmaar. The Dutch government considered issuing the patent for the device for 'seeing faraway things as though nearby', but hesitated because it regarded the device as presented as too simple to warrant a patent. Instead, the government awarded both inventors funding for the development of a binocular instrument, which was considered far more useful.

Meanwhile, news of the invention of the monocular telescope spread like wildfire across Europe. Within months it was possible to buy spy-glasses with a magnification of ×3 in Parisian spectacle-makers' shops.

The first significant application of the telescope came in August 1609, when the Englishman Thomas Harriot observed the Moon through an instrument with a ×6 magnification. In the same month Galileo built his own version of the telescope, with a ×8 magnification, and presented it to the Venetian senate. He made major improvements to the instrument, building one after another, and within a few months he was looking at the sky through a telescope with a ×20 magnification.

This was the major breakthrough, and with this improved instrument Galileo was able to see the mountainous landscapes of the Moon and identify the satellites of Jupiter. The modern science of astronomy was under way.

THE INVENTION OF LOGARITHMS

(1614)

LOGARITHMS REPRESENT THE practical limb of mathematics, a route for practical applications in engineering and astronomy. Logarithms are a short cut in calculation, in the same way that multiplication is a short cut for several additions: five plus five plus five plus five represents three calculations, whereas four times five is only one calculation. Logarithms are a short cut for exponents or powers. Five to the power of two is twenty-five. A logarithm is the power to which a given number, the base, has to be raised to yield another given number. The logarithm of 100 to the base 10 is 2, and can be written down in the following form: log 10 100 = 2.

Although others have been acclaimed as the inventors of logarithms, and it is true that several people helped to develop the idea, the main innovator was John Napier.

Napier went to St Andrew's University in Scotland and left without taking a degree, which was common practice at that time. He is believed to have travelled in Europe in the late 1560s, and stayed for some time in Paris. In 1571 he returned to his estate in Scotland, where he devoted a great deal of his time to the religious controversies of the day. Napier was a committed Protestant and considered his most important publication to be *The Plaine Discovery of the Whole Revelation of St John* (1593), in which he attempted to explain the Book of Revelation. His idea was that the symbols were all numerical and therefore the solution to the book was mathematical.

Religion was Napier's main concern. He considered mathematics to be nothing more than a pastime. His discussion of logarithms was published in 1614 in a work entitled *Mirifici logarithmorum canonis descriptio.*

Napier's logarithms were slightly different from those in use today; he defined his logarithms as a ratio of two distances in a geometric shape rather than the modern way of looking at them as exponents. Napier's motive in inventing logarithms was to save astronomers time in their calculations, and reduce the number of mistakes they made. His idea was that his logarithms, by 'shortening the labours, doubled the life of the astronomers'.

Later, in 1617, Napier described a novel way of multiplying by using rods with numbers marked on them. This was the earliest attempt at a calculating machine. Napier had a very fertile mind, and he produced a surprising range of innovations. He devised revolutionary methods for ploughing and fertilizing arable land. He also devised several 'secret inventions' that were designed to defend England and Scotland from attack by the Spanish. One of these was a heavily armoured circular chariot with guns mounted on it, clearly the forerunner of the tank invented in the First World War. Another was an underwater ship, self-evidently a submarine. He also invented a gun that could mow down a whole battalion of soldiers at once; this was an early attempt at a machine gun. It is said that he tested his gun on a flock of sheep and after seeing the appalling result he swore never to make another gun and never to reveal how to build one.

John Napier will be mainly remembered for inventing logarithms, though. Several other mathematicians adjusted and developed them. Henry Briggs, an English mathematician, was the first to compile a table of logarithms. Briggs wrote to Napier immediately after Napier's ideas on logarithms were published, to ask why he did not create tables using base 10 and log 1 = 0. Napier answered that he too had had that idea, but had been unable to devise the tables owing to illness. This led on to a collaboration; Briggs visited Napier in the summer of 1615 and for a month they worked together devising logarithms using the base 10. In 1685, John Wallis discovered that logarithms could be defined as exponents; the same property was discovered independently by Johann Bernoulli in 1694. By this stage, the concept of logarithms had been fully developed.

Napier's logarithms were an original invention, but they were foreshadowed by the comparison of arithmetic and geometric series. An arithmetic series uses an ascending scale of 1, 2, 3, 4, etc. A geometric series uses an ascending scale in which each term is multiplied by a constant. A geometric series might proceed by doubling (1, 2, 4, 8, 16, 32). In 1620 Joost Bürgi published his work on arithmetic and geometric series in Prague, though this went unnoticed amid the turmoil of the Thirty Years War.

Napier died in 1617 and it was in that same year that Henry Briggs published his tables of logarithms to base 10, to fourteen places of numbers.

The invention of logarithms greatly increased the speed with which mathematicians could compute solutions. This in turn meant that engineers and astronomers could find mathematical solutions faster, as Napier intended. Logarithms facilitated and prepared the way for the great strides that were made from the time of Newton, later in the seventeenth century, to the age of computers in the twentieth. They also made possible the leap forward in engineering that was a major component of the Industrial Revolution.

THE INVENTION OF THE SLIDE RULE

(1614)

THE SLIDE RULE looks superficially like a ruler, but it functions very differently. It consists of two finely divided scales or rules, a fixed outer pair and a movable inner one. There is also a sliding window called a cursor. Before the invention of the pocket calculator near the end of the twentieth century, the slide rule was the most commonly used instrument for calculation in engineering and science. The use of slide rules continued to grow even in the 1960s, when digital computers were gradually introduced. By the 1970s, when the electronic scientific calculator was invented, the market for the slide rule collapsed. With the invention of the calculator the slide rule became instantly obsolete.

The slide rule performs mathematical operations by using distances along non-linearly divided scales. Among the simpler operations the slide rule can perform is straightforward multiplication. If there was subsequent dispute over who invented the telescope, there was a major contemporary dispute, in the seventeenth century, over who invented the slide rule.

One contender was Edmund Gunter (1581–1626). His major publication was *Description and Use of the Sector*, published (in English) in 1623. It was later to be described as 'the most important work on the science of navigation to be published in the seventeenth century'. The sector, in this context, was a mathematical instrument. It consisted of two hinged arms bearing engraved scales that could be used to assist with calculations. The sector was not a slide rule. What made Gunter's sector innovative was that it was the first mathematical instrument to be inscribed with a logarithmic scale to help in solving problems in calculation. It was used in conjunction with compasses, and in practice the points of the compasses tended to damage the engraved scales, and this reduced the accuracy of the instrument.

Another contender for the title 'inventor of the slide rule' is William Oughtred (1574–1660). He was a clergyman as well as a mathematician. Oughtred is credited with inventing the symbol '×' for multiplication in 1628; it was published in 1631 in his book *Key to Mathematics*, which was to become a key textbook. Newton read and was strongly influenced by Oughtred's *Key*.

Today it is generally agreed that William Oughtred was the inventor of the slide rule. In 1632 he wrote *The Circles of Proportion and the Horizontal Instrument*, in which he described both straight slide rules and circular rules, but there is evidence to suggest that he had invented the slide rule some years before and had simply not got round to publishing it.

A third figure in this seventeenth-century quarrel was Richard Delamain (1600–44). He was a maths teacher (he taught Charles I) and originally one of William Oughtred's students. Delamain's book *Grammelogia* contained the first recognizable account of a slide rule. The model he described was a circular slide rule, consisting of two metal discs held together in the middle with a pin.

Delamain and Oughtred argued over the years, in their various books, over which of them was the inventor of the circular slide rule. Their colleagues inevitably became entangled in the quarrel, which became heated. Delamain was a great advocate of mechanical aids to calculation and a great enthusiast of the slide rule. For Oughtred the slide rule was more incidental. For him 'the true way of Art is not by Instruments, but by Demonstration; and that it is a preposterous course of Artists, to make their Schollers only doers of tricks, as it were Juglers . . . That the use of instruments is excellent, if a man be an Artist, but contemptible, being set and opposed to Art.' Oughtred's position was that he had invented the slide rule, but hadn't considered it a priority to publish it.

When the quarrel was at its height, Delamain and Oughtred each accused the other of stealing the invention. Oughtred wrote in *Circles of Proportion* (1632), 'I borrowed and perused that worthless pamphlet [Delamain's *Grammelogia*], and in reading it (I beshrew him for making me cast away so much of that little time remaining to my declined years) I met with such a patchery and confusion of disjointed stuffe, that

I was stricken with a new wonder, that any man should be so simple as to shame himselfe to the world with such a hotch-potch.'

What appears to have happened is that Oughtred and Delamain invented the circular slide rule independently, probably at the same time, though Delamain was the first to publish what he called his mathematical ring. Oughtred's instrument on the other hand was more sophisticated. His circle of proportion was more detailed and more versatile. There is little, if any, doubt that William Oughtred invented the rectilinear (straight) slide rule.

THE INVENTION OF
THE MICROMETER

(1636)

THE SCIENCE OF the seventeenth and eighteenth centuries is sometimes known as the Newtonian age, because it was dominated by Newton. But there were many important inventions by people other than Newton which made new discoveries possible. Many of the errors in scientific calculation before the seventeenth century can be put down to the crudeness and inaccuracy of the instruments that were being used. Indeed, some scientists had not even appreciated how important exactness was, so there had been little interest in improving instruments.

It was in astronomy above all that accuracy was most important. Small inaccuracies in an instrument could be magnified markedly by the enormous distances involved in the observations. Tycho Brahe, who lived and worked in the sixteenth century, seems to have been the first astronomer who used instruments that show great care in their engineering. Once the telescope had been invented, in the first decade of the seventeenth century, and astronomers could see very long distances with some clarity, precise observation became more and more important. The telescope was greatly improved by Galileo, and then by the Huygens brothers, who built a model with a focal length twelve feet long.

It was the Huygens brothers and their collaborators who first applied the micrometer to the telescope.

The micrometer was invented by William Gascoigne of Yorkshire in 1636. The micrometer as it is used in telescopes allows observers to measure very small angular distance accurately. Before the telescope was invented, the only angles that could be measured were those that were distinguishable with the naked eye, and then only approximately assessed. Even Tycho Brahe, who was concerned about accuracy, could only measure with fair precision. Once Gascoigne's invention was applied to the telescope, almost perfect accuracy was possible straight away. It was one of those step changes in science that tend to go unnoticed outside the specialist field.

The principle that lay behind the micrometer was the concept of two pointers lying parallel to one another, and pointing to zero. These pointers were arranged so that the turning of a screw separated them at will. The angle so formed could be determined with absolute accuracy.

The micrometer that Huygens used in his telescope was a tapering slip of metal inserted into the telescope's focus. By noting at what point this exactly covered the object under examination, the Moon perhaps, and knowing the telescope's focal length, he was able to deduce the angular width of the object. Huygens found that an object placed in the common focus of the two lenses of a Kepler telescope appears very clearly defined, and the micrometers invented by Malvasia, Auzout and Picard were developed from this discovery. Malvasia's micrometer, described by him in 1662, was made of fine silver wires placed at right angles at the focus of his telescope.

As telescopes developed and increased in power of magnification, even the finest wires were too thick for accurate observations. They also obliterated the objects the astronomers were trying to measure. The spider-web was devised that is still in use in micrometers. By that time, though, the wires of the early micrometers had already revolutionized astronomical observation.

The jolt of innovation was too great for some of the older generation of astronomers. Johannes Hevelius (1611–87), the 'founder of lunar topography', refused to use the micrometer and other new mechanisms because he knew they were going to invalidate all his earlier observations. He had devoted many years to gathering data in the old way and could not bear to think of all that time and effort wasted.

THE INVENTION OF THE CALCULATING MACHINE

(1642)

BLAISE PASCAL (1623–62) and his father collaborated in a variety of experiments that led to the invention of the barometer, the hydraulic press and the syringe. The idea of mathematical probability evolved out of a correspondence between Blaise Pascal and Pierre de Fermat about the division of stakes in games of chance.

In 1647, when he was still only nineteen years old, Pascal invented and patented a calculating machine. He had built the device to help his father with his work, which involved computing taxes in Rouen. The machine added and subtracted using wheels numbered from nought to nine. It had an ingenious ratchet mechanism to carry the tens of a number greater than nine. As in other areas of endeavour, Pascal's calculating machine was a prototype that others would later come along and develop further. His work can now be seen as a major stepping stone to future developments.

In 1692 the German philosopher and mathematician Gottfried von Leibniz (1646–1716) devised a more advanced calculating machine. This could multiply by repeated addition. It could also divide, as well as the basic adding and subtracting that Pascal's machine could do. Leibniz used a stepped drum to mechanize the calculation of trigonometric tables.

It is easy to see, with hindsight, how this in turn led on to the development of Charles Babbage's analytical engine in 1833. This was a large-scale digital calculator. Babbage managed to get hold of some money from the British government, but he had to sink £20,000 of his own – a huge sum – to develop his calculating machine. The government pulled out of the project in 1842. The prime minister Robert Peel joked, 'How about setting the machine to calculate when it will be of use?'

The steadily evolving calculating machine was the forerunner of the electronic computer of the twentieth century – and it began with Pascal's adding machine.

THE INVENTION OF THE BAROMETER

(1643)

A BAROMETER IS an instrument for measuring the pressure of the air. Since low pressure systems like depressions and hurricanes bring wind, cloud and rain, and high pressure systems bring calm, cloudless and dry conditions, the pressure of the air is obviously a key characteristic. Being able to measure the atmospheric conditions was self-evidently useful, and being able to gauge whether the pressure was rising or falling meant that weather forecasting became possible. The invention of the barometer marked the beginning of a more scientific approach to weather forecasting.

When it was first invented, the barometer answered one of the fundamental questions in natural philosophy at the time – 'Does air have weight?'

The barometer was consciously invented by Evangelista Torricelli of the Florentine Academy, although Gasparo Berti made one inadvertently some years before while trying to make a vacuum. A Genoese citizen by the name of Baliani wrote to Galileo in 1630, telling him that he had designed and built a siphon over a low hill about seventy feet high, but that the water could not be made to rise over the hill. He wanted to know why. A suction pump was supposed to create a vacuum which would suck the water up, the water rising to fill the vacuum. At that time it was believed that there was no height limit on this effect. Baliani's experience in Genoa showed that that was not so. Galileo was intrigued, investigated the problem and reported that there was a working limit to a suction pump. He declared that the pump could not lift water higher than thirty-three feet. Beyond that, the vacuum was not strong enough to pull the water up.

Galileo sensed that there was something more behind this problem and passed it over to his disciple Torricelli. Evangelista Torricelli (1608–47) was a physicist and mathematician. He was born in Faenza and it was as late as 1641 that he moved to Florence to assist Galileo. He was more than a mere assistant, though; he was a pioneering scientist in his own right. He wrote, among other things, about logarithms, the motion of fluids and proposed a theory of projectiles. He explored determining the value of gravity by observing the motion of two weights joined by a string passing over a pulley.

After discussion with Galileo, Torricelli in turn set up an experiment which was carried out by one of his students in 1643. What Vincenzo Viviani built under Torricelli's supervision and direction was a mercury barometer. Torricelli had earlier built a water barometer, but this had required a very long – sixty feet long – glass tube. Mercury had the advantage of being thirteen times denser than water, and so the length of the tube could be shortened to about thirty-five inches.

The first barometer was a long-necked glass tube. It had a closed bulb at the end and was filled with mercury. The tube was then inverted into a basin that was full of mercury. The mercury did not run right out of the tube, but its level fell so that its surface hovered about thirty inches above the level of the mercury in the bowl. At that level it remained steady, only fluctuating a very small amount. Later it became apparent that these small fluctuations were caused by variations in the atmospheric pressure over the mercury in the basin, pushing sometimes more, sometimes less mercury back up into the tube. The small fluctuations in the mercury level were in effect measuring atmospheric pressure.

As we have seen in other branches of science and technology, other people, elsewhere, were thinking along similar lines. It is possible that ideas and information were swapped. Gasparo Berti, a mathematician and astronomer, undertook rather similar experiments a few years before. He was also working on Galileo's problem. Berti's instrument consisted of a huge lead tube filled with water and sealed at the top. When the lower end of the lead tube was opened into a bucket of water, not all of the water in the tube ran out. The space above the water was, as Berti proudly claimed, an artificially created vacuum. What Berti saw was that he had invented a device for making a vacuum. The experiment is so similar to Torricelli's experiment that it seems likely Torricelli knew about it, copied it, and drew an additional inference from it.

The French scientist and philosopher Descartes described an experiment to find atmospheric pressure back in 1631, but it seems unlikely that he succeeded in building a working barometer. He was, on the other hand, the first to devise a scale for quantifying readings of pressure. In 1647, after Torricelli had invented the barometer, Descartes wrote to Marin Mersenne, 'So that we may also know if changes of weather and of location make any difference to it [atmospheric pressure], I am sending you a paper scale two and a half feet long, in which the third and fourth inches above two feet are divided into lines; and I am keeping an exactly similar one here, so that we may see whether our observations agree.'

In 1648, Blaise Pascal put forward the important theory that air pressure dropped with height above sea level. If air had weight, then on mountain summits there must be less air pressing down on the air at ground level and the air pressure must be lower there. Pascal got his brother-in-law to carry a barometer up to the summit of the Puy-de-Dôme in the Massif Central. Perier was startled to see that the mercury column had fallen dramatically by the time he reached the mountain top, at about 4,900 feet above sea level. The column was 3.6 inches lower than when he had measured it at sea level. Perier's observation confirmed Pascal's theory. An important property of the lower atmosphere, the lapse rate, had been explained. The temperature of the air regularly decreases with altitude as a direct result of the regular decrease in pressure.

Enormous progress was made with the barometer and its revelation of the secrets of the atmosphere in that first decade. After that, the barometer developed more slowly. Robert Hooke made a major improvement in 1665, when he invented the wheel barometer. He added a circular scale and a dial to the mercury barometer. This made taking readings from the barometer far easier. After that, the barometer was progressively refined and improved.

It seems to have been the English scientist Robert Boyle who gave the instrument the name barometer. In 1669, Boyle wrote a manuscript entitled *Continuation of New Experiments*; in it he described a design for a barometer that would be easier to carry.

The next major development in the barometer was the aneroid barometer. This was to be a barometer that operated without liquid, and was first discussed as a theoretical possibility. In about 1700 Gottfried Leibniz proposed the possibility of an aneroid barometer that would operate on sealed bellows, but he did not build one. The first working prototype of the aneroid barometer was made in 1843, the invention of the French scientist Lucien Vidie. By removing the fragile and easily broken mercury column and the mercury reservoir, this new invention was a major advance. The aneroid barometer was an easily portable instrument, and it was to become standard equipment for field researchers and explorers. Sailors and farmers bought them to help them predict the weather. Barometers soon found their way into middle-class homes; with no expertise at all, ordinary people were able to see whether the pressure was rising or falling, and therefore whether rain was more or less likely in the next few hours.

The metal aneroid cells have today been largely replaced by sensitive electronic sensors. These, used in conjunction with microprocessor chips, have made possible the miniaturization of barometers; the pocket barometer (and altimeter) is now a reality.

THE INVENTION OF THE AIR PUMP

(1650)

IN 1650 THE German physicist and engineer Otto von Guericke invented the first air pump. He was experimenting with and improving air compressors. Guericke found that he could use his air pump to create a partial vacuum. This was a preliminary to experiments on the role of air in respiration and combustion.

The first compound air compressor was patented in 1828. This performed staged compression in a succession of cylinders.

In 1872, the efficiency of air compressors was improved by having the cylinders cooled by water jets. This led to the invention of water-jacketed cylinders. Probably the best-known compressed-air device is the pneumatic tyre.

In the first half of the twentieth century the pneumatic tube was a popular device in large office buildings for transporting messages and small objects from one office to another. The cash carrier based on the same principle sent money in small tubes from one location to another round department stores. The first of these devices was invented by D. Brown in 1875. Another device, invented by Samuel Clegg and Jacob Selvan in 1940, was a wheeled vehicle on a track inside a tube; the vehicle was in effect squirted along by compressed air.

In 1865, a pneumatic train subway was invented and built by Alfred Beach. The subway ran for a short time in 1870 in New York, for just one block, but it was New York's first underground train.

THE INVENTION OF THE PENDULUM CLOCK

(1656)

THE CLOCK IS one of those devices that did not have a single inspired inventor. It gradually evolved over a very long period, with different inventors adding improvements from time to time. As long ago as 3000 BC, people in Britain and Ireland were building time markers into their monuments. The sunrise position moves gradually backwards and forwards along the eastern horizon during the year, and people noticed that it stopped at its southernmost position (in the south-east) in the middle of winter and at its northernmost position (in the north-east) in the middle of summer. These standstill moments were the natural turning-points of the year, markers for the calendar. We still celebrate one of them, at Christmas, and in antiquity, long before Christ, the megalith builders installed devices in their monuments to signal when they arrived. At the Newgrange passage grave in Ireland, a decorative slot like a large letter box was created above the stone door, so that on 21 December, and only on that day, the rising sun would send a beam of light right to the heart of the tomb. At Stonehenge, two rough pillars were set up outside the earth circle at the north-east entrance, and on midsummer morning, 21 June, and only on that day, the rising sun would be seen from the circle's centre just floating free of the horizon exactly between the two pillars.

These were the major time markers for people living in middle or high latitudes in the prehistoric period, the pivots of the turning year. The daily pattern of life for thousands of years depended on sun time. When there was light, people could work. When it was dark they could not; then it was time to relax by a fire and sleep.

Sundials were invented in Egypt in about 3500 BC, and these enabled people to trace the passage of time through the day. By 1500 BC, the Egyptians had devised a portable sundial, which we can see as the ancestor of the modern wristwatch. Another type of clock invented by the Egyptians was the water clock. This operated by the slow dripping of water into a container, which contained a floating marker that told the hour.

The first mechanical clock with escapements was invented in 1280. The escapement is the mechanism that makes clocks tick, so this was very recognizably a modern clock. Another development that happened shortly afterwards was the chime. The first clock to strike the hours was set up in Milan in 1335, but it was not very accurate.

In 1520, Peter Henlien of Nuremberg devised a clock powered by a spring. This made the clock run at an even speed, and was therefore more accurate, but there was still the problem that it ran slower as the spring unwound. Clocks of this period still only had hour hands, as it was not possible to mark time any more accurately than that. It was not until 1577 that Joost Burgi invented the first clock to have a minute hand. The poor accuracy of the clock did not really justify adding it, but it shows that there was an interest in making time-keeping more precise. The reasons were the obvious ones. If you were in any kind of business, especially in a town or city, where business depended on working with other people, it was important to make contact in an organized way. Today we take for granted that we can arrange business meetings at specific times, and being specific means that we can use our time efficiently. As commerce became more complex, the need to fix times more exactly became more pressing.

Galileo's observations in the 1580s regarding the regular timing of the pendulum swing were to be crucial to the development of a more accurate clock. Galileo saw this possibility, and made drawings and models of a pendulum clock together with his son Vincenzo. Galileo himself died before this could be turned into a working reality, but his son continued the work and produced a working pendulum clock in 1649.

The Galileo clock was perfected by Christiaan Huygens in 1656. Huygens invented the first weight-driven pendulum clock. This device made it possible for the first time to keep time accurately. The Huygens clock was still fitted with an hour hand only. It was in 1680 that clock makers started fitting minute hands as a regular feature, and some shortly afterwards optimistically tried fitting second hands, but many weight-driven pendulum clocks built in the first quarter of the eighteenth century still had hour hands only. The fine old grandfather clocks that are now seen as highly collectable antiques were all built as weight-driven pendulum clocks. At one time, practically every office and upper- and middle-class home in Europe and America had one.

Time was now precisely marked. Commercial and social engagements could be kept, ships could be loaded and unloaded in a more organized way, stage coach services could be properly coordinated.

THE INVENTION OF THE REFLECTING TELESCOPE

(1668)

IN THE EARLY days of the telescope, it was generally assumed that the object (lower) lenses of telescopes were only subject to the normal engineering errors involved in creating accurately curved surfaces. With Newton came all sorts of disturbing revelations. One was his discovery of the properties of light refraction, in particular the different refrangibility of light of different colours. This meant that images seen through the lens were likely to suffer from a slight blurring effect.

An English surveyor, Leonard Digges, is said to have built a reflecting telescope as early as the sixteenth century, though there is no proof. An Italian Jesuit astronomer, Niccolo Zucchi, is believed to have built a reflecting telescope in 1616 and discovered the belts of Jupiter while using it in 1630. In 1652, Zucchi wrote a treatise describing his work, entitled *Optica philosophia*. This may have been the inspiration for the later books by James Gregory and Sir Isaac Newton. Gregory's *Optica promota*, published in 1663, described the images produced by lenses and mirrors. He mentioned the slightly distorted images produced by some lenses and mirrors, pointing out that if the curves of the glass surfaces are conic sections the so-called spherical aberration is corrected. Gregory was conscious that earlier attempts at perfect telescopes had failed, even by using lenses of various types of curvature, so he was fully aware of the significance of his proposal. The new reflecting telescope that he described was called the Gregorian telescope. Gregory was a theorist, and did not have the practical skills to build his telescope. Nor could he find a lens manufacturer prepared to turn his idea into reality. James Gregory had invented a refined, fully workable reflecting telescope, but only in his head. It was left to Isaac Newton to build it; he is the first person who is known for certain to have built a reflecting telescope.

In 1666 Newton discovered that light of different colours is refrangible to different degrees. He realized at once that the defects of the refracting telescope were due far more to this varying refrangibility than to the 'spherical' shape of the telescope's lenses. Some of Newton's work at this time was too hasty, and he was too quick to assume that no further improvement could be expected of the refracting telescope. But this dismissal of the refracting telescope made him turn his attention to the construction of a reflecting telescope. He carried out exhaustive experiments on the material for his specula (telescope mirrors) and settled on an alloy of tin and copper. When it came to grinding the specula, he did not attempt the parabolic shape recommended by Gregory, because it was too difficult to do. He was also convinced that it was not so much the curvature of the lenses and mirrors that had caused the problems as the chromatic aberration, the effects of light refraction.

Newton's new telescope was astonishingly successful. The images he could see through it were much clearer than he had been able to see through a refracting telescope. Using his new reflecting telescope, he was able to see the moons of Jupiter and the horns of Venus. The prototype was so successful that he built a second telescope, one that would enable him to achieve a magnification of ×38, and he presented this to the Royal Society in December 1672.

A French scientist, Cassegrain, also built a reflecting telescope in that year, but there was no advance in the design of reflecting telescopes until 1723. That was when John Hadley presented to the Royal Society a much more powerful instrument on Newton's model but enabling magnification up to ×230. When scientists looked at Hadley's telescope they appreciated more fully than before what a great advance Newton's invention had been. The manufacture of telescopes on this new design began. Two London opticians, Hearn and Scarlet, started making them on a commercial scale.

It was James Short of Edinburgh who eventually built James Gregory's telescope. Short was, in about 1732, given space in the rooms of the Professor of Mathematics at Edinburgh University to carry out work on the telescope. Short at first made his specula of glass, as proposed by Gregory, but switched to making them of metal, like Newton. He successfully made the specula truly parabolic in shape; he gave them a high polish and very sharp definition. James Short went on to make Gregorian telescopes on a commercial scale, and became rich in the process.

THE INVENTION OF CALCULUS

(1666)

THE INVENTION OF calculus is one of the most famous disputed inventions of all time. Was it Newton or Leibniz who invented calculus? Newton started working on calculus in 1666, whereas Leibniz started on his version eight years later and published his first paper using calculus in 1684. Like Oughtred before him, Newton delayed. He did not publish his view of calculus until 1693 (partially) or 1704 (fully). These dates suggest that Leibniz was quicker off the mark than Newton, but the water is muddied by the fact that in 1676 Leibniz visited London, where he was shown at least one of Newton's unpublished manuscripts. It is not certain, but maybe Leibniz was given valuable information in these notes, information that enabled him to develop his calculus.

This was the murky beginning of a major academic debate about the origins of calculus. The controversy was simmering by 1700, and broke out in full fury in 1711.

Leibniz wrote in his notebooks that he made an important breakthrough on 11 November 1675. On that date, he said, he used integral calculus for the first time to find the area under the function y = x. He used a range of notations that are still used today, including an elongated S (for the Latin *summa*) as the integral sign, and d (for the Latin *differentia*) to represent differentials. Leibniz had a difficult life, and his final years, from 1709 until his death in 1716, were soured by quarrels with John Keill, Isaac Newton and other people about whether he had really invented the calculus independently of Newton. His opponents and detractors accused him of just inventing a different notation for ideas that he had cribbed from Newton.

Newton was a difficult, quarrelsome man and he fuelled and

manipulated the controversy. Newton claimed that he had already developed his method of fluxions at the time when Leibniz started working on the calculus. There was no proof of this, and Newton had published nothing that could be produced as evidence, yet curiously no one ever expressed the slightest doubt that he was telling the truth. The nearest thing to evidence that he had invented something special was a calculation of a tangent which was accompanied by the note, 'This is only a special case of a general method by which I can calculate curves and determine maxima, minima and centres of gravity.' He did not explain this general method until twenty years later, by which time Leibniz had published his version of the calculus, and Newton could have copied that. Newton's notes were found after his death, but they could then no longer be dated, so they shed no light on the matter.

The infinitesimal calculus could be expressed in one of two forms of notation. One was Newton's – fluxions. The other was Leibniz's – differentials. The earliest use of differentials can be traced to 1675 in Leibniz's notebooks, and he used this notation when he wrote to Newton in 1677.

There are several arguments in support of Leibniz. He published his method years ahead of Newton. He always referred to calculus as his invention and no one challenged it for a long time. He always behaved as if he had acted in good faith. His notes show that he developed calculus in a completely different way from Newton. He was ready to work in collaboration with Newton.

Some arguments have even so been levelled against Leibniz. He saw some of Newton's papers on the subject in manuscript. He may have obtained the basic idea of the calculus as a result of seeing those papers.

It is possible, as we have seen in some other inventions, that the two men arrived at the calculus coincidentally at the same time. The fact that Leibniz used a different method, strongly suggests that he was not merely plagiarizing Newton. The situation was complicated by the fact that in addition to the formal publications there were letters circulating, there were meetings – all sorts of informal exchanges of information. By these informal routes, it became clear to both Leibniz and Newton that the other was a long way along the path to the calculus. Leibniz actually mentioned it, but only Leibniz was pushed by the situation into publishing.

In 1849, when a researcher was going through Leibniz's manuscripts, he uncovered extracts from one of Newton's papers, copied in Leibniz's handwriting. These notes were probably made in May 1675, when it is known that a copy of Newton's manuscript had been sent for Tschirnhaus to look at. As Tschirnhaus was collaborating at that time, it is very likely indeed that the manuscript was shown to Leibniz. Other scientists, Collins and Oldenburg, seem to have had access to Newton's manuscript in 1676, and because Leibniz was collaborating with them too he may have had a second opportunity to make notes on the manuscript then. Leibniz mentioned that someone else, Collins, had shown him some of Newton's papers, but implied that they were of little or no use to him.

Newton went out of his way in 1711 to make a case for Leibniz having seen his, Newton's, notes on the calculus, but by then he had a vested interest in arguing this. In 1704, someone published an anonymous review of one of Newton's works in which the reviewer implied that Newton had plagiarized the idea of the fluxional calculus from Leibniz. He also implied that there was no question that Leibniz had invented the calculus independently of Newton. This was the spark that brought the controversy out into the open, made the integrity of the two great scientists a matter of public debate. The case against Leibniz was published by Newton's friends and supporters in 1712 as *Commercium Epistolicum*. Newton was behind this attack. Poor Leibniz had no claque of friends and supporters ready to do the same for him. Johann Bernoulli wrote a letter in 1713 making a personal attack on Newton, but the charges he made were false and, when challenged, he weakly denied having written it.

Newton wrote privately to Bernoulli, 'I have never grasped at fame among foreign nations, bit I am very desirous to preserve my character for honesty, which the author of that epistle, as if by the authority of a great judge, had endeavoured to wrest from me. Now that I am old, I have little pleasure in mathematical studies, and I have never tried to propagate my opinions over the world, but I have rather taken care not to involve myself on account of them.'

In the face of the appalling public controversy, Leibniz retreated into silence. He wrote in a letter in 1716, 'In order to respond point by point to all the work published against me, I would have to go into much minutiae that occurred thirty, forty years ago, of which I remember little. I would have to search my old letters, of which many are lost. Moreover, in most cases I did not keep a copy, and when I did, the copy is buried in a great heap of papers, which I could sort through only with time and patience. I have enjoyed little leisure, being so weighted down of late with occupations of a totally different nature.'

Leibniz may not be wholly innocent. He did not actually acknowledge that he had made notes on an unpublished work of Newton's; that only emerged in the nineteenth century. In addition, Leibniz more than once deliberately altered important documents.

While Leibniz's death put a temporary stop to the controversy, the debate persisted for many years. He altered documents that he quoted in publications. He also falsified a date on a manuscript, from 1675 to 1673, so that it would seem to be in advance of other publications.

If the world of science had operated in 1700 as it operates today, then the date of publication would be the decider, and Leibniz would be regarded as the sole and undisputed inventor of the calculus. But at the time, the general presumption that Newton must be the inventor of the calculus prejudiced any real debate. The Royal Society set up a committee to pronounce on the priority dispute but that committee never invited Leibniz to give his version of events. It was this committee that published *Commercium Epistolicum* in 1713. It came out in favour of Newton. It was not surprising – the document was written by Newton himself!

Newton was not the pillar of honesty the Royal Society wanted him to be. John Flamsteed helped Newton with his *Principia*, but subsequently held back information from him. Newton then seized all of Flamsteed's work and tried to get it published with the aid of Flamsteed's enemy, Edmond Halley. Flamsteed had to resort to a court order to block the publication of his own work, and only just succeeded. In retaliation, Newton had the acknowledgement of Flamsteed's help deleted from future editions of his *Principia*.

This sad and rather shabby episode in the history of inventions shows its negative side. Too often the ego of the inventor, or would-be inventor, takes over and truth is left far behind.

THE INVENTION OF THE PRESSURE COOKER

(1675)

THE PRESSURE COOKER is a hermetically sealed pot, an airtight saucepan with a tightly fitting lid, which produces steam to cook food quickly while retaining its nutritional value. The pressure cooker was invented in 1679 by the French physicist Denis Papin. The generation of steam contained inside the cooker raises the pressure. The temperature inside the pot is raised as high as 130 degrees Celsius, well above the normal boiling point of water, 100 degrees.

Papin demonstrated his so-called 'steam-fume chamber' at the Royal Society in London. The pressure cooker was marketed as Papin's Digester. The lid or cover was fitted with a safety valve. This allowed steam to escape when a safe upper level of pressure had been reached and prevented the cooker from exploding.

Sir Christopher Wren, a member of the Royal Society, encourged Papin to write a booklet about the invention. Denis Papin saw the device being used in the kitchen, but his Papin's Digester found its first commercial use in the world of industry, where it was used for sterilizing.

It was really only in the twentieth century that Papin's cooker found its inventor's first intended use. It became a popular kitchen utensil in American and Europe during the Second World War, when people realised how much fuel they could save by using it. A casserole could be cooked in half the time on the same heat. People also found that because the cooking time was shorter, the food kept more of its flavour.

In the twenty-first century, with the campaign about carbon dioxide emissions and growing concerns about the use of fossil fuels, the pressure cooker looks set for a second kitchen revival.

THE INVENTION OF THE STEAM PUMP

(1698)

THE STEAM PUMP was invented by Thomas Savery, who was born at Shilstone Manor neat Modbury in Devon in about 1650. Savery was a born inventor. His first interest was in military and naval applications of engineering, and one of his inventions was a somewhat premature design for a paddle wheel. Then he became interested in pumps. Pumps were of increasing importance in fuel supply. As deforestation progressed in Britain, wood as a fuel source was becoming increasingly scarce. The replacement fuel was coal. In the exposed coalfields this could be quarried at or near the surface, but as soon as miners began to follow seams of coal down into the Earth they encountered ground-water. To mine deeper, there had to be effective pumps to keep the mines dry. The same was true of the Cornish tin mines. It was a clear case of necessity driving invention.

In 1698, Thomas Savery invented an early steam engine. It was based on Denis Papin's 1675 pressure cooker. He applied for and was given a patent for it and then, in 1702, he published details of his machine in the book *Miner's Friend*, claiming that his engine could be used to pump water out of mines. Thomas Savery's pump did not have a piston. Instead it used a combination of atmospheric and steam pressure to raise the water. The machine consisted of a closed vessel filled with water. Steam under pressure was introduced, forcing water upwards and out of the mine shaft. Then a cold water sprinkler cooled the closed vessel down, to condense the steam in the closed vessel. This created a vacuum, which in turn sucked more water up from the mine shaft by way of a valve at the bottom.

The atmospheric action was limited to raising a column of water about thirty feet high, as Galileo's correspondent had already commented. This figure could be raised to fifty feet by the use of steam pressure, but the stress put on the boiler rendered it unreliable. This meant that in spite of Savery's claim his pump was not really capable of lifting unwanted water from a mine. Virtually the only known working versions of Savery's pump were used to supply water in London. An attempt was made to use the steam pump to clear water from the Broadwaters Mine at Wednesbury in the Black Country. It just did not work.

If there was a single starting point for the Industrial Revolution, the invention of the steam pump by Thomas Savery was it. It may not have been successful, but in terms of the thrust to develop it, in terms of its intention, in terms of its capacity to develop into something vitally useful, it set the character of the age. Above all it was a steam engine, and the Industrial Revolution was built on steam.

THE INVENTION OF THE PIANO

(1700)

THE PIANO, A very versatile instrument that has been classified variously as a percussion, string or keyboard instrument, was invented in 1700 by Bartolomeo Cristofori.

The piano was a great invention. It can hold its own as a solo instrument and huge numbers of composers have written pieces for it, from Mozart, Beethoven and Chopin to Scott Joplin, George Gershwin and Bill Evans. It is also widely used as an accompanying instrument for a solo singer or for a solo instrument such as a violin or clarinet. It can be used as a continuo for a chamber ensemble, holding the group of players together. It can also contribute colours of its own to a symphony orchestra. The piano may not sing like a violin or stir like a trumpet, but it is undoubtedly the most versatile and useful of all Western musical instruments. Perhaps its one shortcoming is its lack of portability, though with the advent of high-quality electronic keyboards, even that problem may be overcome.

The piano evolved out of the harpsichord. Its formal name, pianoforte, comes from the final phrase of its full original Italian name: *gravicembalo col piano e forte*, meaning 'harpsichord with soft and loud'. The harpsichord, involving plucking strings with quills, had a narrow dynamic range. The piano, which had padded hammers striking the strings with varying degrees of force, had a very wide dynamic range.

There were earlier attempts at keyboard instruments with struck rather than plucked strings, but the first successful instrument of this design was built by Bartolomeo Cristofori, an instrument maker from Padua in Italy. The exact date when he built his first piano is not known, but an inventory made for the Medicis, who employed Cristofori, shows that a piano existed in 1700. Today, there are just three pianos left that were built by Cristofori in the 1720s.

The design of the piano was not totally revolutionary. As with many other inventions, the inventor made some significant changes to something that already existed. Cristofori was an experienced and expert harpsichord builder, and it is no accident that his pianos looked exactly like harpsichords. The revolutionary new instrument did not look revolutionary. It had a harpsichord keyboard and case, and it stood on three legs; it was known that this keyboard was playable and that this sort of case worked well acoustically. Where Cristofori used great originality was inside, in designing the striking mechanism. The problem he had to overcome was very precise. The hammer had to strike the string, but must not stay in contact with it. The hammer had to bounce off it to let it sound properly. On the other hand, it had to return to its rest position without bouncing too far back otherwise a rapid repeat of that note would be impossible. Cristofori's brilliance lay in solving these two apparently contradictory needs and the hammer action he invented was closely followed in all later pianos.

What Cristofori invented was an instrument of far greater range and potential power than the harpsichord, and of far greater expressiveness when required. Cristofori had invented something truly remarkable, but initially there was little response and little enthusiasm. Then, in 1711, an Italian writer called Scipione Maffei wrote an article about Cristofori's piano, enthusing about it and including a diagram of the mechanism. The article was widely distributed and widely read. Instrument makers across Europe were intrigued and tried their hand at building the new instrument.

One of these experimenters was Gottfried Silbermann, an organ builder, who added the damper pedal which lifts all the dampers from the strings at once and allows the sound to resonate. Silbermann showed one of his early pianos to Johann Sebastian Bach, but Bach did not care for it. He thought the high notes were too soft to be heard clearly. Silbermann was irritated by this criticism, but it seems he took it to heart and modified his design: Bach gave his approval to a later model in 1747. Bach was even ready to become an agent, helping to sell Silbermann's pianos.

Piano making developed during the eighteenth century. There was a Viennese School of piano makers, which included Johann and Nannette

Stein. The Viennese School pianos had wooden frames, two strings to each note to thicken the tone, and leather-padded hammers. These were the instruments for which Mozart wrote his piano sonatas and piano concerti – some of the greatest work ever written for the instrument. These 'Mozart' pianos had a softer, clearer, more bell-like tone than modern pianos.

In the nineteenth century, the piano was developed to create a more sustained and powerful sound. The Industrial Revolution played its part in this transformation. It had by then become possible to produce very high-quality steel for the strings (piano wire), and precision casting to produce cast-iron frames. It became the norm to use three strings to each note instead of the previous two. These changes in materials added enormously to the power and weight of the sound, as we can hear in pieces like the Brahms piano concerti from the end of the nineteenth century. By this stage, the upright piano, grand piano and concert grand piano we know today had reached their full development.

THE INVENTION OF THE SEED DRILL

(1701)

JETHRO TULL WAS educated at Oxford, where he studied law. Later he travelled on the mainland of Europe, where he noted a variety of farming practices. On his return, Tull inherited land in southern England, and there he was able to put into practice some of the new ideas on agriculture he was developing. He became a member of a group of 'improving farmers' who set up the Norfolk system, a pioneering effort to put farming in Britain on a more scientific basis.

Jethro Tull invented the seed drill in 1701. Until that moment, seed had been broadcast by hand. This involved the rather uneven and therefore uneconomical scattering of seed over the surface of a field. Because many seeds remained on the surface, they did not take root. A great deal of seed was being wasted. The device Tull invented, the prototype of which he constructed out of organ pedals from the local church, allowed farmers to sow seeds in well-spaced rows at specific depths. Tull's drill worked. Using the seed drill meant that a higher proportion of the seed germinated. Crop yields increased.

The invention of the seed drill was the first of a whole sequence of agricultural improvements through the eighteenth century that led to improved efficiency, higher yields per acre and greater overall food production. Just as Savery's steam pump was the start of the Industrial Revolution, Tull's seed drill was the start of the Agricultural Revolution.

Jethro Tull was a visionary as well as a farmer. Not only did he invent the seed drill, he also invented the horse-drawn hoe, and an improved version of the plough. In 1731, he published his ideas on agriculture in *The New Horse Hoeing Husbandry, or an Essay on the Principles of Tillage and Vegetation*. It was to be a book that had enormous influence on landowners all over Britain in the eighteenth century.

THE INVENTION OF IRON SMELTING USING COKE

(1709)

IRON WAS A key resource in Britain from around 500 BC onwards. There were reserves of it in moderate quantities at several locations around the country. Small quantities were quarried and smelted during the Roman occupation. During the Middle Ages and the Renaissance, ever-increasing amounts of iron ore were quarried out and smelted to meet the growing demand. Iron was used in the home for nails, window catches, locks, keys, hinges, cutlery, firebacks, cauldrons and other utensils. It was used in industry for making a wide range of tools. It was used in agriculture for making ploughs, harrows, hoes, rakes, scythes and sickles. It was used in the shipbuilding industry for nails, braces, staples, hooks and anchors. It was used in the army and navy for guns, cannons and shot and (converted into steel) for swords and armour.

In other words, well before the explosion of the Industrial Revolution, there was a very important iron industry. Most of the early iron foundries were small, and scattered across the countryside, such as the Forest of Dean and the Weald of Sussex and Kent, close to sources of wood for charcoal and the ore for iron. As the economy became more diverse, the demand for iron grew. As the forests dwindled, the need to find an alternative fuel for smelting grew.

An improvement in technique had taken place in 1496, with the invention of the blast furnace in the Weald. The blast furnace technology had spread to other parts of the country by 1700. Change was in the air.

Abraham Darby I was born in about 1678 near Dudley in Worcestershire. He was the son of a Quaker farmer. As a youth he was apprenticed to a malt-mill maker. Then in 1704 he went on a visit to Holland. He came back to England bringing with him several Dutch brass founders, who helped him to set up the Baptist Mills brass foundry in Bristol. But Abraham Darby was not content with that. He wanted to make another change too. He started thinking of iron. As a cheap substitute for brass, iron cooking utensils would find a large and ready market.

Darby patented the technique of sand casting and took a lease on an old iron furnace at Coalbrookdale in Shropshire. There he went on working away at the process of iron making. The shortage of charcoal drove him to make the revolutionary and world-changing experiment of switching to coke as a fuel. Charcoal was running out. There was also a problem because of its mechanical weakness. Charcoal worked well enough in a small furnace, but in a big furnace it became crushed under the weight of the ore. What was needed, as the scale of the furnaces increased, was a mechanically stronger fuel. Coal was an obvious raw material, because there was so much of it in Britain. Nineteenth-century French school children were taught that Britain was an island made of coal. Coal could be improved by being roasted first, to get rid of undesirable impurities such as sulphur, which could spoil the quality of the finished iron.

Abraham Darby I crucially invented and pioneered the use of coke. He was not the very first person to use coke as a substitute. About 100 years earlier, Dud Dudley (1599–1684), the illegitimate son of the 5th Earl of Dudley, had tried it out at his father's ironworks, but the experiment had not been a success. It was Abraham Darby who, in reinventing it in 1709, made the process work, and on an industrial scale.

The age of charcoal smelting was over. That meant a significant switch in location. There was no point in having the furnaces in the dwindling, thinning forests when the coal was tens or hundreds of miles away in the coalfields. New furnaces were set up on the coalfields of South Wales, the Scottish lowlands, Yorkshire, Staffordshire and the small Shropshire coalfield which was the Darby's headquarters, the hub of this fast-expanding new industry. Darby set a magnificent example by producing iron of the highest quality at his works in Coalbrookdale.

The Darby men were not to be long-lived. Abraham Darby died in 1717 aged only thirty-nine. His son, another Abraham, was to die at only fifty-two. Abraham Darby II, born in 1711, was to become another industrial pioneer.

Abraham Darby II invented the process for making wrought iron. Keeping crucial trade secrets like this was all-important to the success of a family business in such times, but when he died Darby did not pass on the closely-guarded details of his process. It was left to an iron founder of the next generation, Henry Cort (1740–1800), to reconstruct from scratch the Darby technique for puddling and rolling so that the brittle pig-iron straight from the furnace could be turned into high-quality refined bar-iron. He acquired the patent for Darby's process in 1785.

Puddling was the removal from pig-iron of the carbon that made it brittle, in a special furnace called a reverberatory furnace. This produced a malleable iron, an iron that could be hammered and rolled. Henry Cort, who became known as The Great Finer, set himself up in business as a founder at Gosport close to Portsmouth naval dockyards; his aim was to supply iron to the Navy. Cort was let down by his financial backer, who embezzled Navy money. When that came out, Cort was bankrupted, stripped of his patents, and died poor. His ideas were exploited by other iron founders.

Meanwhile, there were still Darbys at Coalbrookdale. Abraham Darby II's son, Abraham Darby III (1750–91), was also short-lived by modern standards, dying at forty-one. The Darby works at Coalbrookdale continued producing iron, maintaining a national reputation for high quality. Abraham III's main achievement was the construction of the world's first cast-iron bridge, which he prefabricated in the factory in 1779. His famous iron bridge still stands and, with admirable simplicity, the place has become known as Ironbridge. Coalbrookdale produced, among other things, cast-iron boilers for Newcomen steam engines, which by the second half of the eighteenth century were in big demand for draining coal mines. The demand for coal was also stepping up because of the requirements of the ironworks. In the early years of the nineteenth century Richard Trevithick would go to Coalbrookdale to order the boilers for his high-pressure steam engines.

The first phase of the Industrial Revolution was now well under way, with massive growth in the iron industry and the coal mining industry. A key characteristic of the Industrial Revolution was also emerging strongly – one industry expanding on the back of another, growing by a kind of mutual feeding.

THE INVENTION OF THE STEAM ENGINE

(1712)

THE STEAM ENGINE is one of those inventions that came about piecemeal. There was evidently a prototype steam engine in the ancient Greek world. Hero of Alexandria described a machine he had designed to open the doors of a temple. Water was heated and steam was expelled through two nozzles that turned the turbine and so opened the doors. But in modern times, the power of steam to do useful work was not harnessed until Thomas Savery, the English engineer, invented a steam pump in 1698 designed to empty water from the shafts and galleries of coal mines.

Savery's pump had been a symbolic landmark, but not a very useful machine. His pump was very inefficient, able only to raise the water about twenty feet, when the mineshafts were much deeper than that. Thomas Savery worked for the Sick and Hurt Commissioners. His work for them took him to Dartmouth, where he probably met Thomas Newcomen, another great pioneer of steam power. In 1701, Savery gained parliamentary permission to extend the lifespan of his patent for another twenty-one years. Then arrangements were under way to develop Thomas Newcomen's more advanced design for a steam engine, which from 1712 was sold, by agreement, under Savery's patent. Newcomen's engine worked by atmospheric pressure, which eliminated some of the dangers involved in using steam under high pressure. Newcomen's engine also used the piston concept invented in 1690 by Denis Papin. These refinements made a huge difference to the effectiveness of the engine – and the pumping. Now, there was a steam engine that was more capable of lifting water out of deep mines, though still not very efficient.

The crucial development was the transformation of Newcomen's engine into an efficient and practical appliance that could be used in factories, and eventually installed in ships and locomotives too. That came from James Watt. It was Watt who perfected the steam engine. It was in 1764 that Watt first became familiar with the Newcomen steam engine, when a working model was brought to him for repair. He analysed the machine and realized that certain key improvements needed to be made. For one thing, the jolting up and down movements needed to be converted into a smoother, rotary motion. For another, the alternate heating and cooling needed to be eliminated. He added a cylinder to the engine which enabled the steam to condense without impacting on the engine's performance.

Watt built a scale model of his own machine, demonstrating it in 1769. Not only did the Watt steam engine operate more smoothly and quietly, it used only one quarter of the amount of fuel. Watt formed a fruitful partnership with the industrialist Matthew Boulton. Watt was the scientist and engineer, Boulton the businessman and financier. They made an ideal team. Their machines were soon powering many different industries and gave the many-sided Industrial Revolution its biggest single thrust.

THE INVENTION OF THE MERCURY THERMOMETER

(1714)

THE FIRST THERMOMETER to resemble a modern thermometer was invented in 1714 by the German physicist, Daniel Gabriel Fahrenheit. He was the first to use mercury in a thermometer. Fahrenheit had earlier, in 1709, invented the alcohol thermometer, which is also still in use today. Sealing a column of mercury inside a glass tube resolved the various problems encountered in the use of water and other liquids. The properties of mercury make it ideal for the purpose. It does not boil as the boiling point of water is approached, and it does not freeze either; nor does it evaporate. Its opacity and silver colour make it easy to see inside the glass tube.

Today, thermometers are calibrated in standard temperature units, either degrees Celsius or degrees Fahrenheit. Before the seventeenth century there were no objective ways of quantifying heat. In 1724, the first temperature scale to be devised was called the Fahrenheit scale, after Daniel Fahrenheit. Fahrenheit's scale puts an arbitrary 180 degrees between the freezing and boiling points of fresh water.

Later an apparently more logical centigrade scale was introduced, using 100 units or degrees to separate the freezing point (00) of fresh water at sea level from its boiling point (100), though Anders Celsius, its inventor, curiously numbered the scale the other way round, boiling = 0, freezing = 100, as if he regarded temperature as a measure of coldness rather than a measure of heat. Anders Celsius was a great man of science, but this was a very strange piece of eccentricity. After Celsius died in 1744 the numbering of the centigrade was reversed. The centigrade scale as we now use it was formally named after Celsius in 1948, and it has become the standard temperature scale in use in the European Union. Celsius was a remarkable figure in eighteenth-century science. Among other activities, he took part in Maupertuis' famous expedition to the north of Sweden in 1736, with the aim of measuring the length of one degree of latitude on the ground along a meridian as close to the North Pole as it was possible to get. A parallel expedition went to Peru to measure the length of one degree of latitude close to the equator. The measurements were significantly different, and they spectacularly proved Sir Isaac Newton's prediction that the Earth would be found to be an ellipsoid: that is to say, not an exact sphere, but a sphere slightly flattened at the poles. So Anders Celsius was closely involved in degrees of latitude as well as degrees of temperature.

The first medical thermometer, used for measuring human body temperature, was invented in 1867 by Sir Thomas Allbutt. All these different types of thermometer have remained in use to the present day, though there are now some alternatives, including some digital versions. In 1984, David Phillips invented an infra-red thermometer. This has two advantages over conventional thermometers. It does not need to be inserted inside the patient's body, only placed in the ear: it also gives readings instantly.

THE INVENTION OF
INOCULATION AGAINST
SMALLPOX

(1718)

INOCULATION AGAINST DISEASE was not invented suddenly, but explored and experimented with in many regions over a long period. It is believed that inoculation was practised in India as early as 3,000 years ago. The Chinese blew powdered smallpox scabs into the nostrils of healthy people after a Buddhist nun discovered that this inoculated people who were not otherwise immune. The principle is that the subject develops a mild case of the disease and is afterwards immune to it. This technique is sometimes called variolation. It carried risks, as up to two per cent of those inoculated died.

The inoculation process spread westwards across Asia to reach Turkey. Lady Wortley Montagu, the British ambassador's wife, arrived in Turkey with her husband at the court of the Ottoman Empire in 1717. While she was there, Lady Montagu heard about smallpox inoculation from Emmanuel Timoni, a doctor from the University of Padua who was connected to the British Embassy in Istanbul. Lady Wortley had the procedure performed on her five-year-old son and four-year-old daughter – a significant act of faith – and they both recovered quickly afterwards. The inoculation was the subject of widespread interest and was reported to the Royal Society by Timoni. The Royal Society published his article in its *Philosophical Transactions* in 1713, and Timoni published it again the following year in Leipzig. Lady Wortley Montagu wrote at great length about her travels:

A propos of distempers, I am going to tell you a thing that will make you wish yourself here. The smallpox, so fatal, and so general amongst us, is here entirely harmless, by the invention of engrafting, which is the term they give it. There is a set of old women, who make it their business to perform the operation every autumn in the month of September, when the great heat is abated. People send to one another to know if any of the family has a mind to have the smallpox; they make parties for this purpose, and when they are met (commonly fifteen or sixteen together) the old woman comes with a nut-shell full of the matter of the best sort of smallpox, and asks what vein you please to have opened. She immediately rips open that you offer to her with a large needle (which gives you no more pain than a common scratch) and puts into the vein as much matter as can lie upon the head of her needle, and after that binds up the wound with a hollow bit of shell, and in this manner opens four or five veins. The Grecians have commonly the superstition of opening one in the middle of the forehead, one in each arm and one in the breast, to mark the sign of the Cross. But this has a very ill effect, all these wounds leaving little scars, and is not done by those that are not superstitious, who choose to have them in the legs, or that part of the arm that is concealed.

The children or young patients play together all the rest of the day, and are in perfect health to the eighth. Then the fever begins to seize them, and they keep to their beds two days, very seldom three. In eight days time they are as well as before their illness. Every year, thousands undergo this operation, and the French Ambassador says pleasantly that they take the small pox here by way of a diversion, as they take the waters in other countries. There is no example of any one that has died in it, and you may believe that I am well satisfied of the safety of the experiment, since I intend to try it on my dear little son. I am patriot enough to take the pains to bring this useful invention into fashion in England.

Shortly after this, in 1721, a smallpox epidemic broke out in London. The royal family were alarmed. They read about Lady Wortley Montague and her success, and they wanted to try inoculation for themselves. Their doctors warned them it was dangerous and they decided to try it out on condemned prisoners first. The inoculated prisoners all recovered in a couple of weeks. The royal family went ahead with their inoculations and assured the British people that inoculation was safe.

There were still those who argued against it, conspicuously clerics who claimed that inoculation was a tool of the Devil and smallpox was God's way of punishing the wicked. This type of objection only encouraged Lady Montague to campaign even harder. By 1723, inoculation was very common, though some doctors held out against it, notably Pierce Dod, a Fellow of the Royal College of Physicians.

In 1721, smallpox arrived in Boston, Massachusetts, on a ship from Barbados. Cotton Mather, a local preacher, asked his slave, Onesimus, if he had ever had smallpox. Onesimus gave the oddly judicious reply, 'Yes and no'. He explained that in his home country, possibly Sudan, there was a practice in which pus from an infected person was deliberately rubbed into a scratch or cut on a healthy person and that this gave immunity. This remedy from seventeenth-century Africa was evidently a precursor to inoculation.

Cotton Mather happened to be conducting a campaign to promote inoculation, but there was strong resistance to the idea from religious zealots. There was even a moment when a lynch mob threatened to hang Mather. When six patients died as a result of the inoculation procedure, Cotton Mather was called a murderer. But at the end of the smallpox epidemic in 1722, he was hailed as a saviour. Seven per cent of the population of Boston had died from smallpox. Of the 300 people who chose inoculation, only two per cent died. The statistics showed that inoculation was the safer option.

The *Gentleman's Magazine* in 1750 printed an article supporting smallpox inoculation, and twenty years later it was considered odd not to be inoculated. But there were still problems with it. There was still a two per cent mortality rate, and while subjects were sick they had to be kept in isolation. George Washington understandably hesitated before ordering his troops to be inoculated. He knew that if the British knew what they were doing they would take advantage of the temporary incapacity of the Revolutionary Army. But the strength of the smallpox outbreak led him to give the order anyway for the inoculation of all troops and new recruits who had not already had the disease.

In 1757, an English boy was inoculated against smallpox. He suffered rather badly for a month before recovering. He resolved after that that when he was older he would try to find a better way of preventing smallpox. His name was Edward Jenner.

Jenner was born in 1749 in Berkeley in Gloucestershire. He became a country doctor in his home village, where he worked and eventually died in 1823. Most of Jenner's patients were members of farming families. They lived close to livestock. In 1788 there was an outbreak of smallpox in the village. While treating his patients, Jenner noticed that those who worked with cattle and had come in contact with cowpox, a much milder disease than smallpox, never went down with smallpox. It seemed that exposure to cowpox was the equivalent of variolation. The problem was proving it. In 1796 he got his opportunity. A milkmaid, Sarah Nelmes, went to him with sores like blisters on her hands. Jenner saw that she had caught cowpox from the cows she was milking. He extracted liquid from her sores, then extracted liquid from the sores of a patient with a mild bout of smallpox.

Jenner's theory was that if he could infect someone with cowpox, they would then recover from it and so acquire immunity to the smallpox that he would inject into them later. Jenner was taking a huge risk, and he must have known it. He approached a farmer called Phipps and asked if he might inoculate his son James against smallpox. He explained to Mr Phipps that if he was right, the boy would never catch smallpox. It is hard to imagine why, but Mr Phipps agreed. Jenner then made two small cuts on James Phipps's left arm, poured on the cowpox liquid and bandaged them up. As expected, James was mildly unwell with cowpox, but recovered again. Six weeks later, Jenner performed the second, far more dangerous, operation, infecting James with smallpox. If the boy had died, Jenner would probably have been arrested for murder. If the boy did not catch smallpox and remained healthy, a remedy for smallpox had been discovered. It was an all-or-nothing gamble that worked. The simple experiment neatly proved that inoculation with cowpox gave people immunity from smallpox. This was a momentous medical breakthrough, and one that has saved hundreds of thousands of lives.

After carrying out more tests, Jenner published his findings in 1798. He called his newly invented technique vaccination, after the Latin *vaccinia* (cowpox). It was Jenner who introduced the word virus too.

Inevitably, like Darwin a few decades later, Jenner had to put up with a lot of ridicule. There was a cartoon showing cows sprouting from various parts of people's bodies supposedly after they had been vaccinated by Dr Jenner. The cartoonist Gillray had some fun at Edward Jenner's expense.

In the end, as more and more doctors found that Jenner's vaccination really did work, the enormous benefit dawned on people. By 1800 most doctors had come round to it. The British Parliament awarded Jenner £30,000 to continue his work on vaccination. The death rate from smallpox dropped dramatically in Europe and America. Onesimus meanwhile had long since been released from service by Cotton Mather, not for helping the cause of inoculation but because Mather found him an unsatisfactory slave – he was disobedient.

THE INVENTION OF THE DIVING SUIT

(1715)

THE DIVING SUIT is one of those inventions that is surrounded with legends and folk tales. This may be partly to do with a fantasy that people had of walking about on the seabed and harvesting the treasure from wrecked ships. The great medieval innovator and scholar, Roger Bacon, has been credited with inventing the first diving apparatus. But this is no more than a tradition, an apocryphal story that might be told about a magician and alchemist. The earliest reliable report of a diving bell comes from the middle of the sixteenth century. It tells of an 'aquatic kettle', which was demonstrated by two Greeks for the benefit of the Emperor Charles V at Toledo in Spain. The aquatic kettle was just like the modern diving bell, except that it had only the finite supply of air contained within the bell. There were no means for replenishing the air, so the divers could only stay under water for a short time. Once the air had become too stale to breathe, the divers signalled that they wanted to be pulled up.

The diving bell was only useful for visiting the seabed, or for looking at structures underwater. It was very difficult to do any actual work underwater from within a completely rigid structure. Dr Halley invented a diving bell in the reign of George I. It was constructed of wood, like a barrel, and loaded with lead along its lower edges to keep it upright and stable. It also had thick glass windows fitted into the top. The water supply was renewed by sending down weighted casks of air that could be released into the bell. As the air casks were lowered to below the lower edge of the bell, the divers reached out, pulled them in and turned them over to let the air escape into the bell.

Starting in the eighteenth century, a variety of inventions addressed the various requirements for diving. Divers needed to be able to descend to significant depths for salvage purposes and for bridge construction. This led to the development of a rigid diving suit that could withstand the pressure. Divers also needed some freedom of movement in order to be able to work under water. This led to the invention of the autonomous diving suit.

In 1715, John Lethbridge designed a completely enclosed diving suit, just large enough to contain one person. It was a reinforced, leather-covered, air-filled barrel. It was fitted with a glass porthole so that the diver could see out, and two armholes with watertight sleeves.

In the same year, an English inventor called Becker gave a demonstration of his version of the diving suit. Becker's suit was a full leather suit, as rubber was as yet unknown, with a large spherical metal helmet with a window. Three tubes ran from the helmet up to the water surface; one was for exhaled air and the other two were for fresh air pumped down by several large bellows. During the demonstration, the diver stayed under water for an hour, which was a convincing display of the suit's fitness for purpose, though there is no record of the depth of the water.

Later in the eighteenth century, in 1771, a Frenchman named Freminet devised a brass diving helmet which had eyeholes. The air was pumped with bellows into a small air reservoir, then pumped down to the diver.

In the 1770s the great engineer John Smeaton constructed a diving bell which was apparently the first to be continuously supplied with air pumped from the surface. It is known that in 1779 Smeaton used the diving bell to repair the foundations of a bridge in Northumberland. In 1788 he worked on the construction of Ramsgate harbour, for which he built an iron diving bell weighing two and a half tons.

In 1797, Klingert invented a diving suit that was very similar to Becker's. It consisted of a jacket and trousers made of waterproofed leather, a round helmet with a porthole and a couple of air pipes leading up from the helmet to the surface.

In 1813, John Rennie used a diving bell that was twice the weight of Smeaton's Ramsgate diving bell. It weighed five tons, and worked on the same principle. This was really the first of the modern generation of diving bells. From then on, through the nineteenth century, engineers built larger and larger diving bells. The one built to fix the large foundation blocks for the North Harbour Wall at Dublin was twenty feet square and weighed over eighty tons. The workmen entered it from above through an access shaft and an air lock; by this stage, the diving bell had become a caisson.

The German inventor August Siebe designed the first diving suit with weighted feet in 1819. His suit too had a spherical metal helmet, though this time with three round portholes, so that the diver had 180-degree vision. Like the earlier models, Siebe's diving suit was supplied with air by pipes and a pump up at the surface. By dispensing with the cumbersome leather suit, Siebe's design gave the diver considerable freedom of movement. The problem was it only worked while the diver remained vertical. The moment the diver bent over, water could enter the helmet under its lower edge and the diver would drown. Siebe's 1819 diving suit was really only a miniature diving bell. Siebe improved on his initial model in 1837 by adding a waterproof garment, bearing close similarities to the earlier English diving suits. This diving suit, which incorporated all the best features of previous diving suits, remained in used for 100 years.

Joseph Cabirol designed his own version of the diving suit in 1855, but really it directly plagiarized the Siebe suit. The spherical helmet was this time fitted with four portholes. It also had a security system. The air intake hose was attached to the helmet close to the right ear, with a valve that could be operated by hand. There was also a safety hose attached close to the mouth, called 'the whistle'. The Cabirol diving suit was a great commercial success, partly because of the public demonstrations and stunts he performed – he once lowered a convict in his diving suit to a depth of 120 feet – and partly because he put his suits on the market at half the price of British diving suits.

THE INVENTION OF
THE FLYING SHUTTLE

(1733)

JOHN KAY (1704–80) was the last of twelve children born to a farmer at Park, just north of Bury in Lancashire, his father dying before he was born. The boy was apprenticed to a reed maker. Reeds are comb-like attachments for hand looms designed to keep the warp threads separate from one another. Reeds were at that time made out of cane or reeds. John Kay showed his first flair for innovation by designing an improved version in polished wire. It was such a marked improvement that weavers all over England wanted the new wire reeds. John Kay was kept busy, making, selling and fitting his new wire reeds to looms across a wide area.

In 1730, John Kay was back in Bury, where he invented a machine for twisting and cording worsted and mohair. Then, in 1733, he made his major breakthrough, the flying shuttle.

Before John Kay invented his flying shuttle, weaving was a relatively slow process. The shuttle contained a bobbin on which the weft (the yarn going across the cloth) was wound. The shuttle was pushed by hand from one side of the warp (the yarn running lengthways through the cloth) to the other.

When the shuttle was passed backwards and forwards by hand, it placed an upper limit on the width of cloth that could be woven. The width of the cloth depended on the weaver's reach to left and right, so cloth could never be more than six feet wide. Kay's invention was to throw a shuttle mounted on wheels mechanically by pulling a cord attached to a driver. When the cord was pulled to the left, the driver made the shuttle shoot, or 'fly', through the warp in that direction. Pulling the cord to the right made it shoot to the right. The flying shuttle allowed the shuttle to pass backwards and forwards across a much wider bed; looms could be built much wider and cloth width too could be much greater if required.

Above all, the flying shuttle was fast. It speeded up the process of weaving and so improved production rates. By doing that, it enabled cloth to be produced in larger quantities at lower costs. It was the beginning of an industry of mass textile production. Operating a loom with a flying shuttle needed only one person instead of two. Here was the root of the textile workers' grievance.

Kay's flying shuttle was the vital first step in the industrialization of textiles, paving the way for other inventions. In particular, it prepared the way for power looms, although they were not to be invented (by Edmund Cartwright) for another thirty years. John Kay was seen by many textile workers as taking work away from people, as a destroyer of jobs. In 1753, his home was attacked and his looms wrecked by angry textile workers.

The mill owners were keen to adopt Kay's flying shuttle. They were less keen to pay him any royalties for it. In fact they formed a manufacturers' association, agreeing among themselves to pay him nothing. Kay tried to defend his patent, spending all of his money in the futile legal battle. Eventually, in despair, he emigrated to France, where he died poor in about 1764. There is an impressive posthumous monument to John Kay in the Kay Gardens in Bury, which does nothing to compensate for the shabby way in which he was treated when he was alive.

THE INVENTION OF THE SHIP'S CHRONOMETER

(1735)

IN THE EARLY days of navigation at sea, people usually kept within sight of land. It was easy, with a little practice, to recognize the coastline of your own home territory and the coastlines of your neighbours' territories. By learning lists of landmarks, it was possible to memorize long itineraries and find your way, coastwise, round long distances of the shores of Britain. This is what happened in the prehistoric period, and what went on happening to a great extent in the Middle Ages too.

The problems began when sailors lost sight of land. One significant clue to a ship's position was the altitude of the sun in the sky. On 21 March and 21 September, the equinoxes, the sun is overhead at midday on the equator. The angle the sun makes with the horizon is known as the sun's altitude. If the observer was on the equator, latitude 00, the sun's altitude was 900. If the observer was at the North or South Pole, latitude 900, the sun's altitude was 00. It was therefore possible for people to calculate their latitude very easily from the sun's altitude.

Latitude told navigators how far north or south they were, but not how far east or west. To pinpoint their position exactly, which is what they needed to be able to do, they needed this second co-ordinate. They needed to be able to fix their longitude. This could not be read from the sun's position or the position of any other heavenly body either.

This uncertainty about longitude was one reason why early maps were very inaccurate. It was also a reason why ships went aground on reefs.

When Christopher Columbus sailed west across the Atlantic Ocean, he discovered land on the far side of the ocean, partly because of a misconception about the size of the Earth, and partly because

Portuguese sailors had already made the crossing unofficially. But he was at least partly convinced that when he reached the Caribbean he had reached what we now call Indonesia, and that is how the islands of the Caribbean came to be known as the West Indies, to distinguish them from the East Indies, which were the real Indies.

The map of the world could only be drawn accurately if there was an accurate way of determining longitude. Navigation could only be made safe if sea captains had a way of determining their location with two co-ordinates instead of one. During and after the great voyages of discovery of the fifteenth and sixteenth centuries, the absence of a way of measuring longitude became an acute problem. People were sailing long distances out of sight of land, and frequently getting into difficulties. Even with dead reckoning, which was calculating a position by sailing in a known compass direction at a measured speed for a measured length of time, captains could still not be certain where they were in relation to features shown on inaccurately drawn maps.

The situation was brought to a head by a particular incident. In 1707 Sir Cloudesley Shovel was returning home to England in command of a squadron of ships after an unsuccessful attack on the French port of Toulon. Sir Cloudesley was on board his flagship, the *Association*, sailing north under full sail when he ran into rocks near the Isles of Scilly on 22 October 1707. The *Association* was badly holed and those on board the *St George*, one of the other ships in the squadron, watched helplessly as the flagship sank in less than four minutes. The ship sank so quickly that everyone on board, Sir Cloudesley included, perished.

The *Association* and four other ships in the squadron were wrecked on the Gilstone Reef. In all, between 1,600 and 2,000 men were drowned in the disaster. The accident was one of the biggest peacetime disasters in British history and rightly regarded as a national disaster. Up to 2,000 men, five ships and the commander-in-chief were lost. The *Association* was also carrying a vast amount of treasure. There were chests of gold and silver coin and plate that were put on at Gibraltar by British merchants. There were chests containing British government funds for the war in France. There were chests containing Sir Cloudesley's own cash and still others containing regimental funds and silverware. The financial loss alone was colossal.

The squadron had sailed northwards onto a known reef and therefore Sir Cloudesley must have mis-navigated. The visibility was very poor that day and he and his officers had been unable to navigate by eye. He was either further to the east or west than he had believed, or the maps were inaccurate and had misled him. In other words he had not been aware of his longitude, or the longitude of the reef.

The outcome of the wreck of the *Association* was the offer of a huge cash prize for the invention of an instrument that could accurately measure longitude. A British parliamentary committee consulted scientists, Sir Isaac Newton among them, and they advised that a seaworthy clock that could be trusted to keep accurate time would supply the information needed – a chronometer. The level of accuracy the committee required was the calculation of the longitude of the port of arrival in the West Indies at the end of a voyage from England; this meant that the chronometer had to keep accurate time for at least six weeks.

For twenty-three years there were no contestants for the substantial cash prize, apart from the Revd William Whiston and Revd Humphry Ditton. They entered a scheme based on ships kept moored at fixed intervals along the major shipping routes. Each night at midnight, each ship would fire an exploding rocket one mile into the air. This would be heard and seen eighty-five miles away, and it would inform navigators whether they had to correct their shipboard clocks. This ingenious scheme was no good, because the moored ships would each have needed clocks that were as accurate as the desired chronometer anyway. It was a circular solution to the problem.

After a very long struggle, this prize was eventually won by John Harrison, who invented the ship's chronometer. In fact Harrison manufactured several clocks that were so accurate that they lost no more than one second per month, which was more than accurate enough for the purpose. The main problems Harrison had to overcome were those of making the chronometer seaworthy. Because ships swayed and pitched so much, it was impossible to use a conventional pendulum clock. The chronometer had to carry on working regardless of big changes in temperature and humidity too.

In 1730, Harrison met Edmond Halley the Astronomer Royal and also a member of the Board of Longitude. Halley looked at Harrison's plans and saw that, if the chronometer worked, it would solve the longitude problem. Harrison was encouraged to build his chronometer with a view to a trial, and he completed the first one, H1, in 1735. It was given a successful sea trial to Lisbon. The Board was favourably impressed and awarded Harrison £500 to enable him to build an improved model that would take up less deck space. H2, which was completed in 1739, was narrower and taller than H1. H2 did not have a sea trial, and Harrison went on refining and redesigning until he produced H4. This was Harrison's most famous invention. It was a great advance on the earlier chronometers because it was much smaller. It was just over five inches in diameter, like a large pocket watch. It was technologically very impressive, and a beautifully intricate piece of craftsmanship.

Harrison's H4 chronometer was given its transatlantic sea trial in 1761. On arrival in Jamaica it proved to have lost just five seconds, which amounted to an error of one minute of longitude. With this triumph, John Harrison qualified for the award, but the Board of Longitude failed to pay up. George III himself intervened when he saw Harrison's final version, H5, in 1772, and forced the Board to pay Harrison his prize money.

The rigorous testing of Harrison's chronometer had proved its worth, yet almost incredibly Captain Cook set off on his earlier voyages without it. In July 1771, though, James Cook set sail from Plymouth with the *Resolution* and the *Adventure* with a copy of Harrison's H4 chronometer on board. This enabled Cook to make far more accurate charts of his discoveries than were ever previously possible. It also accurately fixed John Harrison's place in history. John Harrison's invention enabled navigators and explorers to pinpoint locations with far greater accuracy than ever before. As a result, the regional and world maps produced in the later eighteenth and nineteenth centuries were far more accurate than the charts Sir Cloudesley Shovel had been using in 1707. It also became possible for ships' captains to navigate their courses more accurately – and safely. The chronometers were difficult to manufacture, and not susceptible to mass production, so their use spread slowly, but by 1850 every vessel in the British Navy was carrying three Harrison chronometers.

THE INVENTION OF THE LIGHTNING CONDUCTOR

(1752)

THE LIGHTNING CONDUCTOR was famously invented by Benjamin Franklin in 1752. The story nevertheless had its beginnings in France. On 10 May 1752, a thunderstorm broke over a French village, Marly-la-Ville. As it did so, an old soldier acting on instructions from Thomas-Francois Dalibard, the naturalist, drew sparks from a tall iron rod that had been deliberately insulated from the ground. The experiment showed that thunderclouds contain electricity, and that lightning is a discharge of electricity. We accept this as a matter of fact today, but in mid-eighteenth-century France the news was sensational. Dalibard's collaborator in Paris, Delor, verified the finding and within a matter of weeks other people all over Europe successfully repeated the iron bar experiment.

Dalibard and Delor reported their results to the Académie des Sciences in Paris, where they acknowledged that they had applied Benjamin Franklin's observations. They were referring to Franklin's pamphlet *Experiments and Observations on Electricity*, made at Philadelphia in America. The publicity gained by the French experiment was important not least because it drew attention to Franklin's paper. By chance it was just a few weeks after this, in June, that Franklin carried out his experiment with the kite and the key, though he at that point had not heard about the Marly-la-Ville experiment.

What Franklin did was to draw sparks from an iron key tied to the conducting string of his kite which was insulated from the ground by a silk ribbon. The French observations and Franklin's own experiments proved the validity of Franklin's supposition that tall metal rods planted in the ground can protect buildings from lightning damage.

This much was science, but there was a problem in overcoming centuries of folklore and superstition regarding thunder and lightning. Christians had grown up with the idea that lightning bolts in particular were hurled by God 'to discipline his servants' and in effect punish people for their sins. Some saw thunderstorms as the work of the Devil or believed the storms were called up by witches. Quasi-religious interpretations like these were encouraged by clergymen and even theologians. Thomas Aquinas wrote, 'It is a dogma of faith that demons can produce winds, storms and rains of fire [lightning] from heaven.' Centuries later, Pope Gregory XIII was in favour of priests 'exorcizing the demons who do stir up the clouds'. Even Martin Luther supported this superstition. He said that winds were good or evil spirits, and that a stone thrown into a certain pond in his home area would create a terrible storm because of the devils that were held prisoner there. Catholics and Protestants alike were gripped by superstitions about storms.

It followed that the only acceptable remedies for protecting churches and cathedrals from lightning strikes were the old sacred remedies – ringing the church bells and reciting prayers. So, when Benjamin Franklin invented his scientific lightning rod in 1752, few Christians were keen to put it on their churches. Benjamin Franklin was a scientist, an arch-heretic, and there was no question of putting his rod of iron next to the holy cross of Christ. Few Christians were ready to give up blaming (or praising) God (or the Devil) for thunder and lightning.

Times have gradually changed, though, and rationalism has crept quietly in. Churches and cathedrals everywhere have discreetly placed lightning conductors running up their towers and spires, to save them from secular storms.

THE INVENTION OF
THE RIBBING MACHINE

(1758)

THE RIBBING MACHINE was invented by Jedediah Strutt (1726–97). He was born at South Normanton in Derbyshire, where his father farmed. At fourteen he was apprenticed to a wheelwright at Findern, not far from Derby. He had lodgings with a hosier called Woollatt, and he eventually married Woollatt's daughter. Naturally there was talk of the problems of the hosiery trade, and William Woollatt his brother-in-law described to him the efforts that had been made, unsuccessfully, to make ribbed as well as plain stockings on the stocking frame.

It was a problem and Jedediah Strutt set about devising a solution. He invented Strutt's Derby ribbing machine. Patents were granted to Strutt and Woollatt in 1758 and 1759. Strutt went to live in Derby, where he and his brother-in-law opened a factory making Derby Patent Ribs, which became popular immediately. In 1762, the two men joined Samuel Need, a Nottingham hosier, and carried on their business in the two towns.

In 1768 they were approached by Richard Arkwright, who had been advised to consult Need about the potential of his cotton-spinning frame. Strutt saw its value straight away and was able to suggest some improvements to it. The new firm of Arkwright, Strutt & Need opened its first cotton mill in Nottingham. In 1780, after Strutt had parted company from Arkwright, he built his own mills at Belper and Milford. The partnership with Need ended in 1773 when the patents expired. But by then Strutt had made a discovery that revolutionized calico; he found that cotton and cotton alone could be used to make it. To house the newly invented gear to make calico in Derby, he built the first fireproof mill in Britain.

Jedediah Strutt was a very characteristic figure of the Industrial Revolution, a self-made man with one eye on technical invention and the other on commercial success. He died in 1797. His son William Strutt (1756–1830) was another great technical and mechanical innovator. It was William Strutt who designed the fireproof calico factory. He was particularly interested in solving the problem of heating houses and went on to invent the Belper stove.

THE INVENTION OF THE SPINNING JENNY

(1764)

THE ACCEPTED HISTORY of the Industrial Revolution tells us that James Hargreaves invented the spinning jenny. The truth may be different, but the official story runs as follows.

James Hargreaves led a simple existence, working in the cottage industry that spinning was in the first half of the eighteenth century. His whole family was involved in spinning and weaving. In the Hargreaves' home, fibre (either cotton or wool) was spun into thread, and thread woven into cloth. What Hargreaves designed was a machine to save himself and his family labour.

In 1764, he built a machine to be operated by one person, which would spin several threads at once. Initially, it had eight spindles, but Hargreaves saw that the machine could handle many more. Hargreaves' machine, supposedly nicknamed after his wife, Jenny, was an instant hit with the merchants and entrepreneurs. Hargreaves was then able to supplement his family's poor earnings by making and selling spinning machines.

Inevitably, others involved in the industry saw the development as a threat to their employment. In 1768, as spinning jennies multiplied, James Hargreaves' neighbours attacked his house and destroyed his machines. Hargreaves took his family to Nottingham, where he founded a small spinning mill. His invention was one of those that transformed the industry, one of the major stepping stones in the Industrial Revolution, yet Hargreaves was overlooked and forgotten, and certainly received no recognition for his contribution to the transformation of the economy. He was even refused a patent for a long time. It was only when he produced a sixteen-spindle version in 1770 that the patent was finally granted. Unfortunately by then Hargreaves had made and sold too many of the eight-spindled machines and competitors had already copied and developed his invention. By the time of Hargreaves' death in 1778 there were more than 20,000 spinning jennies in use in Britain alone.

But there is an alternative view of events. As we have seen in many other situations, it is quite possible for two or more people to come up with an innovative idea independently at the same time. It looks as if someone else invented the spinning jenny, and that someone was Thomas Highs. Very little is known about Highs, except that he was an ordinary workman from Leigh and he designed and built the spinning jenny perhaps a year before Hargreaves. Highs had no money, and therefore could not afford to patent or develop his invention. He built several machines and rented them out, but did not have the resources to take the idea any further. It is not known how Hargreaves came into possession of the design for the jenny. It is possible that he acquired one of Highs' rental machines and copied it.

Hargreaves certainly deserves credit for his perseverance in developing and promoting the use of Highs' jenny, and seeing that it found its way into general use. A great many inventions never see the light of day. Even so, it should be remembered that he was probably making use of a machine invented by Thomas Highs. And Hargreaves' wife was not called Jenny: her name was Elizabeth.

THE INVENTION OF THE CONDENSING STEAM ENGINE

(1765)

IN 1759, JAMES Watt began to think of steam power as a force that might be harnessed, to an extent that went far beyond its current restricted use to drive pumps in coal mines.

In 1763, in the course of his routine paid work as mathematical instrument maker for the University of Glasgow, he was sent a working model of the Newcomen steam engine to repair. This apparently routine repair job turned out to be a historic moment. Watt put the model engine back into working order and while he was doing so he noticed some significant defects in the machine's design. The old Newcomen engine was in fact extremely inefficient, consuming enormous quantities of steam (and therefore enormous quantities of coal) to do relatively little work. He also saw that fitting a separate condenser might be a way of improving it. Later he found additional ways of improving the engine's performance, including an air pump to remove the spent steam, a steam jacket for the cylinder to make sure the cylinder was as hot as the steam entering it and a double-action for the engine.

Watt's design improvements were relatively easily made in his mind, but making them and testing them cost him a great deal of money and he put himself into debt in order to demonstrate the merits of his improved engine. He allowed Dr Roebuck, the founder of the Carron Ironworks, two-thirds of the profit from his invention in return for bearing the costs. An engine was then built at Kinneil near Linlithgow. In the course of constructing it, Watt saw further ways of improving the machine.

In 1768, James Watt met Matthew Boulton, who owned the Soho Engineering Works. Boulton agreed to buy Roebuck's share of Watt's steam engine so that they could work together. In 1774, Watt set up in partnership with Boulton, and the improved steam engine went into production at the Soho Engineering Works, protected by a patent Watt had prudently taken out in 1769. The partnership was an unusually happy one. Watt saw to the engineering side and Boulton looked after the business. Between 1781 and 1785, Watt took out no fewer than six patents for further component devices, including the centrifugal governor, a self-regulating device that ensured an even running speed in a rotating steam engine.

James Watt's improved steam engine soon superseded Newcomen's as the engine of choice for mine and other pumps. For one thing, it used only a quarter of the fuel – a major saving. But Watt's imagination ran ahead. Why should this extremely efficient machine be used just to drain mines? Why not use it to drive carriages along roads or ships across the sea? In 1784, Watt described a steam locomotive in one of his patents, yet for some reason discouraged William Murdoch, his assistant, from experimenting with steam locomotion. What they had in mind was a steam carriage that would travel along ordinary roads. Watt had not thought of using rails, and when that was proposed by others in the last years of his life he would not consider it at all.

His son, also called James Watt, in 1817 fitted his father's steam engine to the first English steamship to leave port, the *Caledonia*. James Watt's outstanding contribution to the Industrial Revolution – and to life in the nineteenth and twentieth centuries – was the improved and more efficient steam engine, the engine that was to transform the nineteenth-century world. Watt envisaged using his improved engine for locomotion. It was largely due to Watt's inventions and the imagination behind them that the steamship era began – and, in spite of himself, the great age of the railway. Watt's steam engine was one of the greatest and most significant inventions of the past three centuries.

THE INVENTION OF VENETIAN BLINDS

(1769)

VENETIAN BLINDS ARE a popular alternative to curtains and in some settings, such as offices, they have been found to be far more practical in controlling and adjusting the amount and quality of daylight coming into a room, from full to none.

As far as British law is concerned, Venetian blinds were invented in 1769. It was on 11 December in that year that Edward Bevan was issued a patent in London. It nevertheless looks as if Venetian blinds already existed before that date. Possibly Marco Polo saw Venetian blinds in Persia, and brought them back to Venice. Because of their exoticness and novelty, they acquired a high profile. It seems they were for a while known as Persian blinds. From Venice, the newly introduced blinds spread from city to city.

The technical design of Venetian blinds has remained unchanged since the late eighteenth century. The earliest surviving technical drawing of a Venetian blind is the work of Roubo, in *L'Art du Menuisier*, published in 1769. The drawing shows suspended multiple slats that are tiltable and can be raised and lowered – exactly like Venetian blinds today.

The blinds existed in America two years earlier, interestingly two years before Bevan took out his English patent. In the *Philadelphia Journal* of August 1767, John Webster advertised:

> *The newest invented Venetian sun blinds for windows, on the best principles, stained to any colour, moves to any position, so as to give different lights, screens from the scorching rays of the sun, draws a cool air in hot weather, prevents from being overlooked, and is the greatest preserver of furniture of the kind ever invented.*

Perhaps the most interesting feature of the Venetian blind is the sheer lack of development over the last 230 years. There have been no quantum leaps. The reason is that the blind had been developing for a long time before its appearance in the West, and that it had long ago gone through the process of refinement and improvement. By the 1760s the blind had by a long process of evolution, perhaps lasting hundreds of years, reached its most efficient form. It cannot be bettered.

THE INVENTION OF COADE STONE

(1771)

COADE STONE WAS a moulded artificial stone with a tough surface finish. It was manufactured and widely used in Britain for making ornamental stonework on buildings and monuments in the late eighteenth and early nineteenth centuries. The stone is no longer manufactured, but it was relatively common 200 years ago. It was a compound that looked very like terracotta, but it was denser and smoother, and ideally suited to resist damage by weathering in a British climate.

Coade stone was made by only one company, owned by Mrs Eleanor Coade and her successors. The Coade family had experienced major financial problems. Mrs Coade's husband had been bankrupted twice, the second time shortly before his death in about 1770. Mrs Coade and her daughter moved house to Narrow Wall, Lambeth, which shortly became the address of the Coade Factory. The younger Eleanor remained unmarried, but confusingly adopted the title 'Mrs' apparently to give her name more commercial gravitas. She took on a business partner, her cousin John Sealey, and she later took on a manager, William Croggon, who eventually acquired the company. The younger Mrs Coade died in 1821. She did not leave the factory to Croggon, as he may have expected; he had to buy it. Croggon continued to run the business until 1833. He went bankrupt in 1833 when one of his clients, the Duke of York, failed to pay his bill.

The Coades set up in business making artificial stone at a good moment. Robert Adam returned to England from his adventures in Rome and at Herculaneum, and used a great deal of Coade stone for the ornaments in his architectural designs. Two of the designs Adam brought back from Italy were copies of the Medici and Borghese Vases. Mrs Coade copied the Borghese Vase and a pair of these copies was installed conspicuously at Kedleston Hall, at the top of the curving steps up to the spectacular garden entrance.

Artificial stone had been made in Lambeth since 1720 by Richard Holt, and the Coades probably deliberately chose this as the local craft. Holt's products were apparently of low quality, with poor design and finish. According to one version of events, the Coades bought Holt's business, but it seems more likely that they bought the artificial stone business from David Picot, who is known to have moved to the address that later became the Coade Factory. An advertisement placed in the newspapers in 1771 by Eleanor Coade makes it clear that the factory was no longer David Picot's.

The stone and the method for manufacturing it are said to have been invented by the elder of the two Eleanor Coades. The younger Eleanor Coade, her daughter, had great commercial ability and was able to exploit her mother's invention. To ensure that their firm had no rivals, and there could be no competition in the production of Coade stone, the two women kept the manufacturing process a closely guarded secret. This made good commercial sense. For a long time it was believed that the younger Eleanor Coade took the secret to the grave, and that now nobody knows how to make Coade stone. This would have been a great loss, as it has turned out to be even more successful, with the passage of time, than the Coades could have realized. Objects moulded in Coade stone are still unweathered, and still as sharply detailed as when they were first manufactured.

In fact the manufacturing process for Coade stone was not entirely lost. Coade stone can still be produced. It is unfortunately used only rarely. One example is a sculpture called The Seated Figure by Mollie Adam and fired by Dogenes Farri of East Sydney Technical College in New South Wales. Modern kilns can easily reach and maintain the temperatures required to make Coade stone, but in the Coades' time the firing methods were fairly crude. Making Coade stone today is relatively straightforward, but making it then was a major achievement.

At first sight, Coade stone looks as if it might be a type of concrete, setting by chemical action, but it is in fact a ceramic, a form of stoneware that has to be subjected to high temperatures for several days in a kiln. Ceramics of this type have been known since antiquity. The famous Terracotta Army dating from around 206 BC shows how old the technology is. The 6,000 or more terracotta soldiers made to act as a bodyguard for the Qin emperor in the afterlife were fired in much the same way as Coade stone.

Mrs Coade's original name for her stone from the middle of the 1780s was Lithodipyra, a composite word made out of three Greek words meaning stone-twice-fire. The material used was sixty per cent ball clay from Dorset and Devon, ten per cent crushed flint, ten per cent sand, ten per cent crushed soda lime glass and ten per cent 'grog'. This grog was finely crushed up material that had already been fired once. The technique of adding grog was not a new one; it was used in Bronze Age pottery and Roman brick, and even in the orange bricks that were used to outline the Long Man of Wilmington in the mid-sixteenth century.

The mix of materials could not be hand-moulded because it was so stiff. Instead, a model had to be made in some other material and a mould made of it. Then the Coade stone mix was pressed into the mould. It was an expensive production method, but once the mould was made, any number of copies could be produced from it. The process was very reliable, and could produce elaborately detailed work with no distortion. The Borghese Vase was modelled in 1771, but copies were repeatedly made, the last known being made in 1827.

Among the many monuments made in Coade stone were the tomb of Captain William Bligh in St Mary's churchyard at Lambeth and the large wall monument to General Bowes in Beverley Minster. Lawrence Castle in Devon has a castellated folly with a large Coade stone figure of General Stringer Lawrence dressed as a Roman general. In 1809, when George III reached the fiftieth year of his reign, Weymouth in Dorset ordered a life-sized statue of the king flanked by a lion and a unicorn. Orders for several more copies of this statue followed. One survives in fragments at Lincoln. In 1810, the Earl of Buckingham mounted it on top of a pillar at Dunston on the Lincoln–Sleaford road, to act as a kind of inland lighthouse. In 1940, George III was pulled down as a danger to low-flying aircraft. The pieces were kept and eventually partially assembled into a bust of George III by John Ivory, a local mason. It is known as *The One-Third*. The original artist was Joseph Panzetta, who brought some incredibly sharp detail to the robes and Garter chain.

THE REINVENTION OF
THE FLUSHING WATER
CLOSET

(1775)

SIR JOHN HARINGTON had invented the flushing water closet at the end
of the sixteenth century. But that was an invention that came and went.
Perhaps it was premature. Whatever the reason, it had to be reinvented.

It was Alexander Cummings who reinvented it in 1775. Cummings
added a detail of his own, the strap. This was a sliding valve positioned
between the bowl and the trap, and it had not been thought of before.

Other designers quickly followed. In 1777, Samuel Prosser was
awarded a patent for a plunger water closet. In 1778, Joseph Bramah
came up with a different design again. Bramah's design had a valve at
the bottom of the bowl which worked on a hinge, a precursor of the
modern ballcock. Bramah's closet was adopted extensively on ships.

Shortly after this came Thomas Twyford, whose firm was to
dominate the manufacture of water closets through the second half of
the nineteenth century and the whole of the twentieth. By 1849,
Twyford was manufacturing both wash basins and closet pans, and
exporting them to many countries, not only in Europe, but further
afield, including America and Australia. The water closet now had a
global reach. In 1885 the manufacture of water closets entered an
important new era when for the first time the toilet was made, trapless,
in a one-piece, all-china design. Twyford had competitors, such as
Moulton and Wedgwood, but the new design was a great advance on
the many other closets on the market, many of them contraptions made
of wood and metal. The new Twyford all-china closet was not only

mechanically effective and efficient, it was far more hygienic than any earlier water closet. Twyford's competitors quickly followed suit.

Developments in North America were only slightly behind developments in Europe. Migrants to the New World had not taken Sir John Harington's water closet with them: only chamber pots. But by the end of the nineteenth century the innovations in water closet design were closely shadowing those in Britain. In the twentieth century the innovation continued, and the US Patent Office received 350 applications for new water closet designs, just in the first third of the twentieth century.

The effect of this invention cannot be exaggerated. It has greatly improved domestic hygiene, reduced the incidence of gastric infections, and improved the general standard of health in the community.

THE INVENTION OF THE SPINNING MULE

(1779)

SAMUEL CROMPTON, THE inventor of the spinning mule, was born in 1753 on a farm. A poor man, Crompton laboured for over five years to perfect the spinning mule. While he was doing this he supported himself by working as a violinist in the pit at the Bolton Theatre. He created the spinning mule in 1779 out of two pre-existing inventions. One was the water frame, at one time thought to have been invented by Richard Arkwright but now believed to have been invented by Thomas Highs; the other was the spinning jenny, at one time thought to have been invented by James Hargreaves, but also very likely invented by Thomas Highs. The new hybrid machine, the spinning mule, was designed to create high-quality textiles quickly.

Before the weaving process can begin, the yarn has to be spun. In the eighteenth century this was done by women and children, but several of these spinners were required to keep one weaver supplied with yarn. This problem had been identified and several inventors had worked towards a solution. In the middle of the eighteenth century they devised several labour- and time-saving machines to allow the yarn to be spun faster. The spinning mule was really a culmination of these efforts. A mule is a cross between a horse and a donkey; the spinning mule was a similar cross-breed.

An important reason for combining the characteristics of the two machines was that the moving carriage of the spinning jenny allowed it to spin faster, while the water frame made use of water power. Combining the two meant that the industry now had a fast machine that could exploit water power instead of man power. Later, the mule would be run on steam power.

The new spinning mule produced strong but fine yarn that was suitable to be made into any kind of textile. This made it very versatile. To begin with, the machine was used to spin cotton, but it later switched to handle other fibres too.

Like Thomas Highs, Samuel Crompton was too poor to apply for a patent to protect his invention. So he sold the rights to David Dale, who made a good deal of money out of the invention. Many of the inventors at the time of the Industrial Revolution had trouble over protecting their rights. There was a considerable amount of resentment among contemporaries that people like inventors should be allowed to control something like an idea or have exclusive rights to build a necessary machine. There was little chance for 'little people' like Samuel Crompton or Thomas Highs to control the future exploitation of their inventions. It was left to rich, powerful and unscrupulous entrepreneurs like Richard Arkwright to do that. It was in fact Arkwright who acquired the patent for Crompton's mule, even though it was not his invention.

A Commons Committee met in 1812 to deal with Samuel Crompton's claims, and reported that, 'the method of reward to an inventor, as generally accepted in the eighteenth century, was the machine should be made public, and that a subscription should be raised by those interested, as a reward to the inventor'. That may have been appropriate in times when inventions did not require much capital to develop them, but the Industrial Revolution had already changed that. The scale and sophistication of the machines coming forward now were such that an ad hoc public subscription was inadequate. Crompton died six years after this Commons Committee reported.

Crompton's mule was an important step towards increased production levels in the textile factories. It was, significantly, too large a machine to fit into most homes, and this in itself was a recognition that the days of textiles as a cottage industry were over. The speeding up of the spinning process was vital to accelerating production in the cotton industry in particular, a key industry in the Industrial Revolution in Britain.

THE INVENTION OF THE HOT-AIR BALLOON

(1783)

JOSEPH AND JACQUES Montgolfier were the sons of a paper manufacturer at Annonay in France, about forty miles from Lyons. The brothers noticed the clouds suspended in the sky and thought that if they could enclose vapour of the same nature as the clouds in a large and light bag it might rise, carrying the bag up into the air.

It sounds like nothing more than a daydream over a lunch-time bottle of wine, but in 1782 they constructed a small balloon which was lifted off the ground by igniting a cauldron full of paper underneath it. The warmed air and smoke inside the balloon decreased in density so that they became more buoyant than the surrounding air. The small experiment was a complete success, and their surmise that it was warmed air that caused many of the clouds to bubble up was vindicated.

The Montgolfier brothers decided to make a larger version of the hot-air balloon and demonstrate it in public. On 5 June 1783, in front of a large crowd of spectators, a huge linen globe a hundred feet in circumference was inflated over a fire fuelled with chopped straw. When the mooring ropes were cast off, the balloon ascended rapidly to a great height, and then came down again when the air inside it cooled down over a mile away. This single momentous event marked the invention of hot-air ballooning.

News of the flight at Annonay spread fast. In Paris, Barthelemy Faujas de Saint-Fond set up a fund to repeat the experiment. The second balloon was made by two brothers by the name of Robert, and the project was supervised by a physicist, J. Charles. Initially the plan had been to use hot air like the Montgolfiers, but Charles wanted to try hydrogen. This balloon too was given a very public audition, and it too obligingly rose rapidly to a height of about 3,000 feet. Rain began to fall as the balloon went up, but the crowd stayed to watch the spectacle, many of them getting drenched.

On 19 September 1783, Joseph repeated the Annonay experiment in Paris, in the presence of Louis XVI, Marie Antoinette and a vast crowd. The inflation with hot air only took eleven minutes, after which the balloon rose to a height of 1,500 feet and floated along in the wind for about two miles before coming down in a wood.

The Montgolfier balloon was painted with ornaments in oil colours and was very eye-catching, looking like a gigantic Christmas tree bauble. The event was beautiful, memorable – and highly significant. Suspended under the balloon was a cage containing a sheep, a cock and a duck; this trio of passengers became the first airborne travellers. The age of air travel had begun.

The first person to ascend in a balloon was Jean Pilatre de Rozier. In October 1783 he made several ascents in a captive balloon; it was anchored to the ground with ropes. He demonstrated that the balloon could rise with people suspended beneath it, and also carry fuel so that a fire could be kindled underneath the balloon while in flight. On 21 November in the same year, Pilatre de Rozier and the Marquis d'Arlandes travelled in a free balloon about two miles from the Bois de Boulogne. They were airborne for about twenty-five minutes. This was the first manned flight in a balloon.

The Montgolfier brothers were unable to develop the balloon technology any further before the outbreak of the French Revolution. Etienne fell foul of the authorities, and Joseph immersed himself in the work of the paper factory.

THE INVENTION OF
THE PARACHUTE

(1783)

THE IDEA OF the parachute was in the air long before the device itself was actually needed. The parachute was imagined, rather than invented, by Leonardo da Vinci (1452–1519). Leonardo's parachute sketch showed a sealed linen cover which was held open by a pyramid-shaped frame made of wooden poles. The original drawing was sketched in a notebook in 1483. Leonardo noted, 'If a man is provided with a length of gummed linen cloth with a length of twelve yards on each side and twelve yards high, he can jump from any great height whatsoever without injury.'

The Croatian Faust Vrancic went as far as making a parachute based on Leonardo da Vinci's drawing and used it when he jumped from a tower in Venice in 1617. Vrancic wrote a book entitled *New Machines*, in which he describes in words and pictures fifty-six advanced machines and structures. Among the fifty-six new machines was the parachute, which he described as *homo volans*, man flying.

The first practical parachute design was invented by Sebastien Lenormand in 1783. The timing was interesting, as it coincided with the Montgolfiers' first balloon flights. If people were going to ascend in balloons, they might well need parachutes to get them safely down to the ground again in an emergency. Parachutes were consciously developed with this emergency need in mind.

Another Frenchman, Jean Pierre Blanchard (1753–1809) seems to have been the first person to use a parachute in an emergency. In 1785, he went up in a balloon and from it he dropped a dog in a basket to which a parachute was attached. In 1793, Blanchard himself descended to the ground using a parachute when he was flying a hot-air balloon

227

which exploded. It was Blanchard who invented the first foldable parachute made of silk; until then all the parachutes had been given rigid frames.

On 22 October 1797, Andrew Garnerin became the first person who is known to have jumped using a parachute without a rigid frame. This was a great risk, as he cannot have known for certain that the parachute would open. Yet Garnerin did it more than once. He jumped from balloons as high as 8,000 feet up. Garnerin added a noteworthy refinement to the parachute, an air vent to allow some air through and so prevent the parachute from oscillating.

Then the inevitable first fatality happened. In 1837, Robert Cocking became the first person to die as a result of a parachute accident.

In 1887, Captain Thomas Baldwin invented the first harness for a parachute, and three years later Paul Letteman and Hathchen Paulus devised the method for folding the parachute and packing it into a knapsack to be worn on the back. This brought the parachute very close to its modern form. It was Paulus who invented the intentional breakaway – the addition of a small parachute which opens first and, in opening, pulls open the main parachute.

Grant Morton and Albert Berry were the first men to parachute jump from a plane – in 1911. The first free-fall jump was made in 1914 by Georgia Broadwick.

In June 2000, Adrian Nicholas dropped from a hot-air balloon at 10,000 feet in South Africa, trusting that a parachute based on Leonardo da Vinci's design would see him safely to the ground. Parachute experts had advised him not to do this, believing that the Leonardo parachute would not work. But it did work. Adrian Nicholas commented that the ride down was smoother than with modern parachutes. He did however cut himself loose at 2,000 feet and from there to the ground he used a modern parachute. He did not want to land on the ground and have the heavy wooden and linen construction falling on top of him. Leonardo's design was vindicated, 500 years after its invention.

THE INVENTION OF
THE POWER LOOM

(1785)

THE POWER LOOM was a steam-powered, mechanically operated version of the earlier hand-operated loom. Its function was simply to combine threads in a systematic way to make cloth.

The first power loom was invented in 1785 by Edmund Cartwright, who set up a cloth-making factory in Doncaster. He was an unusual figure in the Industrial Revolution. Many of the great inventors were jobbing artisans who saw ways of making their own work easier or making the machines they were using more effective. Others were employers or factory owners who wanted to find ways of increasing production. Very few of these men were highly educated or academic and none, I think, except Edmund Cartwright, was functioning completely outside industry.

Cartwright was nevertheless a polymath destined to become a great innovator, and he went on to invent a wool-combing machine in 1789, a steam engine fuelled by alcohol and a rope-making machine. As well as all this, he helped Robert Fulton with his steamboats. Edmund Cartwright was born into a well-to-do family at Marnham in Nottinghamshire. He and his two brothers were all to make national names for themselves in different fields. George Cartwright went into the Army, and then became a fur trapper and explorer in Canada. He earned the nickname 'Labrador Cartwright' and became a celebrity when he brought an Eskimo family to Britain. John Cartwright became a famous radical politician. Edmund Cartwright became famous as the inventor of the power loom, but he was other things besides.

Unlike most of the other major figures in the Industrial Revolution, Edmund was intensively educated. He attended Wakefield Grammar

School and University College, Oxford. He became a Fellow of Magdalen College, where he made a name – another name – for himself as a poet and critic. He wrote a mythic poem called *Armine and Elvira*, which was popular and praised by no less a figure than Sir Walter Scott as 'a very beautiful piece'. He went into the Church, in 1772 becoming curate at Brampton near Wakefield and in 1779 vicar of Goadby Marwood in Leicestershire.

While visiting a young boy who was seriously ill with typhus, he noticed a tub of yeast and remembered that there was an old wives' tale that rotting meat hung over yeast would turn sweet again. He dosed the boy with yeast, and the boy recovered. He treated other parishioners in the same way and cured them too. News of the cure spread and it became widely used by English doctors. But it is the power loom, with all its consequences, that we remember Edmund Cartwright for now. How did a man of the cloth become a man who made cloth? Cartwright himself explained:

> *Happening to be at Matlock in the summer of 1784, I fell in company with some gentlemen of Manchester, when the conversation turned on Arkwright's spinning machinery. One of the company observed that as soon as Arkwright's patent expired, so many mills would be erected and so much cotton spun that hands would never be found to weave it.*
>
> *To this I replied that Arkwright must then set his wits to work to invent a weaving mill . . . The Manchester gentlemen unanimously agreed that the thing was impracticable; and in defence of their opinion they adduced arguments which I certainly was incompetent to answer or even to comprehend, being totally ignorant of the subject, having never at any time seen a person weave.*
>
> *I controverted, however, the impracticability of the thing by remarking that there had lately exhibited in London an automaton figure which played at chess. 'Now you will not assert, gentlemen,' said I, 'that it is more difficult to construct a machine that shall weave than one which shall make all the variety of moves that are required in that complicated game?"*

When the Revd Edmund Cartwright returned home after his holiday in Matlock, he began to work seriously on the problem of designing such a machine, and succeeded in inventing one that worked.

As with other inventions, Cartwright's power loom needed refinements. It was improved upon by William Horrocks and Francis Cabot Lowell, and after that, from around 1820, the power loom went into general use. The first power loom in America was built in 1813 by a consortium of merchants in Boston led by Francis Cabot Lowell.

The power loom had far-reaching effects. One major effect of the fully developed, fully mechanized, looms was that it was possible for women to operate them. A major social change took place, as women took over from men as weavers. Although it raised production levels astronomically, it cannot be said that the power loom added anything to the sum of human happiness. As a clergyman Edmund Cartwright would have been mortified if he could have foreseen the hardship and misery his invention would cause. As the power looms took over, the hand loom weaving industry petered out, causing widespread misery and unemployment.

THE INVENTION OF
THE THRESHING
MACHINE

(1786)

SINCE TIME IMMEMORIAL, after wheat was harvested the stalks and chaff had to be separated from the grain. This was done by a simple but arduous process called threshing. The freshly harvested wheat was laid on a specially prepared hard surface, a threshing floor, and threshed – or thrashed – with a flail. This released the grain from the chaff. After the threshing, the mixture of grain and chaff on the floor was picked up in baskets or trays and tossed repeatedly in the air; the chaff was taken away by the breeze and only the grain fell back into the basket. This process was called winnowing.

Circular threshing floors have been found in Crete dating from the Bronze Age; they were in use as early as 1500 BC. The arduous process of threshing and winnowing by hand went on in much the same time-consuming way right up to the nineteenth century.

In 1830, the wheat crop produced on just five acres of land might take as much as 300 hours of labour to reap, thresh and winnow. The McCormick reaper that was patented in 1834 helped to speed up the harvesting process itself. The threshing was still often done by hand.

In 1786 a threshing machine was invented in Scotland by Andrew Meikle. His father had invented a winnowing machine as early as 1710, but it had not been generally adopted. Perhaps Mr Meikle Senior was ahead of his time. In country areas, new-fangled contraptions were often regarded with deep suspicion. The son was more successful. Andrew Meikle was a millwright at Houston Mill on the estate belonging to John Rennie at a place called Phantassie in East Lothian. Meikle was lucky to have Rennie's support; Rennie helped him to get his machines installed at other mills.

Not every farmer could afford to own a threshing machine. The big farmers could, but the smaller farmers came to rely on a travelling thresher, with its own work team, to do the job for them. The steam engineer set up the thresher either close to the wheat field where the crop was being harvested or among the farm buildings if that is what the farmer preferred. The machine was started up, and a team of bundle haulers toured the field, loading stooks of wheat onto a horse-drawn cart and taking it back to the threshing machine. Men stood on the cart and pitched bundles of wheat into the machine. The grain was separated by rotating cylinders inside the machine and the grain fell to the bottom. Chaff and dust were taken out by a fan, while a conveyor belt took the grain to a grain wagon. The stalks were blown by a fan to make a separate stack.

The threshing machine, or thrashing machine as it was initially known, had huge social and political consequences. It was partly responsible for the Captain Swing Riots. When the Napoleonic Wars ended in 1815, there was an economic slump. Farm workers had to endure more than a decade of high taxation and low wages, as well as poor living conditions. They had had to face large-scale unemployment for years, as a direct result of the introduction of the threshing machine – and also the continuing enclosure of the agricultural landscape. Before the Agricultural Revolution there had been work for scores of men on the land even in the smallest English village. With low wages and no prospect of things ever improving, the introduction of the threshing machine was the last straw. It was seen as the invention that would drive them to the point of starvation. In 1830, all over southern England farm labourers smashed threshing machines, threatened the farmers who owned them and set fire to their haystacks and barns.

The farm rioters were punished very harshly. Nine rioters were hanged; 450 more were transported to Australia. It was never clear who was organizing the riots, and the farm workers named a fictitious 'Captain Swing' as their leader.

The first threshing machines were fed by hand and powered by horses. They were also fairly small – no bigger than a piano. By the end of the nineteenth century they were built bigger, with a considerable range of processes incorporated into the one machine. Rakes, shakers and fans were added, all driven by the same machinery, so that all of the processes could be carried out in an orderly way: threshing, shaking and winnowing. On the medium-sized and larger farms, over 200 acres, steam power was more economical than horse power. The threshing machine was a great asset to the farmer, greatly reducing the cost of processing the wheat harvest.

But there was no advantage to the agricultural workers. As the mechanization process continued through the nineteenth century, fewer and fewer farm workers were needed. This led on, inevitably, to rural depopulation in the late nineteenth and early twentieth century, and a drift to the towns. By 1900, eighty per cent of the population of Britain was living in towns. Although the threshing machine cannot be blamed for all of these changes, it played a very significant part.

THE INVENTION OF COAL GAS LIGHTING

(1792)

GAS LIGHTING WAS one of the most important inventions of the late eighteenth century. It was invented in 1792 by William Murdock, the son of an Ayrshire millwright. Murdock worked with his father for a time, then in 1777 he walked the 300 miles to Birmingham, where he found James Watt and asked for a job. Watt's partner, Matthew Boulton, gave him one and later described him as the finest engine builder he had known.

Watt spent a lot of time in Cornwall trying to persuade tin mine owners to buy his steam pumping engines. In 1779 he became ill, returned home, and Murdock was sent to Cornwall instead to develop the interests of Boulton and Watt, selling pumping machinery and installing it in the tin mines. Murdock worked in Cornwall for nineteen years. During that time he spent a lot of his free time inventing. Among other machines he invented a steam-powered locomotive. He even built a three-wheeler model of this steam carriage in 1786, and ran it late one evening through the streets of Redruth. The people of Redruth must have wondered what it was that had woken them up.

One of the people who saw this steam carriage was a fifteen-year-old local boy called Richard Trevithick, who is now widely regarded as the inventor of the steam locomotive, though he borrowed the idea from Murdock's model. Murdock was keen to develop his invention, and naturally approached Boulton and Watt in the expectation that they would sponsor the building of a full-sized version. Unfortunately for Murdock, Boulton and Watt did not have sufficient foresight. They saw no future for the locomotive, and went so far as to make Murdock promise to abandon the idea.

Murdock did abandon his steam locomotive. He diverted his

inventive imagination into other areas. He invented a new slide valve for the steam engine. In 1791 he took out a patent on a process for making dyes out of coal tar and also extracting a type of paint from coal that would inhibit the growth of barnacles on the hulls of ships.

Murdock's big invention, since he was deprived of the credit for inventing the steam locomotive, was coal gas lighting. In 1792, Murdock roasted coal in an iron retort with a pipe attached. The gas released from the heated coal squirted out of the pipe and the gas jet at the end was ignited to produce a steady flame. Two years later, he developed this into a gas lighting system for his house at Cross Street in Redruth. He built a closed iron vessel in the garden and ran a network of pipes through the house to a series of gas jet burners. This was the first system of gas lighting in the world.

Almost incredibly, Matthew Boulton and James Watt opposed this great invention too. Murdock could not take any more opposition. He knew he had a good idea and that Boulton and Watt were wrong. He parted company with them in 1794 and went back to Scotland. Boulton and Watt realised that, whatever they may have thought of the locomotive or the gas lighting, they had made a terrible blunder in losing the services of William Murdock; they now realized the commercial potential of his gas lighting scheme too. In 1795, they offered him a job managing the Soho Works.

In 1801, Philippe Lebon gave a public demonstration of gas lighting in Paris – another case of near-simultaneous invention – and in 1802 Boulton and Watt agreed to let Murdock install two gas lamps outside the Soho Works. This was the first installation of gas street lighting in Britain. The next year the whole factory was lit by Murdock's gaslight. In 1806 the Phillips and Lee spinning mill in Manchester was lit by Murdock's gas lighting, and from then on it was installed in one factory after another. It was bound to catch on rapidly and it was only a short time before all the big factories were running on gaslight.

Murdock was awarded the Royal Society's Gold Medal for his invention. Murdock himself was only interested in interior lighting for individual buildings, such as houses or factories, each dependent on its own gas production equipment. He was not interested in lighting for whole streets or areas, let alone whole towns by installing gas mains. It was left to others to make that development, which required another imaginative leap.

Gas lighting in the home was a great improvement on the candlelight that most people had depended on during the hours of darkness. But reading or sewing by candlelight was tiring on the eyes, because of the flickering, unless you were sufficiently well off to be able to afford to burn lots of candles. Gaslight opened up the winter evenings for a great many people. Gas lighting in the factories was another matter altogether. It was greatly in the interests of the factory owners to keep the work of the factory going on into the evening, it made more efficient use of the machinery and raised production. But it was not in the interests of the workers, who were suddenly forced to work much longer hours. The introduction of gaslight was, for the ordinary worker, one of the worst developments of the Industrial Revolution, heralding decades of miserable shift work and often working hours that fast approached slavery. It would eventually lead to a raft of reforming legislation that imposed limits on what employers could demand, but often those reforms were ineffective against the enormous and ever-increasing power of the factory owners.

THE INVENTION OF THE COTTON GIN

(1793)

As A BOY, Eli Whitney spent a lot of his free time in his father's metalwork shop taking clocks and watches apart and reassembling them, just to find out how they worked. He had the natural intense curiosity that inventors often have. He also had the drive needed to see an invention through. By the age of fourteen, he had his own nail making business. He manufactured nails using a machine he had invented and built himself.

Then Whitney went to Yale, ran up some debts, and took a teaching post in order to pay them off. This took him to a plantation in Savannah in the state of Georgia, where he listened to the cotton farmers' complaints about the amount of time it took them to clean the cotton crop. The seeds had to be removed and it might take all day to clean just one pound of cotton. Whitney was a keen observer and had an eye for detail. He watched the hand movements of the workers as they separated the seeds from the fibres. One hand held the seed, while the other teased out the strands of lint.

Eli Whitney decided to design a machine that would imitate these hand movements and carry out the tedious and frustrating task far more quickly. He came up with a relatively simple device. It consisted of a metal drum with wire teeth. Raw cotton straight from the field was fed through the cylinder and as it was spun round the teeth passed through narrow slits in a piece of wood. The hook-shaped teeth pulled the fibres of cotton through and left the seeds behind; a rotating brush, moving four times faster than the drum, then cleaned the lint off the hooks. This simple device proved to be fifty times faster than picking out the seeds by hand.

The importance of Eli Whitney's invention, the cotton gin, was far-reaching. It did far more than make the working day of the cotton farmers easier. It stepped up the commercial production of cotton on an incredible scale. In 1793, the year when Eli Whitney invented his machine, about 180,000 pounds of cotton were harvested in the USA. Within two years, with the use of the cotton gin, this had grown to over six million pounds. By 1810, production had reached 93 million pounds per year.

This stepping up of production did much to revive a flagging industry in the Deep South. It also kick-started the Industrial Revolution in America. Once steam engines were harnessed to cotton gins, the process became automated – a fully industrial process. The success of the cotton industry outshone possible rival cash crops, such as tobacco and indigo.

This is an outstanding example of a single-shot invention, an invention made all in one go. It was also an invention that was designed and created very quickly. From the moment when Eli Whitney started working on the problem to the moment when the machine was finished it took just ten days. In those ten days, the face of the Deep South and the character of the American economy were changed out of recognition. It was the single most important invention in the history of the US economy.

THE INVENTION OF THE INTERNAL COMBUSTION ENGINE

(1794)

THE INTERNAL COMBUSTION engine is one of those inventions that has been both a benefit and a menace to humanity. It has made it possible for people to travel far afield for their holidays, broadening their horizons, and made it possible for people to have more choice about where they live in relation to their place of work. It has become easier to transport food, and shops sell a far greater diversity of food than was possible before. In many occupations, a lot of time has been saved by the use of the engine, and therefore more hours of leisure have been generated. Many of the advances in quality of life in the twentieth century were made possible by the internal combustion engine. But it has also been a great polluter of the environment, making villages, towns and cities busier and noisier places.

Experiments with the principle of internal combustion engines began well before the Industrial Revolution got under way, with the Dutch engineer Christiaan Huygens and the French Jean de Hauteville and Denis Papin experimenting with gunpowder – and this was before 1700. But gunpowder was an unsuitable energy source. It was not until other sources were tried that real progress began. Explosive gases, alcohol and distillates of petroleum were far more promising.

In 1794 Robert Street gained a patent in Britain for the first true practical internal combustion engine. It consisted of a cylinder fitted with a piston inside it, linked to a pivoting arm that worked a simple water pump. The cylinder was fitted with a water jacket to keep it cool and it extended into a furnace that was hot enough to ignite a mixture of air and liquid fuel. The fuel dripped in under gravity, and the air was pumped in by hand. Some of the features of the engine are clearly recognizable, and they survived into later models, but there were obvious shortcomings in the design and subsequent inventors worked to resolve them during the following century.

Samuel Brown started building and marketing a gas-burning internal combustion engine in England in the 1820s. At the same time, the French engineer Nicolas Carnot published *Reflections on the Motive Power of Heat*, which contained most of the principles of the modern internal combustion engine, though Carnot did not actually build an engine.

In the 1830s William Barnett designed the first practical two-stroke engine. This was later to find wide application in the development of the smaller petrol engines. Barnett was an important innovator. It was he who invented the pilot light ignition system, which became a very important method for igniting fuel engines until the electric sparking plug was invented, and is still used in gas central heating boilers.

In the 1840s and 1850s inventors went on tweaking the design of the engine, but it remained very inefficient in terms of converting fuel into usable motion. In 1862 Alphonse Beau de Rochas published a paper detailing several potential improvements, including what became the standard sequence for the operation of a four-stroke engine: intake, compression, power and exhaust. Like Carnot, Beau de Rochas was a theorist, not a builder. It was the German engineer Nikolaus Otto who took Beau de Rochas' theory and built it into a working machine. Otto started building engines in 1867. The first ones were under-powered and noisy, but within ten years he had produced an improved design which is very much the internal combustion engine as we know it today. Different numbers of cylinders have been used – two, four, six, eight and sometimes more – but the basic principle remains unchanged.

THE INVENTION OF THE PENCIL

(1795)

THE PENCIL HAD its beginnings in Cumbria in 1564. An unknown passer-by walking along Borrowdale found pieces of shiny black material clinging to the roots of a fallen tree, picked them up and found that they could be used to draw and write on paper. The black material was graphite and the chance discovery generated considerable local interest. Graphite, which became known as 'blacklead', was difficult to exploit as a writing medium because of its brittleness. It needed to be held and supported in some way. Initially, people tried wrapping the graphite in string. It was only later that people inserted it into hollowed sticks. Making the hollow sticks to hold the graphite was laborious, but it was in this way that the pencil was invented.

The first patent for a pencil-making process was granted in 1795, to the French chemist Nicolas Conté. He used a mixture of powdered graphite and clay, fired it to make the 'lead', then placed it in a slot in a cylindrical wooden tube. The thin strip of wood cut from the slot was then replaced to hold the lead in position. From the start, Conté was able to produce a range of leads, some hard, some soft, depending on the way he kiln-fired the graphite. The versatility of his process made Conté's pencils extremely useful to draftsmen, writers and artists.

The first mass-produced pencils were manufactured in Europe and exported to America. But wars interrupted trade, and it was not long before the Americans were manufacturing their own pencils. The early American pencils were unpainted to show off the quality of the wood casing; many still are. Eastern red cedar, native to the eastern USA, was used, a strong wood that did not splinter and the manufacturers were proud of it.

Today, millions of pencils are produced every year. Thanks to Conté's innovation, they come in a variety of lead hardness, and they have become an indispensable part of our everyday lives.

THE INVENTION OF LITHOGRAPHY

(1796)

LITHOGRAPHY WAS INVENTED by Alois Senefelder in 1796. It is a method of printing text or artwork onto smooth surfaces such as paper, using chemical processes to create an image.

The positive area of a plate image might, for example, be a water-resistant chemical such as an oil-based medium, while the negative part of the image would be water. In that case, when an ink and water mixture is introduced to the plate, ink will stick to the positive part of the image, and water will clean the negative part of the image. Both positive and negative parts of the image are on the flat surface. The benefit of this printing method is that it allows the print plate to be flat and allows for much longer print runs than the previous 'physical' methods of imaging, such as engraving, which involved a three-dimensional image.

The lithography process works because of a simple basic principle, the repulsion of oil and water. There is a very large range of oil-based mediums that might be used to make the positive image, but the chosen medium needs to be one that can withstand water and acid. After the placement of the image the plate is coated with acid emulsified with gum arabic. The resulting emulsion creates a salt layer over the area surrounding the image. The salt layer penetrates the surface of the plate, etching it. The printer then removes the oil-based drawing medium with turpentine, leaving the salt layer holding the image's form.

When wetted with water and inked, the plate and paper are run through a press, which produces a finely detailed image.

Lithography was the first new printing process to be invented since Gutenberg in the fifteenth century. It was widely used during the nineteenth century for book texts, but more especially for fine art work and maps. In the early days, only black and white images were produced, but within a few years multi-coloured lithographic printing became possible. By 1850, this had become commonly known as chromolithography. It entailed using a separate plate or stone for each colour, and the print had to go through the press separately for each plate. The main technical problem was keeping the paper exactly in alignment (in register) for each plate.

The colour lithography technique lent itself to large areas of flat colour, and this had a marked effect on the style of poster art.

Modern high-volume lithography is still used to make posters, but also books, newspapers, credit cards, patterned CDs. Almost any mass-produced item with a smooth surface can be printed lithographically. Imaging is assisted by laser technology, which allows images to be transferred directly from a computer file onto the plate.

THE INVENTION OF CAST-IRON BEAMS AND COLUMNS

(1796)

ARCHITECTS STARTED USING cast-iron beams and columns in the construction of factories to reduce the fire risk. There were several disastrous fires in very large timber-floored factories. This led, at the close of the eighteenth century, to the concept of a fireproof building. The fireproof construction relied on the use of strictly non-combustible materials. Interior brick walls, brick arches and slab floors were supported on iron beams and iron columns, which in turn were generally supported by exterior brick walls.

A problem was that it was impossible to calculate the behaviour of any of these materials in the context of a specific fire. Nor was it possible to know how strong a slab floor might be when fire broke out in a building. In fact the safety record of these early fireproof buildings was very good. There were, however, a few catastrophic failures.

Cast-iron beams and columns became available as an option with the massive increase in the output of iron during the Industrial Revolution. In the course of the eighteenth century, cast-iron production in Britain increased tenfold. Wrought iron was only used in buildings for firings such as ties and brackets.

The first fireproof building ever to be constructed was Ditherington mill in Shrewsbury. It was designed by Charles Bage and built in 1796–7, using cast-iron beams and columns supplied by William Hazledine's foundry. The brick floor arches sprang from wedge-section beams, with the same construction repeated at roof level. Inside the building, solid columns with cruciform cross-sections gave the beams intermediate support.

Although Ditherington mill may have been the first thorough-going fireproof building, cast-iron beams had already been in use for a time. It is thought that the architect John Nash was the first to use cast-iron beams to support timber floors over spans that were too great to be supported by timber beams.

In almost all later buildings the cruciform section column was replaced by columns that were hollow tubes, circular in cross-section. These were found to be structurally more efficient. The inverted T shape for the beams used at Ditherington was used in all iron beams for the next thirty years. The beam spans were no more than fourteen feet, which necessitated lots of columns.

Beginning in 1825, Eaton Hodgkinson ran experiments to find the ideal shape of beam, with the support of a Manchester engineer called William Fairbairn. Testing several beams to destruction, he found that beams with a large bottom flange were the most efficient. Hodgkinson devised a formula for the ideal beam shape, which was used from the early 1830s onwards. Its first use was in the design of a railway under-bridge on George Stephenson's Liverpool and Manchester Railway. But progress was sporadic. Some later beams reverted to less efficient shapes. Notably, the iron beams of Sir Charles Barry's 1830s Houses of Parliament were made with a symmetrical I-shaped cross-section; this was recommended by the civil engineer Thomas Tredgold, but it was a poor choice as the I shape is structurally weak.

Fireproof floors of brick arches supported on iron beams were used in several public buildings in the 1820s and 1830s, such as Robert Smirke's General Post Office in London and William Wilkins's National Gallery. Typical multi-storey construction in cast-iron used storey-height column-lengths, fitted together with simple spigot and socket joints.

Some of these industrial buildings have been saved from demolition and refurbished. A landmark case was the fireproof warehouses in the Albert Dock in Liverpool; these were designed by Jesse Hartley and built in 1842.

Fireproof buildings were built not only to withstand fire, but to withstand the enormous weight and vibration of early nineteenth-century machinery. It has been found that these buildings can be saved even after they have been abandoned and neglected for many years. It was for this reason that it was possible to save and refurbish so many nineteenth-century warehouses in the London Docklands. Some were turned into new industrial premises, while others were converted into highly desirable flats, many with fine views of dock basins or the River Thames. Unfortunately the discovery that these fine old buildings could relatively easily be rescued and given a new lease of life came too late for many of them. By 1992, half of the 2,400 textile mills of Greater Manchester had been demolished, most of them in the 1970s and 1980s. During the property boom of the 1980s, mills were being destroyed at a rate of two a week.

Strong though they undoubtedly were, nine British cast-iron buildings collapsed during the nineteenth century and a further four collapsed during the twentieth century. When Gough's mill in Salford fell down in October 1824, seventeen people were killed. Gough's mill fell down when it was one year old because of the failure of a beam supporting the top floor. The beam was defective, and it was also heavily loaded with partition walls and carding engines. The building collapsed from the top, progressively downwards through its six storeys and most of the dead were on the ground floor. The other accidents were similar in nature. The failure of a girder above the students' dining hall at King's College, London in December 1869 and the progressive collapse of the structure could have been catastrophic if it had happened a few hours later, when the hall would have had 160 people in it; luckily it was empty at the time.

The American inventor James Bogardus, born in 1800, was responsible for transferring cast-iron beams and columns from Britain into American architecture. His factory in New York City, which was built five storeys high entirely as a cast-iron structure, was the first building to be constructed in this way in the United States. It was so successful that Bogardus entered the business of building iron-framed buildings in cities throughout America.

THE INVENTION OF
THE TOP HAT

(1797)

THE TOP HAT was invented in 1797. On 15 January that year, the stylish new headgear was worn for the first time in public by James Hetherington. The sight drew a huge and disorderly crowd and created a mini-riot in a London street. According to a newspaper account, some passers-by panicked, some fainted, some screamed; dogs barked; an errand boy's arm was broken in the stampede. Hetherington was arrested and charged with causing a breach of the peace, having 'appeared on the public highway wearing a tall structure of shining lustre and calculated to disturb timid people'. He was fined £50.

James Hetherington was a London haberdasher and he had experimentally introduced a new fashion. He designed it, made it and was the first to wear it in public. In fact his new hat was, like many another invention before and since, a development of something that already existed. The contemporary riding hat had a similar shape, but was covered in beaver fur. What Hetherington did was to give his new hat a narrower brim and a higher crown, and he covered it in black silk.

Although people ridiculed it initially, in the same way that a gentleman was once derided for effeminacy for sporting an umbrella, the top hat gradually gained favour in the upper echelons of English society. Prince Albert gave it his seal of approval in 1850, which effectively ended the controversy. From that point on, the top hat was accepted. The switch from beaver to silk had a massive impact on the beaver-trapping industry in North America: it went into steep decline.

The height of the top hat made it a particularly impractical piece of headgear. This made it an ideal marker for the leisured rich. It was a major social and economic statement on the part of the wearer, and remained so for many decades. The phrase 'high hat' indicated the snobbish attitude of mind and the arrogance of bearing that went with wearing the top hat.

The shape of the hat changed from place to place and from decade to decade. In France, early nineteenth-century dandies known as *Les Incroyables* (the Incredibles) wore outlandishly large top hats, so large that there was no room for them in crowded cloakrooms at the theatre or opera house. So, in 1823, Antoine Gibus invented the collapsible opera hat.

In England, among the commercial classes, the famous stove-pipe hat, worn uncompromisingly vertical, made gatherings of the captains of industry look just like an industrial townscape bristling with factory chimneys. The top hat defined the nineteenth century.

The John Singer Sargent portrait of Lord Ribblesdale, the epitome of the Edwardian aristocracy, shows how English gentlemen wore their toppers in 1902, tilted very slightly forward and very slightly to one side. The top hat defined the English aristocrat's total self-assurance and aplomb.

By the time Fred Astaire made his film *Top Hat* in 1935, the age of the top hat was over, in both Britain and America. But the rakish way he wore it showed how versatile it was. It could express irresponsible raffishness just as well as aplomb, just by varying the tilt.

THE INVENTION OF THE ELECTRIC BATTERY

(1800)

THE BATTERY IS one of those devices that was invented more than once. There is some evidence that crude batteries were in use in Iraq and Egypt as early as the year 200 BC. They were used specifically in craft industries for gilding and electroplating. But it was in the eighteenth and nineteenth centuries that the modern battery, the battery as we recognize it, came into being. It was Benjamin Franklin who coined the word battery – to describe an array of charged plates.

In the 1790s, Luigi Galvani, an Italian anatomy professor, found that he was able to make a frog's muscles contract by touching the nerves with electrostatically charged metal. In his later experiments he found he was able to stimulate muscular contraction with different metals without an electrostatic charge. He assumed from this that the body tissue of animal contained an inherent life-force, which he called 'animal electricity'.

Alessandro Volta disagreed with Galvani about this theory and in 1800 he designed an experiment to disprove it. He built a voltaic pile, which demonstrated that the electricity did not come from the animal tissue but was generated by the contact between different metals in a moist environment. The voltaic pile consisted of alternate discs of copper and zinc with pieces of cardboard soaked in brine between the metal layers. This superficially unpromising device actually produced an electric current. It was the very first wet cell battery.

There is general agreement that the voltaic pile represents the invention of the battery; it was the first device to produce a steady and reliable current of electricity.

Volta's battery led on to the development of electrochemistry, electro-magnetism and the whole range of modern applications of electricity. The volt, the unit of electromotive force, is appropriately named after Volta.

As with many other inventions, it was subsequently improved by other people. In 1836, John Daniel invented the Daniel cell using two electrolytes, copper sulphate and zinc sulphate. The Daniel cell was safer than the Volta cell, and less corrosive. Shortly after that, there were experiments with liquid electrodes. In 1859 Gaston Plante devised the first workable and re-chargeable storage lead-acid battery; this became the standard car battery of the twentieth and twenty-first centuries. In 1881 Carl Gassner invented the first commercially successful dry cell battery. In 1901 Thomas Edison invented the alkaline battery.

It would be hard to exaggerate the importance of the electric battery. We use batteries in a huge range of everyday appliances and tools: torches, clocks, watches, cameras, mobile phones, lap-top computers. The whole texture of our everyday lives depends on the existence of batteries.

THE INVENTION OF
THE METAL LATHE

(1800)

HISTORY IS UNEVEN. It records some things and not others, and the events that are recorded depend entirely on the values of the time when the events happened. There is a contemporary carving of the Persian king Darius (548–486 BC), sitting on a throne which has lathe-turned legs. We know the name of Darius and a great deal about what he achieved, but nothing at all about the person who made the throne, nothing about the craftsman who invented the lathe that turned the throne's legs. The carving is in fact the first indication that somebody had invented the lathe by that date.

The lathe continued in use, in the background, throughout the succeeding centuries. There are medieval illustrations showing the spring pole lathe in use, such as the one in a Parisian manuscript from the middle of the thirteenth century. A seventeenth-century text, Joseph Moxon's *Mechanick Exercises* (1680), describes the design:

> *The pole is commonly made of a Fir-pole, and is longer or shorter according to the weight of the Work the Workman designs to turn. The thicker the pole, the harder the tread mut be to bring it down. Upon the thin end of the pole is wound a considerable bundle of string, so much as will compass the mandrel twice or thrice. The string is made of the guts of beasts.*

The workmen who operated lathes were known as turners. It is clear from Moxon's book that some turners were using their lathes to turn metal as well as wood.

Brasiers that turn Andirons, Pots, Kettles, &c. have their Lathe made different from the Common Turners Lathe. As the common Turners work with a round string made of Gut, as hath been described, the Brasiers work with a Flat Leather Thong, which is wrapping close and tight about the Rowler of their Mandrel.

The modern metal lathe was invented by Henry Maudslay. Maudslay was the son of a joiner who had worked on the wooden frames for cotton machines, but ended up working as a shopkeeper at the Woolwich Arsenal in London. As a boy, Henry Maudslay worked as a powder monkey, making and filling cartridges. He worked in the carpenter's shop, but as he developed an increasing interest in working with iron he was allowed to work in the nearby blacksmith's shop instead. At the age of fifteen he was gaining a reputation for his skill and accuracy as a worker. Joseph Bramah, the famous lock-maker, was looking for an apprentice who could make the precision tools needed for manufacturing complex lock mechanisms. Maudslay was apprenticed to Bramah and eventually became his foreman. While he worked for Bramah, Henry Maudslay made a padlock which Bramah put in the window of his shop, offering a reward to anyone who could pick it. The lock went for fifty years before being picked; it took a locksmith called A. C. Hobbs sixteen days to open it in 1850. He got his reward. Maudslay meanwhile married Bramah's housemaid in 1791 and in 1797 asked Bramah for a pay rise to enable him to support a family. Bramah refused, so Maudslay left to set himself up in business on his own, building machinery.

Maudslay invented the slide rest lathe. This was a major innovation, a machine that greatly improved the precision of the lathe, and with it Henry Maudslay had invented the first modern machine tool. He was the first to see the importance of making machine parts that were standardized, so that they would be interchangeable and easily replaceable. His lathe allowed the turning of large pieces of metal, including the components of weapons, with great precision. In doing so, it also changed the nature of warfare and contributed to Britain's dominance at sea. The lathe was also used in the manufacture of many musical instruments.

In 1810, Maudslay moved his premises and took on a partner, Joshua

Field. The new firm of Maudslay and Field sold improved versions of Maudslay's lathe, and Maudslay went on to invent more machine tools. He also built machinery for minting coins. He invented a tool that could shear and punch holes in boiler plates, a task that had until then been done by hand. As a result of this invention, he was contracted by the Royal Navy to make plates for ships' tanks.

Maudslay made advances in the technique of screw-cutting, inventing the first screw-cutting lathe. Characteristically, he worked towards ensuring the uniformity and standard pitch of screws, as part of his interest in standardization. Maudslay developed a system for the number of threads on a screw. Prior to Maudslay, there had been no system at all, and every nut and bolt was unique. The problem was that parts were not interchangeable.

Henry Maudslay had a great reputation as a master craftsman. But he was also well known as an affable person. He was also a great teacher. Young men arrived from all over England to be taught by him. He trained a number of youths who went on to make big names for themselves as engineers: Joseph Clement, James Nasmyth and Sir Joseph Whitworth.

Richard Roberts was another highly inventive engineer, who had learnt his trade with Maudslay before starting his own machine-manu-facturing business in Manchester. He too built lathes, and to standards of which Maudslay would have been proud. Roberts built improved machines for the textile industry, including a more efficient power loom and a mule for spinning yarn. Henry Maudslay did not just invent a useful machine; he invented a whole approach to machines and machine tools that made many other inventions possible in the nineteenth century. Maudslay's career was one of those necessary steps that made many other advances possible in the decades that followed.

IV

THE NINETEENTH-CENTURY WORLD

THE INVENTION OF THE JACQUARD LOOM

(1801)

THE JACQUARD LOOM was invented in 1801 by Joseph Marie Jacquard. His loom used a revolutionary new system for creating patterns in woven fabric. Many different aspects of textile manufacturing, from the spinning process onwards, had been mechanized during the Industrial Revolution. What Jacquard did was to introduce mechanization to one of the most difficult and intricate aspects of cloth-making, which was weaving intricate designs. Jacquard did not invent the technique completely from scratch, but built on earlier inventions by Basile Bouchon, Jean Falcon and Jacques Vaucanson, manufacturers of textiles in the early eighteenth century.

Jacquard's method used pasteboard cards with holes punched in them in a grid pattern that was eight holes by twenty-six holes. Each punched card carried encoded in its punched holes the sequence of threads for just one row of the pattern. To make a design on the cloth involved many rows and therefore many cards. The rectangular cards were strung together in order, edge to edge, to make a continuous roll that rotated loosely round a drum. Each hole corresponded to a hook which could either raise or lower the harness that guided the warp thread, running across the fabric, and so determine whether the weft thread, running along the fabric, would lie above or below it. It was the sequence of raised and lowered threads that made the pattern.

The Jacquard loom was not only the first loom to use punched cards to control it: it was the first machine of any kind to be controlled by punched cards. The device was purely mechanical, involving no electronics of any kind, yet it marked an important first step in the development of computer technology. The rational procedure of stepwise, predetermined decision-making was the precursor of the computer program, which is based on the same procedure. When Charles Babbage came to design his early computer, the analytical engine, he planned to use punched cards to store its programs, seeing the card system as a reliable way of storing and reactivating instructions for the machine. Babbage did not in the end, as we shall see, live to prove that the card system would work in computing. But computer developments in the twentieth century showed that punched cards really did work.

The punched card system as applied to the Jacquard loom paved the way for the mass production of pattern woven fabric. This made production cheaper, and offered customers greater variety. Modern Jacquard looms are controlled by electronic computers, not by punched cards. This has allowed for many more variations in the pattern to be introduced.

THE INVENTION OF
THE SUSPENSION
BRIDGE

(1801)

FOR MANY HUNDREDS of years bridge technology was fairly static. The stone bridges built in the fifteenth and sixteenth centuries in Europe were based on the same stone arch principle as the bridges built by the Romans over the Tiber. Things began to change in the mid-eighteenth century. In 1747 in Paris a specialist engineering polytechnic was founded, the Ecole des Ponts et Chaussées. It devoted a great deal of study time to bridge design, with the result that the massive arch bridge design inherited from Roman times was pared back to something lighter but with the same strength.

By the late eighteenth century, when the Industrial Revolution was well under way in Europe, iron presented itself as a natural alternative material for bridge building. Some new bridges were built of stone reinforced with iron (truss bridges), and this enabled engineers to design bridges with spans up to 200 feet wide. Longer spans enabled bridges to clear small and medium-sized rivers in one leap, and architects and engineers looked for ways of making the spans even longer.

The suspension bridge represented the great leap forward in bridge design. The principle is simple. The central part of the bridge is held up by cables anchored at each end of the structure and passing over two or more high towers to create the upward pull. This technique allowed for much longer spans. The longest truss bridge is 1,200 feet long. The Verrazano Narrows Bridge in New York has a main span of 4,260 feet and the overall suspension span is 6,690 feet.

The first suspension bridge was built in 1801 by James Finley. He used wrought-iron chains to hold up a seventy-feet-long twin tower bridge at Uniontown in Pennsylvania. It was a small beginning. Thomas Telford made a much larger suspension bridge over the Menai Strait connecting Anglesey to mainland Britain in 1826. This was 580 feet long. It has twin stone towers and wrought-iron chains supporting a wooden deck. The size and the dramatic setting of Telford's bridge quickly made it world-famous. It is still functioning, after being virtually rebuilt in 1940.

Later suspension bridges were longer and more ambitious. Steel cables made that possible. The famous Brooklyn Bridge was begun in 1869, under the direction of the engineer John Roebling. Roebling died in an accident shortly afterwards, and the project was seen through by his son Washington Roebling. He, too, almost died, when he suffered from the bends after spending too long in compressed air conditions inside a caisson. The finished Brooklyn Bridge is both practical and beautiful. Most suspension bridges make a major impact on the land-scape because of the height of the superstructure. The suspension bridge design is a striking example of an engineering solution that is both practical and elegant.

THE INVENTION OF THE STEAMBOAT

(1802)

THE INVENTION OF the steamboat is often attributed to Robert Fulton, though others had already designed and built working steamboats before him.

Fulton was an American, born in Pennsylvania. He showed an early aptitude for technical drawing, and local gunsmiths went to him for designs for their guns. He moved to Philadelphia, where he set himself up as a portrait painter and miniaturist. After four years, he decided to move to England to study under the artist Benjamin West.

When he reached England, Fulton was bowled over by what he saw there – the Industrial Revolution in full swing. Coal mines were being opened up all over the country, networks of canals were being dug, new bridges were being engineered. Fulton was so excited by it that he put aside his plans to become an artist and settled instead for being an engineer. As a boy of fourteen he had put together a design for a steam-powered paddleboat. Now he wanted to build one.

Fulton applied to the British government for permission to buy a steam engine and take it to America. His request was refused. Fulton went on trying for three years until, in 1803, he was able to buy an engine from the firm of Boulton and Watt. Then he had to wait another three years before he could ship it to the States. Now in partnership with Robert Livingston, Fulton set about designing a paddle steamer. They disagreed about the design, with Livingston favouring a single rear paddle wheel (the design chosen for the Mississippi stern-wheelers), but Fulton eventually had his way, with a paddle wheel on each side of the boat's hull.

They installed their Boulton and Watt steam engine into a ship called the *Clermont*. She was 100 feet long and on 17 August 1807 she made her maiden voyage as a steamboat on the Hudson River. Shortly after this, Fulton put the ship on a regular schedule and more and more passengers used it. It was a breakthrough in promoting a new mode of transport, but Robert Fulton had not invented the steamboat – he had merely popularized it.

The real inventor of the steamboat was William Symington. He was English, born in Leadhills and became a mechanic in the Wanlockhead mines. He was also an innovator. In 1787, he took out a patent for a road locomotive, the same idea that Richard Trevithick later worked on. In that same year, in America, James Rumsey drove a boat on the Potomac River at 4 mph. It was powered, rather ineffectively, by a stream of water forced through the stern of the boat by a steam-powered pump. At the same time, John Fitch was experimenting with an oar-driven steamboat on the River Delaware.

Perhaps William Symington heard about these experiments, because in the following year, 1788, he switched his interest from land transport to water transport. He mounted his road locomotive engine, which was a direct-action steam engine, onto a paddle boat. The paddle boat Symington designed together with Patrick Miller, another inventor and steamboat pioneer, was just twenty-five feet long. It had twin hulls, like a catamaran, with paddle wheels mounted between them. The boat was mainly Miller's design, but Symington supplied the engine.

Miller and Symington launched their unusual vessel on Dalswinton Loch at Miller's estate not far from Dumfries. The paddleboat worked well enough, and the paddles were the most efficient device so far invented, yet this landmark invention attracted no commercial attention whatsoever.

In 1801–2, William Symington built the *Charlotte Dundas* at Grangemouth on the southern shore of the Firth of Forth. The *Charlotte Dundas* was one of the earliest practical working steamboats ever to be built. Symington intended that she should work as a tugboat, and she certainly proved her capabilities by undertaking towing work on the Forth Clyde Canal. Unfortunately for Symington, vested interests stopped the *Charlotte Dundas* from being taken into regular service. The specious argument raised against her was that she would stir up too much wash and erode the banks of the canal.

Because he failed to capture any commercial interest in his (and Miller's) invention, William Symington died in poverty. Yet there is no doubt that he was one of the greatest innovators of the Industrial Revolution. He made the first working steamboats, and prepared the way for the building of ever-larger steamships through the second half of the nineteenth century. Robert Fulton saw the *Charlotte Dundas* in action, and he was impressed by her effectiveness. He went back to America to mount a Boulton and Watt engine into an existing ship, the *Clermont*. He designed neither the ship nor the engine, and he did not invent the concept of a steamship either.

THE INVENTION OF THE STEAM LOCOMOTIVE

(1804)

RICHARD TREVITHICK, A Cornishman, invented a plunger pole pump in 1797 for keeping deep mines clear of water. The next year he applied the principle of the pump to the design of a water-pressure engine. In 1800, he built a high-pressure non-condensing steam engine, which competed with James Watt's low-pressure steam engine. Trevithick was working in parallel with a number of other inventors, who were all thinking of mounting a steam engine on a carriage and running it along roads. The idea of steam locomotion was in the air.

As things turned out, steam locomotion was never successfully to work on roads. Once steam locomotives were invented and mounted on railway tracks, that was universally accepted as the way forward. But in 1800 the railway did not seem like an inevitable future: just one of several possibilities. James Watt did not believe in railways, and from 1800 until 1815 Trevithick built several experimental steam road carriages.

Trevithick also built the first steam railway locomotive, which was clearly the prototype for George Stephenson's commercially successful locomotives, and it is hard not to see Stephenson as something of a plagiarizing entrepreneur. He took Trevithick's invention, made money out of it – and took all the credit for it as well. On the other hand, Trevithick himself had borrowed the locomotive idea from William Murdock, whose road locomotive Trevithick had seen fifteen years earlier trundling through the streets of his home town of Redruth one evening in 1786.

It was on Christmas Eve 1801 that Trevithick's road locomotive, nicknamed *Captain Dick's puffer*, carried the first load of passengers ever to be transported by steam locomotion. The age of steam locomotion had begun in earnest, in the first year of the new century. In March the following year Trevithick and Andrew Vivian applied for a patent for steam-propelled carriages. In 1803, another steam carriage was driven through the streets of London.

In 1804, Trevithick worked on a tramroad locomotive for industrial use at Pen-y-darran, where it proved to be capable of carrying twenty tons of iron. This, the first railway locomotive, proved to be the truly ground-breaking invention, the one that Stephenson would copy. He built a similar engine for the Wylam colliery at Newcastle in 1805. In 1808 he built a circular railway track near Euston Square in London. To demonstrate the new technology, he took members of the public round the track on a steam locomotive at speeds up to 15 mph. Trevithick was sometimes called the Giant of Steam. He was a huge, powerful man who could hurl a sledgehammer over an engine house.

He went on inventing and innovating, but it was Trevithick's 1804 invention of a steam railway locomotive that was destined to transform the map of Britain, the industrial geography of Europe, and reshape the transport systems in several continents. With that 1804 invention, the age of the railway had begun.

THE INVENTION OF THE REFRIGERATOR

(1805)

It was William Cullen of Glasgow University who invented the first modern refrigeration technique, as early as 1748. He cooled air by the evaporation of liquids in a vacuum, but made no attempt to put the technique to any practical use. In eighteenth-century Britain, there was a perceived need to keep fresh foods cool in order to stop them from going bad. Houses were often fitted with larders that had slate or marble slabs, and these kept food cool. The wealthy had ice houses built in their grounds. These were partly subterranean brick-built igloos, in which snow and ice collected from the fields, rivers and lakes in the winter, or brought down from mountains, were stored, packed in straw. Buckets of ice could be taken to the house when needed to preserve meat or chill drinks and desserts. Some ice stored in this way might be sold to butchers and fishmongers or distributed to the homes of tenants. Ice preserved in ice houses lasted several months, on into the spring and summer.

The word refrigerator was invented by Thomas Moore in 1800. The device Moore invented was what we would now call an icebox rather than a refrigerator. It consisted of a cedarwood tub lined with rabbit fur and filled with ice, surrounding a container made of sheet metal. Moore's icebox was designed for transporting butter from Maryland to Washington DC.

The first refrigeration machine was designed in 1805 by the American inventor Oliver Evans. In 1844, an American doctor called John Gorrie made a machine based on Evans's invention, specifically to cool the air and give relief to his yellow fever patients. So the refrigeration process was already being applied to a simple form of air conditioning. In 1902, Willis Haviland Carrier demonstrated the first purpose-designed air conditioning system. All through the nineteenth century, different inventors had been making small improvements to the refrigeration technique. Useful contributions were made by James Harrison, Charles Tellier, David Boyle and Raoul Pictet.

The first commercial refrigerator for chilling food was manufactured in 1911. In 1915 Alfred Mellowes developed a more self-contained model, which nevertheless had to be hand-made. What was needed was a model that might be mass-produced for both commercial and domestic use. Alfred Mellowes was bought out by W. C. Durant, president of General Motors, three years later, and Durant founded the Frigidaire Company to mass-produce refrigerators for the American domestic market.

In 1900, around half of the households in the United States used melting ice and an icebox to keep food cold: the other half had no provision for cooling food at all. Only in the homes of the very rich were any serious attempts at refrigeration to be found. One of the earliest domestic refrigerators was installed at the mansion of an oil tycoon, Walter Pierce.

The French inventor Marcel Audiffren was keen to develop refrigerators for domestic use. His patents were bought by the American Audiffren Refrigerating Machine Company. Machines were manufactured in Indiana. The first Audiffren refrigerator was sold in 1911, for about 1,000 dollars, which was twice as expensive as a car. A feature of these early domestic refrigeration systems is that while they had a cold box located in the kitchen, the machinery (including a motor and a compressor) had to be housed in an adjacent room or basement, so a fair amount of space was required.

The retail price gradually fell so that more and more American families could afford to buy them. They also became more compact. In 1923, Frigidaire brought out the first self-contained unit. Also in the 1920s, freezer compartments were introduced, so that it became possible to produce ice cubes for drinks.

The first refrigerator to be widely sold was the General Electric Monitor-top refrigerator, which was a simple white icebox on short legs, with its mechanism mounted on top in a white drum-shaped housing. This, really the first modern 'fridge', was introduced in 1927.

Over a million of them were sold, and some of them are still in working order today.

From then on, the refrigerator saw refinements rather than any revolutionary changes. The freezer units became more and more popular during the 1940s. In the 1950s and 1960s, features such as automatic defrosting were introduced. In the 1980s there were increases in efficiency and environmental concerns led to the banning of the use of CFC coolants in sealed systems. CFCs, we tend to forget, were adopted because the earlier refrigerants were extremely dangerous. Refrigerators right up until 1929 used three toxic gases: ammonia, methyl chloride and sulphur dioxide. There were several fatal accidents in the 1920s when methyl chloride leaked out of refrigerators.

Now nearly every home in the more economically developed countries of the world has a refrigerator. This makes it possible for people to store and preserve food products in a fresh state for much longer than in the past. With anything up to ninety per cent of the population living in cities, few people have gardens large enough to grow their own fresh food. With the pace of modern life fewer people have the time to tend a garden on a regular basis. This increases dependence on shops and supermarkets. Increasingly, people buy food in greater bulk than in the past, and store what they don't immediately require. The overall result has been an enrichment of diet, with more people eating a far greater range of foods than they did fifty or eighty years ago. The astonishing variety of foods available in supermarkets is the visible proof of this. Another major result of refrigeration is improved health, with fewer cases of food poisoning.

THE INVENTION OF CARBON PAPER

(1806)

CARBON PAPER WAS invented twice over, in the same year, by the Italian Pellegrino Turri and the Englishman Ralph Wedgwood. The spur to this invention was the huge expansion of the world of commerce that accompanied the Industrial Revolution. The larger scale of industrial production and the increase in sales and other transactions meant that more and more paperwork was generated. More invoices, more receipts, more contractual agreements – and more copies. Carbon paper meant that it was possible to write a document and generate a second copy without writing it out a second time. The specific motives impelling both Turri and Wedgwood were nevertheless nothing to do with business; both men were trying, independently and at the same time, to devise a means by which blind people could write without having to use a quill and ink.

It seems that James Watt made an earlier attempt at copying paper in the 1770s. He did not trust his clerks to hand-copy business letters with total accuracy, so he invented a form of carbon paper. Tissue paper moistened with certain fluids was pressed onto the original letter, which had to be written using special ink. He was ready to exploit his invention commercially, but found there was no market for it.

Ralph Wedgwood was granted a patent for his stylographic manifold writer, as he called it, in 1806. He saturated sheets of thin paper with printer's ink, then dried them between sheets of blotting paper. The inked carbon paper was then sandwiched between a sheet of tissue paper and the sheet of regular paper for the top copy. A metal stylus was then used to scratch the writing onto the tissue paper. This created an image that read correctly on the regular paper for the top copy and another that was a mirror image on the back of the tissue paper. Reading the mirror image was fairly easy, as it could be seen in reverse through the back of the tissue paper. The metal stylus was necessary, as the quills that were in regular use at the time could not be pressed hard enough. In about 1820 the process was refined to the extent that only one side of the carbon paper was inked, and an indelible pencil might be used instead of the metal stylus.

In the 1820s Wedgwood ran a profitable business, selling carbon paper from his shop at 4 Rathbone Place in Oxford Street, London. The new invention was not an immediate runaway success, though, as many businessmen distrusted it. They feared forgery and much preferred originals written in ink. There is still the same fear of photocopies today, for the same reason.

In 1823, true carbon paper was invented by Cyrus P. Dakin. His copying paper was coated with oil and carbon black. For the next three decades it continued to be a struggle to sell the product to businesses. It was really only with the invention of the typewriter in 1867 that carbon really gained acceptance. Typewriters produced a clearer and more even second copy as well as a clearer original. In the 1860s Lebbeus H. Rogers made his carbon paper by placing sheets of paper on a stone table and coating them with a mixture of carbon black (soot) and naphtha. He went on to invent a machine that would apply the coatings. The technique for making carbon paper has remained fundamentally the same ever since.

Carbon paper made it possible for people in business, on however small or large a scale, to keep accurate records of all their correspondence. This made business and legal transactions more transparent, reduced the number of disagreements, made both suppliers and customers more accountable, and put the Western economies on a sounder commercial footing. The negative side of the invention has been the proliferation of paperwork, the endless increase in the volume of records and archives, and the associated increase in bureaucracy. It has also accelerated the consumption of paper, and therefore accelerated deforestation.

THE INVENTION OF THE TIN CAN

(1810)

THE ENGLISH MERCHANT Peter Durand invented the tin can in 1810, and was issued with a patent in that year. In 1813, the first commercial canning factory was opened, in England, by John Hall and Bryan Dorkin. At first the process of manufacturing tin cans was slow, but it was developed until 1846, when Henry Evans invented a machine that could make cans at a speed of one per minute, ten times faster.

The first tin cans were made of heavy-gauge metal. They were so thick that they had to be hammered open. Experimentally, thinner metal was used and found to be just as effective at preserving food. The thinner cans could also be opened more easily with specially designed can openers. The first can opener was invented by Ezra Warner of Waterbury in Connecticut in 1858, and American troops used it during the Civil War. In 1866 another kind of can opener, the key opener, was invented by J. Osterhoudt in 1868; this is the design that is still used on tins of sardines. The style of can opener that is used in most kitchens today, cutting with a roller wheel round the can rim, was invented by William Lyman in 1870. An electric version of this type of opener was developed in 1931.

In the nineteenth century, the intention was to keep food fresh by enclosing it in a completely airtight casing. The tin can was entirely successful in achieving this. In 1935, cans were used for the first time to keep drink fresh. The first canned beer, Krueger Cream Ale, was sold in January that year in Richmond, Virginia.

In the days before the existence of domestic refrigerators, canned food was indispensable. Every home in the Western world had its larder stocked with tinned food: tinned fruit, tinned fish and tinned meat.

Before refrigeration, the only safe way of getting Argentinian beef to consumers in North America and Europe was to can it: hence the invention of corned beef. Refrigeration has made much of this canning redundant, but there is a kind of inertia in the marketplace, and of course canned food can be stored on open shelves outside the refrigerator.

THE INVENTION OF
THE METRONOME

(1812)

THE METRONOME WAS invented in 1812 by the craftsman and inventor Diedrich Nikolaus Winkel (1773–1826). Winkel was born in Germany, but went to live in Amsterdam in 1800, where he made high-quality musical instruments.

Winkel made the first usable metronome by adding two weights to a pendulum. The larger weight was hung low down on one side, while the lighter weight was mounted on a slide on the other side. There had been earlier attempts at time-beaters, but they had come up against the problem that slow beats needed pendulums over six feet long, such as the metronome built by Etienne Loulie in 1696, which has a claim to being the earliest (as well as largest) metronome.

The modern metronome is still reliant on the same principle as Winkel's initial design, which is on display in the Municipal Museum in The Hague.

The German instrument maker and music teacher Johann Maelzel saw Winkel's metronome in 1816. He pirated Winkel's invention, took out patents on it in London and Paris, and successfully presented it to the world as his own. The device became known as Maelzel's metronome, and was referred to in nineteenth-century music scores as 'M.M.' alongside the composer's recommended setting.

Winkel also invented the componium, a barrel organ. This was an automatic organ that depended on two separate barrels turning simultaneously. The tunes on the barrels were arranged so that one barrel would give a few bars, then a few bars would come from one of ten tunes on the other barrel, randomly selected. There was a kind of random variability, so that endless permutations of music would come out. The componium is remembered as a mere toy. Winkel deserves to be remembered for his successful invention of the portable metronome, which changed the way composers annotated their scores, and ensured performances that were more faithful to their intentions.

Before the metronome was invented, composers were limited to rather generalized indications of tempo, usually given in Italian, graded from slow to fast: *adagio, adagietto, andante, andantino, allegretto, allegro* and *presto*. These could be modified or qualified by adding *molto*, but they were still only relative. Ultimately, it was not certain what a long-dead composer might have intended by *andante* (at walking speed), because some people walk briskly and others dawdle, and therefore it was not certain what any of the other *tempi* were supposed to be. Once composers had access to a standard instrument for measuring beats, they could indicate to performers and conductors precisely what they had in mind. This was done by writing above the stave a crotchet symbol with the desired metronome setting beside it. A passage in Tippett's Concerto for Double String Orchestra is marked 'quaver = 180'. Usually, composers still give an indication in words of the mood they want as well, so that Richard Strauss in the score of *Elektra* writes 'minim = 60' but adds *molto tranquillo* underneath just to be sure.

The usefulness of metronome markings has expanded in recent years with the invention of software for composing and printing music, such as the wonderful Sibelius 7, the invention of brothers Jonathan and Ben Finn. With systems of this kind it is possible to write music and play it back at the desired speed, and the way to tell the software what speed is required is simply to add the metronome mark.

THE INVENTION OF
THE MINER'S SAFETY
LAMP

(1815)

THE MINER'S SAFETY lamp was invented in 1815 by Sir Humphry Davy. In 1798 he took a job as assistant to Dr Beddoes in the new Pneumatic Institution in Bristol. Within just four years, Davy had established himself as a major scientist in his own right after his experiments with gases. He became addicted to laughing gas, saying it gave all the benefits of alcohol without any of the disadvantages; he damaged his eyesight in a lab experiment, then almost killed himself by inhaling a newly discovered gas. After that he took a holiday in Penzance.

In 1803 Davy became a Fellow of the Royal Society and later its President. He was knighted in 1812, gave a farewell lecture to the Royal Institution, married a wealthy widow and then took a long holiday in Europe. This holiday was perhaps the most remarkable thing of all: to decide to go on a tour of Europe in the midst of the Napoleonic Wars. In fact his celebrity was such that he was able to get special permission from Napoleon to undertake the trip.

Sir Humphry Davy returned to England in 1815, to be asked if he would design a safety lamp for miners. This project was prompted by a particularly bad accident in 1812 in a coalmine at the Felling Colliery near Sunderland. Miners in the eighteenth century worked exclusively by the light of naked flames, just as they had way back in the Late Stone Age flint mines in Sussex and the Bronze Age copper mines in North Wales. Coal gave off gas. Coal miners in particular were in danger from explosions caused when pockets of firedamp (mainly methane gas) ignited. Explosions of firedamp had caused many injuries and fatalities in coalmines. The Felling Colliery disaster killed ninety-two men and boys and led to demands for safer lamps in mines.

Davy's solution to the problem of supplying illumination in a mine where there were possible flammable gases was elegantly simple, and he arrived at it very quickly. He devised a simple, easily portable, cylindrical lantern nine inches tall. The design of the lamp was based on the simple principle that for a substance to be ignited it must first be heated to its kindling point; if that heating can be prevented, then combustion cannot happen. Davy completely enclosed the flame inside the lamp with a fine copper gauze. Air was able to pass through the gauze to supply the flame with oxygen, but the gauze kept the temperature of the surrounding envelope of air to below that of the methane combustion point. The lamp was tested successfully in the Hebburn Colliery in 1816, and it was widely adopted as standard equipment in the British coal mining industry. Minor improvements were later made to the Davy lamp. Special locks were added to make sure the lamp could not be opened by accident.

Because only oxygen would keep the flame burning, an excess of non-combustible but still dangerous carbon dioxide in a mine would put the lamp out. Coal miners quickly realized that this made the Davy lamp into a carbon dioxide detector. They often placed their lamps near ground level in the galleries, in places where carbon dioxide was likely to collect. This made the lamp doubly useful. Today, other gas-detecting devices are available, and the lighting is supplied by electricity, but Davy's invention transformed the working lives of coalminers for the rest of the nineteenth century.

Davy's motive was entirely philanthropic. He intended anyone and everyone to take advantage of his new safety lamp. He took out a patent on his invention probably to prevent someone else from doing so and exploiting it commercially; there is more than one reason for patenting an invention. After this triumphantly successful invention, Davy was made a baronet (in 1818) and then President of the Royal Society.

The Davy lamp is one of those exceptional inventions that result from brilliant insight and a perfect understanding of the nature of the problem to be solved. Sir Humphry came up with the solution to a serious problem, by inventing a technically simple device that does precisely what is required of it.

THE INVENTION OF
THE MACADAM ROAD
SURFACE

(1815)

IT WAS JOHN Loudon McAdam, a Scot born in 1756, who invented a reliable method for giving roads a hard, all-weather surface. The structure consisted of three layers of stones laid on a slightly convex or crowned surface drained by two side ditches. The first two layers of stone were angular hand-broken stones no bigger than three inches across, laid tightly together to a depth of about eight inches. They were covered with an upper layer about two inches thick of smaller stones that were no more than one inch across. Each layer was compacted with a heavy roller, which made the angular stones lock together.

McAdam discovered that the best material for surfacing roads was stone that had been mechanically broken or crushed, then graded to give a uniform size of chippings. Naturally produced gravel, like beach shingle or river gravel, was useless for this purpose; the process of attrition gave the stones smooth rounded surfaces that made them slippery and too mobile. The gravel therefore had to be artificially produced.

McAdam's technique was sometimes later called water-bound macadam, to distinguish it from later developments using tar. His method used a lot of manual labour, but the technology was simple and it produced a strong, free-draining and very serviceable road surface. Roads that were surfaced in this way were described as macadamized.

John McAdam's road surface was probably the greatest advance in road design since the paved roads of Roman times. A great advantage of macadam roads is that they did not involve any high technology and were therefore relatively cheap to make.

The road network of Britain was transformed. The biggest difference was felt in the clay lands of southern England. There, in winter, the roads often turned into ribboned pools of thin mud and became completely impassable, while in summer, they solidified leaving deep ruts and ridges made by carts and carriages, as hard as concrete. Macadam turned these difficult roads into all-weather roads and greatly improved communications. McAdam became known as the Colossus of Roads.

When motor cars appeared at the end of the nineteenth century, dust rising from the macadamized road surface became an increasing problem. The vacuum created underneath fast-moving vehicles sucked dust up from the road surface, whirling it up into dust clouds that were both unpleasant and reduced visibility. Motor vehicles also put greater stress on the structure of the macadamized surface, literally pulling it apart. The solution was to spray tar on the surface. This created tar-bound macadam, tarmacadam or tarmac. Later the tar was mixed with aggregate to make a distinct surface layer on the road. Nearly everywhere, the old water-bound macadam surfaces have been replaced, though some stretches have been preserved, such as the United States National Road.

THE INVENTION OF THE STETHOSCOPE

(1816)

IN THE EARLY nineteenth century, a Paris doctor, Rene Laennec, was working with tuberculosis patients when he made his great invention. Someone asked him to press his ear against a piece of wood. When he did, he heard the amplified sound of a pin being scratched across the wood grain. He immediately saw an application for this, made a paper cone and used it to listen to the sounds inside his patients' chests. To his amazement, the simple device worked and after a few autopsies he realized that the amplified sounds he was hearing were genuinely significant.

Laennec went on to make a cylindrical wooden stethoscope on a lathe. It was, in principle, similar to the ear-trumpets that deaf people had been using for some time. Using his stethoscope, Laennec was able to distinguish between conditions that were minor ailments and those that were life-threatening. His reputation as a doctor soared.

He worked to refine his stethoscope, setting up a small workshop at his home. His final design was a lathe-turned wooden trumpet. One end of it was shaped to fit against the doctor's ear. The other end flared into a cone. This could also have a brass cylinder fitted inside it for listening to the heart. The first documented use of the stethoscope was in March 1817, when Laennec used it to examine the heart of a forty-year-old woman. Unfortunately Laennec himself contracted tuberculosis, probably from his patients, and died.

After Laennec's death, further improvements were made to the stethoscope. In 1828, Pierre Piorry added an ivory earpiece and chestpiece. In 1855, a binaural stethoscope was introduced, an instrument with two earpieces. This is the model that is used in most parts of the world today.

THE INVENTION OF THE KALEIDOSCOPE

(1817)

THE KALEIDOSCOPE WAS invented in 1816 by the Scottish scientist, Sir David Brewster. He devised the name for his invention from three Greek words, *kalos, eidos* and *scopos,* meaning beautiful, form and watcher. The kaleidoscope was a tube containing a collection of loose pieces of glass and other coloured items. Also inside the tube was a set of mirrors or glass lenses fixed at angles to one another. The idea was that the reflected images of the objects would create patterns when viewed through the end of the tube. The mirrors created axes of symmetry, so however the loose pieces of glass were shaken about inside the tube they always made a symmetrical pattern. Usually there are three axes of symmetry, so the impression given is frequently of a multi-coloured snowflake.

In the 1870s, an American called Charles Bush made improvements to the kaleidoscope, went in for mass production and generated a kaleidoscope craze. Brewster took out a patent for the kaleidoscope in 1817, but Bush was granted patents for his versions of the kaleidoscope in 1873.

What Brewster invented was a typically charming and innovative Victorian toy. It was a great favourite with both nineteenth- and twentieth-century children. One of the attractions was that the kaleidoscope not only made attractive patterns, but with a tap of the finger, the pattern could be made to change slightly. With repeated taps, the pattern could be made to grow and evolve like a living thing. There was a revival in kaleidoscope art in the 1970s, and today there are hundreds of people making them.

THE INVENTION OF THE SEAGOING IRON SHIP

(1821)

SOME PEOPLE BELIEVE that there may have been an iron ship as early as 1203. There are Chinese archives that credit General Qin Shifu with inventing a ship shaped like a falcon; it was allegedly sheathed in sheet-iron armour and fitted with an iron ram. The problem is that this sounds like the stuff of legend, rather than objective reporting. There are also records dating from 1413–15 which tell of Korean turtle ships, but it is only at the time of the 1592–98 Imjin War with Japan that the chronicles specify that the turtle ships were iron-clad and even then it looks as if the cladding was on the deck rather than the hull.

In spite of these reservations, it looks as if the idea of nailing protective sheets of metal onto wooden-hulled warships was around for a very long time. It is difficult to be certain when it began.

The idea of building a ship's hull out of solid iron was a very much later concept. In a way, that could not have been thought of until iron was produced in sufficient quantities to supply such a project. In the sixteenth century people would no more have thought of building an entire ship of iron than of gold. There just wasn't enough of it.

It seems that an iron vessel was built as early as 1787 in Staffordshire, about as far from the sea as it is possible to get in England. It was apparently built by John Wilkinson of Bradley Forge. The evidence is contained in a letter that he wrote on 14 July 1787: 'Yesterday week my iron boat was launched. It answers all my expectations, and has convinced the unbelievers, who were 999 in 1000. It will be only a nine days' wonder, and afterwards a Columbus's egg.' Wilkinson's iron barges plied on the River Severn for a time, but the idea of iron vessels did not catch on. After all, people reasoned, iron sinks in water, so it must be an inappropriate material from which to build a ship. After Wilkinson's experiment, other boat-builders and ship-builders carried on using timber.

It would be twenty years before anyone else tried to follow Wilkinson's lead. In 1810, Onions and Son of Brosely made the decision to build some more iron river barges, also for use on the River Severn. In Liverpool in 1815, a Mr Jervons built a small iron barge for use on the Mersey. All of these vessels were built for use in sheltered inland waterways; none were seagoing.

A braver venture was the iron paddle steamer *Aaron Manby*, which was built in 1821 under the direction of Charles Manbury at the Horsley Works in Staffordshire. The ship was ordered by Charles Napier and his son, also called Charles. The *Aaron Manby* was prefabricated at the Horsley Works, and then assembled on the Thames at Rotherhithe. There was a general expectation that this iron ship, 120 feet long, would sink. In fact, when complete and floating in the Thames, she floated unexpectedly high in the water, drawing about a foot less water than the wooden steamships anchored round her.

There are, peculiarly, no surviving plans, drawings or paintings of this important and innovative vessel. It is known that she had a flat-bottomed hull made of quarter-inch iron plates, paddle wheels twelve feet in diameter mounted on each side and a tall smoke stack forty-seven feet high. She had a single wooden deck and a bowsprit.

After trials, in May 1822, the *Aaron Manby* made the voyage from London to Le Havre under the command of Captain Charles Napier, later Sir Charles Napier. This ship crossed the Channel with a cargo of linseed and iron castings, reaching Le Havre at the mouth of the Seine on 10 June, then continuing up the River Seine to Paris. The *Aaron Manby* was the first iron ship to be built anywhere in the world, and the first to negotiate the waters of the open sea. The *Aaron Manby* plied up and down the River Seine, making regular river trips until 1830, when the Napiers sold her. With her flat bottom, she was more suited to river work than steaming on the open sea. Her new owners used her on the River Loire at Nantes.

But iron was still a long way from going into general use. In 1832,

Maudslay and Field built four iron vessels for the East India Company, and then the tide turned; within a few years iron hulls became general, for both merchant ships and warships.

The great transformation that came about as a result of this change of material was that there was now no limit to the size that ships could be built. So long as there was enough coal to power it, enough men to crew it, enough cargo to justify it, the ship could be doubled or trebled in length. It was not long before the captains of industry realized the economies of scale, that sending large consignments, especially long distances, was more economically done in one large vessel than several smaller ones. This reached its ultimate conclusion in the design of the supertankers of the late twentieth century.

The first large iron ship to cross the Atlantic was the famous *Great Britain*, designed by Isambard Kingdom Brunel, launched in 1843 and still to be seen today in Bristol. The revolution in ship design happened in Great Britain, where many other major innovations and inventions had taken place in the Industrial Revolution. Britain therefore naturally took the lead in the building of this new generation of iron ships, which was a further major boost to the rapidly expanding British economy. In the later nineteenth century, iron ships were made in huge numbers in British shipyards.

THE INVENTION OF THE DIGITAL CALCULATING MACHINE

(1823)

THE DIGITAL CALCULATING machine was invented by Charles Babbage in 1823. Babbage was born in 1791, was educated at Trinity College, Cambridge, and founded the Analytical Society with Herschel and Peacock in 1820. He had a restless mind and was what we would call a polymath. Babbage founded the British Association for the Advancement of Science in 1831 and died in 1871.

Charles Babbage was an aesthete but an unusual one. He saw beauty and interest in odd places. He attended a performance of Mozart's *Don Giovanni*, but became as restless as the opera's anti-hero. 'Somewhat fatigued with the opera, I went behind the scenes to look at the mechanism.' One of the stage hands offered to show him round, and backstage he met actors dressed as devils with long forked tails, who were on their way up via the stage lift and trapdoor to snatch Don Giovanni down to Hell. On another theatre visit, he became preoccupied with the stage lighting effects and devised his own ballet complete with spectacular visual effects designed to show the interior of the Earth, though it was never publicly performed. Babbage seems to have been obsessed with fire. He once had himself baked in an oven at 265 degrees Fahrenheit for 'five or six minutes without any great discomfort'. On another occasion, he had himself lowered into the crater of Mount Vesuvius so that he could see the molten lava.

In June 1823, Babbage met the Chancellor of the Exchequer, who granted him the money he needed to proceed with the building of the difference engine. He began work straight away. Five years later, his grant had been used up and he was faced with financing the completion of the machine. The Exchequer could not remember promising a further grant, and the initial meeting had not been minuted. The costs involved were considerable. We think of calculators as being lightweight devices no bigger than mobile phones, but the machine Babbage was building was made of brass, steel and pewter clockwork, and would weigh two tons when finished.

In 1829, a group of Babbage's supporters lobbied the British Prime Minister, the Duke of Wellington. Wellington went to have a look at a model of the engine and ordered a substantial grant of £3,000. The engineer Joseph Clement was commissioned to construct the engine for the government. Unfortunately Clement was very inflexible, refused to collaborate with Babbage at Babbage's house, Babbage refused to pay Clement and work on the engine stopped. In the midst of this fiasco, Babbage changed the specification of his machine. The initial design was to calculate to six decimal places, and he now redesigned to calculate to twenty. It was probably this foolhardy decision that meant that the project would never see completion.

Babbage tried to interest the government in a new machine at a time when they were heartily sick of hearing about the old one. He went on for nearly a decade trying to get the government to commit itself to the suspended difference engine or start work on the new analytical engine. Finally, in November 1842, Sir Robert Peel told Babbage that the government had spent enough money on the project and that it was being abandoned. Babbage was furious. Peel had mixed feelings about the machine. He wanted the project wound up but he was also aware that Babbage had gained nothing at all from the years spent on it. Peel was generous enough to offer Babbage a baronetcy in recognition of his work. Babbage, still the man of fire, was so angry that he turned it down. Meanwhile, the great idea, the first digital computer, remained an idea and was never built.

THE INVENTION OF THE ELECTROMAGNET

(1825)

AN ELECTROMAGNET is a device in which magnetism is generated by an electric current. The electromagnet was invented by William Sturgeon (1783–1850), an English physicist and auto-didact. Sturgeon was born at Whittington in Lancashire, was apprenticed to a shoemaker and then, in 1802, joined the army. William Sturgeon taught himself mathematics and physics and eventually, in 1824, became a science lecturer at the Royal Military College at Addiscombe in Surrey.

In 1825, Sturgeon exhibited the first electromagnet. The device consisted of a horseshoe-shaped lump of iron, wrapped loosely with several turns of coil. When an electrical current was passed through this coil it became magnetized; when the current was switched off the coil became demagnetized. Three years later he built the solenoid for which Ampère had put forward the theory. In 1836 he founded the journal *Annals of Electricity* and in the same year invented a galvanometer.

The use of a magnet that could be switched on and off was obvious, to Sturgeon and everyone who saw his demonstration. Switched on, the electromagnet could lift a piece of iron. Switched off, it released it again. Sturgeon's electromagnet was relatively weak, but it showed great potential. It was the beginning of the period when electricity could be used for making machines easy to regulate and control. It also laid the foundations for large-scale electronic communications.

Five years after William Sturgeon invented the electromagnet, an American called Joseph Henry made a more powerful electromagnet. This showed the potential of Sturgeon's invention for long-distance communication. Henry sent a current along a mile of wire to activate an electro-magnet, which made a bell strike. It was the birth of the electric telegraph.

THE INVENTION OF THE RAILWAY

(1825)

THE RAILWAY WAS invented by George Stephenson (1781–1848) in 1825. The steam locomotive, the vehicle for running on the rails, was invented by Richard Trevithick, but Trevithick thought in terms of running his locomotive on roads. Rails, too, had already been invented; they were in use for manoeuvring wagons in coalmines. The development of the railway depended simply on combining these two inventions or concepts, and it was Stephenson who did that.

Stephenson was born in the mining village of Wylam near Newcastle in 1781. He was the son of a colliery mechanic and was self-educated, like a great many of the innovators of the Industrial Revolution. He became the chief mechanic at Killingworth Colliery. He first became interested in steam in 1813 when, in his professional capacity, he looked at one of the 'steam boilers on wheels' designed by one of the mine managers, John Blenkinsop. It was used at several collieries for carrying coal. George Stephenson saw that the machine had the potential to do rather more than colliery work.

In 1814, Stephenson started building his own steam locomotive at Killingworth. He named it *Blücher* after the great Prussian field marshal. Blenkinsop's engine had cogged wheels which engaged with cogs in the sides of the rails. Stephenson's had the design that all modern trains have: rimmed wheels that ran on smooth rails, giving a faster and smoother ride. Stephenson had trouble with *Blücher*, which kept breaking down. He worked for months to improve the engine.

Then Stephenson made a major breakthrough: the steam blast technique. This worked by directing exhaust steam into the chimney through a narrow blast pipe. The expelled steam sucked air in after it,

ABOVE: *A wooden Viking chariot, showing the wheels fitted with wooden pegments, circa AD 800.*

ABOVE: *An ancient Arabic astrolabe which would have been used for celestial observations.*

ABOVE: *An early microscope designed by Robert Hooke, 1665. Hooke (1635–1703) also made original contributions to many other fields of science.*

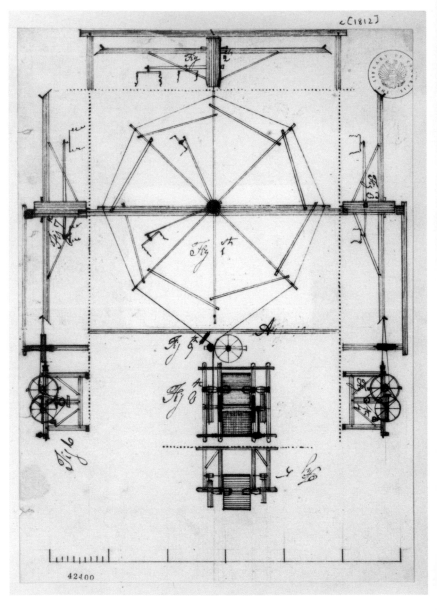

ABOVE: *A circa 1750 sketch of Spinning Jenny which was used for spinning thread and invented by James Hargreaves.*

ABOVE: *An original Singer sewing machine dated 1854 and still in working order.*

ABOVE: *A cash register on an old store counter, and a string spool for tying up packages, circa 1800s.*

ABOVE: *A robotic model showing the machinery workings of the body, circa 1928.*

ABOVE: *A man removing an instantly developed photograph from a Polaroid camera, circa 1948.*

increasing the draught in the furnace, raising the temperature, providing more power and speed.

The first railway train had its first outing on 27 September 1825, with its inventor, George Stephenson himself, at the controls. The little locomotive called *Locomotion* with its tall chimney steamed along, pulling thirty-two open wagons carrying more than 300 people, along with twelve coal wagons with more passengers perched on top of the coal. This first train journey followed the only railway line then in existence, a stretch from Shildon Colliery to Darlington and from there on to Stockton. It was just twenty miles in all, but long enough to prove that the railway was a practicable and viable new form of transport.

In the early days, the trains steamed along at little more than walking speed, and a horseman rode in front carrying flags to warn of the train's approach. As trains reached fifteen miles per hour, the horsemen were replaced by faster scarlet-coated huntsmen, also on horseback. But even a coach-and-four, then the fastest transport known, was outpaced by the speeding train.

When *Locomotion* reached Stockton, 40,000 people had gathered to see it arrive, while a brass band played the national anthem. Everyone who saw it knew that a historic event had taken place that September day in 1825 – a revolution in transport that would transform England. In fact its effects were much farther-reaching than anyone, George Stephenson included, could have foreseen. A replica of Stephenson's *Locomotion* still pulls wagons along the tracks at the Beamish North of England Open-Air Museum in County Durham.

Stephenson's success with *Locomotion* led to his appointment as the engineer of a passenger and freight railway that was to run between Liverpool and Manchester. This came in 1826. For this project, George Stephenson and his son Robert designed and built a new locomotive, the famous *Rocket*. This engine was built at their Newcastle workshop. They fitted it with a multi-tubular boiler in which the water was turned into steam by coming into contact with twenty-five copper tubes that were heated from the firebox. The *Rocket* is now on display at the Science Museum in London.

The Liverpool and Manchester Railway opened on 15 September 1830. Over 50,000 people gathered at the starting-point of the journey to Manchester, the engine yards at Liverpool. When a cannon was fired, not one but eight locomotives set off. They were led by Stephenson's latest and most powerful engine, the *Northumbrian*, which Stephenson himself drove. The *Rocket* followed, now already superseded. Stephenson had some distinguished passengers in the *Northumbrian*'s carriages. One was the Duke of Wellington, now prime minister. Another was Prince Esterhazy, the Austrian ambassador. Also present was the Tory MP for Liverpool, William Huskisson.

The triumphal procession moved on out of Liverpool without incident. There was no problem until the *Northumbrian* stopped about eighteen miles along the line at Parkside, to take on water. Two of the following locomotives, the *North Star* and the *Phoenix*, overtook the *Northumbrian* on a parallel track. Huskisson and Prince Esterhazy decided to get out and stretch their legs by walking along the track beside the train. Wellington waved to Huskisson and opened the door of his carriage. Huskisson hurried forward to shake Wellington's hand and chat. Suddenly the *Rocket* steamed into view along the adjacent track. The Prince, who was small and light, was easily pulled up out of harm's way into one of the carriages. Huskisson was tall, partly paralysed down one side, and far from nimble. To make matters worse, he changed his mind about what best to do. He tried to climb up into one of the carriages, but fell back into the *Rocket*'s path and had his thigh badly crushed. With great presence of mind, Stephenson ordered all but the leading carriage of the *Northumbrian* uncoupled. He put Huskisson, who was in agony and bleeding badly, into the carriage and went at full speed to Eccles, fifteen miles away on the outskirts of Manchester. He managed to reach Eccles in twenty-five minutes, but Huskisson died of his injuries in Eccles vicarage that evening. He was the first victim of a railway accident. Huskisson's death was in no sense the result of any failure in the technology of the locomotive; it was simple human error, the result of a bad decision. But it was also the shadow side of the Industrial Revolution, a glimpse of the misery and suffering that the new technology could bring in its wake.

Nothing could halt the relentless advance of technological and industrial progress. Commerce was all. The morning after William Huskisson's death, the first fare-paying passengers left Liverpool for Manchester.

By the year 1840, there was a network of 1,500 miles of railway covering Britain. Then the web of iron spread out across Europe. By the 1890s, railway networks were developing across North and South America, Australia, southern Africa and India. The railway age was a truly global phenomenon.

Stephenson, regarded as the Father of the Railway, worked as a consultant on many railway schemes in Britain and mainland Europe. He retired in 1838 to Tapton House, from where he could watch his trains go by on the North Midland Railway. He spent his final decade tending his garden.

THE INVENTION OF THE FRICTION MATCH

(1826)

THE FRICTION MATCH was invented by John Walker (1781–1859). Walker was a chemist and apothecary, with a shop in the High Street in Stockton-on-Tees. Walker had a reputation as an amateur botanist. He was also interested in geology and minerals, and spent a lot of time experimenting with chemicals. In 1826 he accidentally invented the friction match while mixing chemicals – potassium chlorate and antimony sulphide in particular. He found that by coating sticks with the mixture along with gum and starch, and then letting the coating dry, he could start a fire by rubbing the stick along any surface.

He started making and selling his matches straight away. He called them Congreves after the Congreves rocket invented in 1808. Walker's first customer was a Stockton solicitor called Hixon. Walker did not apply for a patent for this important invention, so he made very little money out of it, but he also refused to tell anyone what his matches were made of.

He was unmarried, living with his niece. He was precise and set in his manner of dressing, always wearing a tall beaver hat, brown tail-coat, white cravat and grey stockings.

Undefended by any patent protection, Walker was bound to lose his invention. A man called Samuel Jones spotted Walker's Congreves and decided to go into business in competition. Jones marketed his matches under the trade name Lucifers. They became popular, especially with smokers. In 1830, Charles Sauria, a French chemist, made friction matches using white phosphorus. Unlike Jones's Lucifers, Sauria's matches did not create an unpleasant smell, but they did make people unwell: white phosphorus is poisonous.

One problem with the early friction matches was that they could be struck on any surface, and that meant that they could be struck accidentally, and cause unintended fires. The invention of the safety match by Johan Lundstrom in 1855 was an important improvement. Lundstrom added red phosphorus to the strip of sandpaper stuck on the outside of the matchbox and the remaining ingredients on the match head. This removed the problem of the phosphorus poisoning and created a match that could only be lit by striking it on the prepared strip. In 1889 Joshua Pusey invented matchbooks, which he called Flexibles.

Matches have remained much the same ever since. They are among those basic, everyday objects that we have around us all the time and depend upon – and yet take entirely for granted.

THE INVENTION OF BRAILLE

(1829)

LOUIS BRAILLE, THE inventor of the reading system for the blind, became blind himself at the age of three. He was in his father's leather shop when one of his father's knives slipped and struck him in the face, blinding him. Louis was an intelligent and determined boy and did everything he could to overcome his disability. When he was ten years old he was given a scholarship at the National Institute for Blind Children in Paris. He was musical, playing both organ and cello.

While the young Louis Braille was at school, he was introduced to a system of reading for blind people invented by the Institute's founder, Valentin Hauy. This system involved running fingertips over letters that were embossed on paper. It was well meant but slow, and Louis and other users of the system found it tedious and time-consuming. Another system was invented by Captain Charles Barbier, called night writing. The captain was a soldier and he developed the night writing system to enable soldiers to read messages at night. Braille encountered the Barbier system when he was fifteen and worked on ways of improving it for the use of the blind.

The system Captain Barbier developed was based on a set of twelve raised dots that were arranged in various permutations to represent all the letters of the alphabet. Braille pared this down, developing his own system that only used six dots. The letter A was one dot, B was two dots one above the other, C was two dots side by side. By using fewer dots, Braille made reading much faster, twice as fast as reading the Barbier system, bringing a blind reader's speed up to about half that of a sighted reader.

Braille spent a great deal of time refining and perfecting his system, and by the time he was twenty it was published. It was also informally adopted by the Institute, where he had, not surprisingly, become a teacher.

The Braille reading system was very slow to catch on, and by the time Louis Braille died of tuberculosis in 1852 it had still not gained general acceptance. Other reading systems for the blind were devised. In the 1860s a New York Point system was devised in America. In Boston an adaptation of Braille's system was introduced by another blind teacher, who called it American Braille. In the end, the simplicity and logic of Braille's original system prevailed and it became the preferred system all over the world. In 1932, it was adopted at an international conference as the official language of the blind.

A stylus was developed that enables the blind to create the symbols on paper themselves, and so write in Braille.

The significance of a usable and efficient reading and writing system for the blind is wide-reaching. It was not uncommon for totally blind people to be dumped in asylums or put to work at crafts that did not require sight. Braille had the effect of empowering blind people in exactly the same way that literacy has empowered the rest of mankind, enabling them to play as full a part in the life of the community as sighted people.

THE INVENTION OF THE LAWNMOWER

(1830)

THE LAWNMOWER WAS invented in 1830 by a self-employed engineer, Edwin Beard Budding.

The lawnmower, a device for cutting grass blades down to a uniform height, was a great saving in time and skill. For hundreds of years there were skilled countrymen who could wield a scythe so expertly that they could trim lawns to bowling-green smoothness. But the work was laborious in the extreme, and not everybody could do it.

Edwin Budding had been a carpenter at Chalford in Gloucestershire, and he may have borrowed his idea for a rotary grass cutter by seeing a rotary cutter used to shear the nap off woollen cloth at Brimscombe Mill. The rotor trimmed the wool far more evenly than it could ever be done by hand shears. As with many other inventions, this one was made by thinking 'what if . . . ?' and transferring a technique from one area of activity to another, apparently unconnected to it. It is making that un-anticipated connection that sets inventors apart from the rest of us.

Budding designed a rotary lawnmower to cut the lawns of playing fields and large gardens. His 1830 patent described the lawnmower as 'a new combination and application of machinery for the purpose of cropping or shearing the vegetable surfaces of lawns, grass-plots and pleasure grounds.' Budding saw that people might actually enjoy the exercise they would get while mowing their own grass; 'country gentlemen may find in using my machine themselves an amusing, useful and healthy exercise'.

John Ferrebee bought from Budding the right to manufacture and sell lawnmowers, as well as license others to manufacture them. An early Budding and Ferrebee lawnmower was used in Regent's Park in London in 1831. After another ten years of development, Budding produced a larger machine that could be pulled by a horse or donkey. It took sixty years before a steam-powered mower was attempted.

Commercial manufacturing of lawnmowers started in the 1850s. In 1862, Ferrebee was manufacturing eight different models. The largest had a roller thirty-six inches across. By the time he stopped manufacturing lawnmowers in 1863, Ferrebee had made more than 5,000 machines. It was in the 1860s that the first modern grass boxes for catching and storing mown grass were fitted.

In 1859, the first chain-driven mower, called the Silens Messor, was built by Thomas Green. By 1900 the biggest name in lawnmower manufacture was Ransome, and the best-known machine was the Ransome's Automaton.

The socio-economic shifts that generated more leisure for the lower-middle and working classes led directly to the rise in popularity of a variety of sports. The spread of lawn tennis, croquet, cricket, football and rugby all increased the demand for lawnmowers. Alternatively, it could be that the availability of the lawnmower made the spread of such sports possible.

There were two great advances inherent in the invention of the lawnmower. One was that it saved an enormous amount of time. The other was that it was no longer necessary to employ someone to cut your lawn: you could do it yourself. The adoption of the lawnmower was an important step towards major social and economic changes in the countries of the West. Upper- and middle-class families in 1800 regularly employed servants to carry out a whole range of domestic tasks, including gardening. By the twentieth century, people carried out most of these tasks themselves and dispensed with servants. Today, most of us mow our own lawns – thanks to lawnmowers.

THE INVENTION OF THE ELECTRIC MOTOR

(1831)

THE FIRST ELECTRIC motor was built by Michael Faraday. Faraday's career as one of the greatest scientists and innovators of the nineteenth century had humble beginnings. He started as an errand boy for a bookbinder in London. But he was very intelligent and had boundless curiosity. He read a great deal, thought a great deal and, on his own admission, had been forced by reasoning to reject many of his own ideas.

> *The world little knows how many of the thoughts and theories which have passed through the mind of the scientific investigator have been crushed in silence and secrecy by his own severe criticism and adverse examination; that in the most successful instances not a tenth of the suggestions, the hopes, the wishes, the preliminary conclusions have been realized.*

Michael Faraday was a serious and rigorous critic of his own thinking, and this was the root of his success as a pioneering scientist.

Faraday attached two wires through a sliding contact to a copper disc. When he rotated the disc through the poles of a horseshoe magnet, he found that he produced a continuous direct electric current. In this way Faraday invented the first simple generator. As a direct result of these experiments, Faraday became the inventor not only of the generator, but of the transformer and the electric motor as well. He succeeded in making the first electric motor in 1831.

As it happened, Joseph Henry was working on a similar motor at the same time. As is often the case with major breakthroughs of this kind, the first motor was far from efficient, and several other engineers improved it over the succeeding decades.

Most of the early motors were direct current motors. It was not until 1887 that Nikola Tesla introduced the alternating current motor. In 1882, Tesla identified the principle of the rotating magnetic field, and exploited the rotary field of force to operate machines. He used this to design a new two-phase induction motor in 1883. It has been said that the introduction of Tesla's motor in 1887 marked the beginning of a Second Industrial Revolution because it made possible the efficient generation and long distance transmission of electricity. Before the invention of the rotating magnetic field, electric motors worked by passing a conductor through a stationary magnetic field. Tesla proposed that the commutators from a machine could be removed and that it could operate on a rotary field of force. Tesla's teacher, Professor Poeschel, said this was akin to building a perpetual motion machine. Tesla obtained a US patent for his electric motor, a classic alternating current electro-magnetic motor, in December 1889.

Today, there are many different types of motor: single-phase, two-phase and three-phase alternating current motors, induction motors, synchronous motors, torque motors, stepper motors, brushless DC motors, coreless DC motors and linear motors.

Some electric motors are immediately apparent around us today, in the form of electric cars and trains. There are thousands of other electric motors that are less obvious, powering a great many devices that we take for granted, such as washing machines, ventilation and extractor fans, and pumps for fountains and swimming pools. One great virtue of electric motors is that they produce no atmospheric pollution, at least where they are used, and they are quiet. These attributes make them ideal for use in the home and in the urban environment generally. The electricity has to be generated, but that may be done from a number of energy sources, green or otherwise, 100 miles or more away from the consumer.

THE INVENTION OF THE MECHANICAL REAPER

(1831)

ROBERT MCCORMICK, A Virginia farmer, made several unsuccessful attempts to design a mechanical reaper. He gave up in 1831, and in that year his twenty-two-year-old son Cyrus took up the challenge. Cyrus studied the technical problems carefully and came up with a design before the end of the year. The machine was built immediately and used to bring in the late harvest on their farm in 1831.

Cyrus did not get round to applying for a patent until 1834, by which time another reaper designed by Obed Hussey had been patented. There was fierce rivalry between the two men, first to settle the legal rights to the invention, then to manufacture enough harvesters to meet the huge demand. In 1847, McCormick moved to Chicago and began manufacturing there. In the end the McCormick Harvesting Machine Company emerged as the leading manufacturer. McCormick was not only a successful inventor, he had the entrepreneurial skills to market his product as well. By 1856 he was selling 4,000 reapers every year by offering them for sale by instalments.

The mechanical reaper, transformed into the combine harvester, made enormous changes to the Prairies. Now huge fields of grain could be harvested at high speed, which meant that best use could be made of spells of good weather. Food production was stepped up enormously. Thanks to harvesters, the Prairies regularly produced huge grain surpluses, which the US could export, and which were to acquire profound political importance in the Cold War. The Soviet Union frequently suffered from crop failures and needed the US grain surplus, which gave the US political supremacy.

The combine harvester, invented by Hiram Moore at about the same time as the McCormick reaper, was (and still is) a large and complex machine that harvests, threshes and cleans the grain by shaking it, and may be used for crops of wheat, soybeans, flax, oats, barley or rye. The harvester produces and separates grain and loose straw, which it bales and throws back onto the harvested field. The early combines were pulled along by horses or mules and employed a bull-wheel to power the machinery. Later combines were pulled along by tractors and used independent diesel engines to power the grain separation. The big modern combines are completely self-contained machines, self-propelled and they operate like miniature grain factories, from the 1970s working from on-board computers.

THE INVENTION OF
THE SEWING MACHINE

(1832)

THE MODERN SEWING machine was invented in 1832 by the New York inventor Walter Hunt. The new machine had a needle with an eye at its point that pushed a thread through cloth to interweave with a second thread carried by a shuttle. Hunt did not take out a patent on his invention. When he suggested to his fifteen-year-old daughter Caroline that she should go into business making corsets with his new machine, she protested that it would put needy seamstresses out of work. Like so many inventions of the eighteenth and nineteenth centuries, the sewing machine brought both benefits and disadvantages. It saved human labour, but thereby created unemployment.

In 1843, the sewing machine was invented all over again by a twenty-seven-year-old Boston machine shop apprentice called Elias Howe. His machine used two threads to make a stitch that interwove by means of a shuttle. Howe made his invention without any knowledge of Walter Hunt's machine, and patented his reinvention in 1846; there were widespread infringements of Howe's patent in subsequent years. In any case tailors and clothing manufacturers were afraid to introduce it in case they antagonized their workers.

The Singer sewing machine, invented by the mechanic Isaac Singer in 1850, was destined to become the machine that took over the market. Singer watched some Boston mechanics trying to repair a primitive Howe sewing machine and could see ways of improving it. The Singer machine was patented in 1851. Isaac Singer went into partnership with his New York lawyer, Edward Clark, in order to mount the strongest defence against patent suits from Howe. Howe won his lawsuit, which did not prevent Singer from producing and selling his sewing machines, but ensured a royalty payment for Howe, who was to make a fortune, not from his own, but from Singer's machines. In 1858, Isaac Singer offered the earlier inventor of the sewing machine, Walter Hunt, 50,000 dollars in five annual payments to clear up any claims he might have had on the machine design. Hunt died a few months later without collecting even the initial payment. Walter Hunt was one of those unlucky inventors who was unable to make any money out of anything he invented – not just the sewing machine, but the breech-loading rifle and his Globe stove.

THE INVENTION OF THE HANSOM CAB

(1834)

THE HANSOM CAB was a light, two-seater, closed horse-drawn carriage. It first saw service in Liverpool in the 1830s but came to be widely used as a taxi in Western cities in the later decades of the nineteenth century. Perhaps its most distinctive feature was the elevated seat for the driver behind the cab; the driver spoke to his passengers through a trapdoor in the cab roof. The two passengers entered from the front through a folding door and sat side by side on a seat positioned above the carriage's single axle.

The hansom cab was named after its inventor, Joseph Hansom, who patented it in 1834. It was initially called the Hansom Safety Cab, and the idea was to combine speed with safety by having a low centre of gravity. Joseph Hansom was an architect from Hinckley in Leicestershire. The design was bold, original, stylish and extremely practical. Having just one axle and one pair of wheels, the cab could turn round in its own length, which made it extremely manoeuvrable in the crowded streets of London. The hansom was, even so, not particularly easy to drive.

The hansom cab was as widely accepted in mid-nineteenth-century London as the taxi is today. It replaced the hackney carriage as a vehicle for hire in London. It was popular because it was fast and could escape relatively easily from London's traffic jams. It was also popular because it could be pulled by a single horse, and was therefore significantly cheaper than travelling in a bigger four-wheeled carriage. Once the clockwork mechanical taximeters were installed to measure the fares, the name was changed from hansom cab to taxicab.

The hansom cab was very popular in London, but it was also adopted by other European cities, such as Berlin, Paris and St Petersburg. A few decades later, it was exported to America, where it was widely used in New York. The hansom found its way to Cairo, Sydney and Hong Kong. It went on being very popular until as late as the 1920s, when motor vehicle transport took over, along with more reliable mass-transport systems such as the London Underground.

THE INVENTION OF THE REVOLVER

(1836)

THE REVOLVER WAS invented by Samuel Colt in 1836. The revolver was a handgun with a relatively long barrel and a revolving cylinder that contained five or six bullets. Samuel Colt also fitted his new gun with a new cocking device. The revolver was a huge advance on the previous generation of handguns, which consisted of one-barrel or two-barrel flintlock pistols. The Colt revolver, named after its inventor, gave the possibility of six shots rather than two. This was a highly significant increase in fire power.

Samuel Colt's invention played a major part in shaping American history. It was said that Abraham Lincoln may have made all Americans free, but Samuel Colt made all Americans equal. It was a distinctive component in the American gun culture. Since Colt's revolver was patented, over thirty million firearms bearing the Colt brand name have been sold. In the second half of the nineteenth century, the Colt revolver was the best-known firearm in the United States, and also in Canada and Mexico.

Although the revolver made a great deal of money for Samuel Colt and his heirs, and provided employment for those engaged in manufacturing the weapons, it is hard to see how this invention has improved the world. The argument can be made that people have guns to defend themselves and that the Colt revolver could be used in self-defence. It is, even so, sobering to contemplate how many people must have been killed or injured by revolvers, and how much crime was empowered by revolvers, and how much needless fear they have engendered.

THE INVENTION OF
THE SCREW
PROPELLER

(1836)

THE IDEA OF using a screw as a propeller for ships is one with a shadowy beginning. Several of the major inventors and engineers of the first Industrial Revolution – Watt, Bramah and Trevithick – all mentioned or described the screw. Trevithick devised several screws of different shapes, though did not actually use any of them.

The man who must be credited with the invention of the screw propeller is Francis Pettit Smith, who was born at Hythe in Kent in 1808. As a boy he had a passion for building model boats. When he grew up he took to sheep farming on Romney Marsh, later moving to Hendon. In 1834 Francis Smith built a boat with a wooden propeller driven by a spring. He demonstrated that it worked, and those who saw it thought it remarkable. It is not known where Smith got his idea from. Perhaps the screw was hovering in many people's minds independently at the beginning of the nineteenth century.

Francis Smith concluded from his experiment that his method of screw propulsion was far better than the paddles that were at that time considered the best method. In 1835, Smith built another, better, model and carried out a number of experiments with it at Hendon. In 1836 he took out a patent for propelling vessels by means of a screw revolving under the water at the stern. After that, he openly displayed his invention in London. Sir John Barrow, Secretary to the Admiralty, was impressed by the way the screw propeller functioned.

An invention of this magnitude needed financial backers, and Smith acquired two backers, Wright and Caldwell, who funded the building of a miniature ship, in effect a scale model, of ten tons and six horsepower. It was fitted with a screw that consisted of two whole turns. On 1 November 1836, the little vessel was tried out on the Paddington canal and then the Thames, which she successfully steamed up and down for nearly a year.

During one of these trips, the propeller hit something underwater and about half of the length of the screw was broken off. Far from impairing the boat's performance, it shot ahead and reached greater speeds than it had ever reached before. As a result of this chance discovery, the vessel was fitted with a new screw consisting of a single turn, which worked much better. The vessel had worked well in sheltered water, so Francis Smith decided to try her in the open water of the English Channel. He took her down the Thanes to Gravesend, took on a pilot and continued to Ramsgate and Dover. Between Dover and Folkestone he encountered rough weather, but the boat continued regardless at more than seven miles per hour. The boat returned home safely, the object of great interest the whole way.

Reason says that at that point in history the paddle should have been given up in favour of the screw. But people's minds, and a certain amount of vested interest, were for the time being locked into the paddle wheel as the right method of propulsion.

At about the same time that Francis Smith took out his patent, John Ericsson, a Swede living in England, also invented a screw propeller. Ericsson had earlier invented a steam locomotive, the *Novelty*, which had competed with Stephenson's *Rocket* in the locomotive contest at Rainhill in 1829. The *Novelty* had been beaten, but it was acknowledged by *The Times* that the *Novelty* had come a close second. Ericsson's fertile imagination now came up with the screw propeller. Like Francis Smith, he started by testing his spiral propeller on a model boat; it drove it along at three miles per hour. He patented it in 1836. To demonstrate it he built a boat forty feet long with two huge propellers, each five feet in diameter. The *Francis B. Ogden* shifted along at ten miles per hour and showed a lot of pulling power. Ericsson perceptively saw that screw propulsion was particularly suited to warships. Being tucked out of sight under the ship, the propellers would be far harder for the enemy to put out of action than the over-conspicuous paddle wheels of conventional steamers.

He tried to interest the Lords of the British Admiralty in an excursion, being towed by his experimental boat. The Admiralty barge was duly filled at Somerset House with celebrities including Sir Charles Adam, Sir William Symonds and Captain Beaufort. The trip on the river was a huge technical success, but it had no effect on their lordships' preference for paddle steamers. Writing to Ericsson afterwards, they objected that because the screws were in the stern, the boat would certainly be impossible to steer, yet this was after they had seen for themselves that this was not so.

An American naval officer, Captain Stockton, took a different view and ordered two iron boats with screw propellers for him to take back to the USA. One, the *Robert F. Stockton*, was seventy feet long. It left for America in April 1839, and needless to say its inventor soon followed, to become a huge success as a great New World inventor.

'Screw' Smith, meanwhile, was stuck with the fact that in Britain the paddle boat was seen as the epitome of the steamer. In 1838, the Lords of the Admiralty asked Smith to allow his vessel to be inspected. This time, they were more ready to be impressed. They now wanted to see the screw propeller tried on a vessel of at least 200 tons. Obviously Smith could not afford to build one but financiers appeared, sensing that there might soon be a breakthrough with the Admiralty, followed by some big orders.

A Ship Propeller Company was formed to build the new test ship. She was to be the *Archimedes*, a wooden ship of 237 tons fitted with a screw of one turn. She had her maiden voyage in May 1839. The Admiralty had said that if her speed was around four miles per hour they would be satisfied. The *Archimedes* reached nine miles per hour. The *Archimedes* passed test after test. She was shown to mariners and engineers in every port in Britain in 1840. After touring Britain, she was sent to Oporto, and did the trip in the fastest time recorded. Only favourable comments were made. But still there were delays in adopting the new method of propulsion.

The establishment resistance to the screw propeller was remarkable, given the incredible range of industrial innovations that had erupted during the previous century. It took intervention by Isambard Kingdom Brunel to cut across this resistance. He was building a big new steamer at Bristol for passenger traffic between England and America. She was the *Great Britain*. He had intended to make her a paddle steamer, but he heard about the *Archimedes*, inspected her, was impressed and recommended to his directors that they should opt for screw propulsion.

The vessel was modified to take screw propulsion. The *Great Britain* reached a speed of ten miles per hour, even though the wind was against her. The Admiralty were still reluctant to shift. Eventually a screw steamer, the *Rattler*, was pitted against a paddle steamer, the *Alecto*, in a straightforward contest. The *Rattler* won and that meant, at last, that Francis 'Screw' Smith had won too. But Smith had by this time lost a lot of money in the struggle to persuade the Admiralty. His company lost about £50,000 in promoting the invention, and then, in 1856, the patent right expired. Following that, scores of naval vessels were fitted with propellers.

Poor Screw Smith was left with very little, though the British establishment recognized that he had been badly treated. He was granted a Civil List pension and given a knighthood shortly before his death in 1871. The triumph of the screw propeller was due, not to its superiority as a technical device, as it should have been, but to the extraordinary tenacity and stamina of its inventors, Ericsson and Smith, and their unshakable belief in the worth of their inventions.

THE INVENTION OF PHOTOGRAPHY

(1838)

THE CAMERA DEVELOPED out of a certain interest in recording scenes accurately. In particular, it developed out of an earlier invention, the camera obscura. The phrase camera obscura means 'dark chamber', which is what it was. It might be a fixed room in a tower, or a portable structure something like a bathing machine or a gipsy caravan. Light entered the camera obscura through a tiny hole, which could be focused through a lens onto a table or onto the wall. There, the image of the landscape might be traced – in perfect perspective – and then used by an artist as a basis for a painting. The camera obscura was also a safe and comfortable way of observing solar eclipses.

The camera itself arose out of an understandable desire to preserve the projected image. One pioneer was a lithographer called Joseph Niepce, who experimented with photographic techniques from 1813 onwards. In 1822, he used bitumen combined with lavender oil to make a primitive photograph. A plate exposed to sunlight through a transparent engraving responded in a significant way; the areas of bitumen struck by the sunlight turned solid and so became fixed, whereas the areas that were exposed to less light did not and were washed away. So a kind of photograph of the engraving was created. Niepce later, in 1826, used this process in a camera obscura and, with an exposure of eight hours and camera containing a treated pewter plate, created a photographic image of the courtyard of his house. It was the first photographic image in history, but he was unable to fix it and it gradually faded. Niepce called his photograph a heliograph or sun drawing.

At the same time, a theatrical scene-painter called Louis Daguerre was developing a similar interest in photography. He had used the camera obscura a great deal in the course of his work. He had also carried out experiments with methods of fixing the images, in particular with using the tendency for silver salts to darken on exposure to sunlight. When he heard of Niepce's success, he proposed that they should work together. Niepce had reached an impasse, aware that he was unable to find a way of developing a holographic image on paper, so agreed to collaboration.

Niepce died in 1833, and after that Daguerre carried on developing the technique alone. In 1838, he discovered that an image exposed on a copper plate coated with a layer of silver iodide became visible once it was exposed to mercury vapour. This was the turning point. Then all that remained was for Daguerre to discover a way to stop the whole picture from turning dark, and he found that dipping the plate into a solution of salt removed the remaining (unexposed) silver iodide from the plate permanently fixed the image.

By chance, photography was invented simultaneously in England, by William Henry Fox Talbot. In 1839, Talbot announced that in the previous year he had invented a photographic technique, which he called photogenic drawing. Talbot's technique had the advantage of enabling the photographer to produce multiple copies of an image, and being on paper they were far more versatile in the way they might be handled. Daguerre's daguerrotypes were on glass plates and could not be reproduced. This limited their usefulness, but the images were much sharper than in Talbot's photogenic drawings.

Fox Talbot's technique involved making prints on silver chloride paper. In 1841 he patented another type of photograph, the calotype or Talbotype. This was the first process allowing the creation of a photographic negative from which any number of prints could be made. He also invented a third photographic technique in 1851, the creation of instant photographs using illumination by an electric spark – the first flash photography.

Photography had far-reaching effects. By the end of the nineteenth century, photography was in regular use for the accurate recording of the appearance of places and people, and this released painters from the need to produce likenesses. Photography allows us to see exactly what the elderly Duke of Wellington or Rossini looked like. The revolutionary art movements of the twentieth century, such as expressionism, futurism and cubism, were 'permitted' by photography. Photography also showed what people, places and events were really like and this had a dramatic effect on the transmission of news, which became franker. Portrait photography started as soon as the technique was available, so for the first time people could see exactly what celebrities looked like; indeed it could be said to have led to a particular cult of celebrity, depending on certain iconic photographic images. There are such iconic images of James Dean, Marilyn Monroe and Che Guevara.

It was also possible to take photographs of events, and this marked the start of modern news coverage. Photographs of the American Civil War and the Crimean War showed war as unglamorous and bloody; they meant that it became harder for politicians to deceive people about the nature of warfare, or particular incidents. Eventually it was possible to capture on photographs historic events such as the assassination of John F. Kennedy and the attack on the World Trade Center. The immediacy of these images has had the effect of making people more engaged with world events than ever before.

THE INVENTION OF THE ELECTRIC TELEGRAPH

(1838)

THE TELEGRAPH WAS a long time evolving before it became a usable invention. Sir William Watson, as early as 1747, showed that an electric current could be transmitted through quite a long wire. In 1758 someone suggested that an electrical telegraph system could be developed. 'C.M.' – whoever it might have been – proposed anonymously to the *Scots Magazine* that a separate insulated wire might be used for each letter of the alphabet. At the receiving end of each wire a pith ball was to be suspended above a piece of paper marked with the appropriate alphabet letter. As a charge passed along a wire, the pith ball would attract the paper beneath, so that words could be spelt out. The ingenious idea sounds rather like an ouija board, and it was tried out by Georg Le Sage in Geneva in 1774.

Several similar systems were tried out over the next two decades. Many scientists and engineers were at work on the idea, but the one who was to make the biggest impact was Samuel F. B. Morse. Morse was an unlikely inventor. He graduated from Yale in 1810, and set sail for Europe to study art. He returned to America in 1813, and became one of America's finest portrait painters. One of his sitters was Eli Whitney, the inventor of the cotton gin. But Morse always had an interest in science. In 1832, Samuel Morse was travelling back to America after a trip to Europe, on the packet ship *Sully*, when he fell into conversation with other passengers about recent work on electromagnetism. Joseph Henry had created an electromagnet that was capable of sending an electrical impulse down a wire; indeed the previous year Henry had succeeded in sending an impulse along a wire a mile long and ringing a bell at the other end. Fired with a dawning new idea, when he got home Morse started making designs for a telegraph recording instrument and an outline scheme for a dot-and-dash code that would utilize the duration of the electrical impulses transmitted. He made an experimental model. Fired with his new enthusiasm, Morse abandoned art in 1837.

In 1838, Morse gave a demonstration of his telegraph for President Martin Van Buren and was granted a US patent for his invention. It was Morse's assistant, Alfred Vail, who devised the Morse code, using combinations of dots and dashes to represent letters of the alphabet; really we should call it the Vail code. This replaced an earlier system using numbers to represent letters. The receiving system initially consisted of a roll of paper and a stylus to punch out the dots and dashes. This was then replaced by an ink version. Vail was later to find that the electric dots and dashes could be made audible, and that with a little practice the receiving operator could interpret the patterns of sounds as dots and dashes. Suddenly rapid progress towards a fully operational system was being made.

Six years later, the American Congress made 30,000 dollars available to Morse, to enable him to build an experimental telegraph line thirty-seven miles long between Washington and Baltimore. Morse was given help in this enterprise by Ezra Cornell and the businessman Hiram Sibley; together they built the world's first long-distance telegraph line.

On 24 May 1844, Samuel Morse transmitted his first public telegraph message. It was sent from the US Supreme Court Room in the Capitol at Washington DC to his assistant Alfred Vail at the Mount Clare railway station at Baltimore. Morse sent the question, 'What hath God wrought?' To show that message had been received, Vail sent the question back again. But progress was not always smooth. Many ordinary people were alarmed at the idea of electric currents passing through the landscape and feared for their safety.

With the help of his financiers, Ezra Cornell and Hiram Sibley, Morse set up more telegraph lines. In 1848, a line connected New York and Chicago. In 1856, Western Union was set up by Cornell and Sibley to amalgamate the smaller telegraph companies; in 1859, Morse agreed to the formation of the North American Telegraph Association, creating a virtual monopoly.

By this stage it was clear that it was possible to communicate by telegraph across an entire continent. Why should it not be possible to communicate across an ocean too? The first submarine cable was laid on the seabed between England and France in 1850. The first attempt to lay a transatlantic cable in 1857 was a failure. The Channel floor had been shallow and smooth, but the Atlantic was very deep and there were previously unsuspected submarine mountain ranges with volcanoes in the middle. The transatlantic cable broke in water that was two miles deep and could not be recovered. The Atlantic was spanned between Newfoundland and Ireland in 1858, though the cable later parted in deep water. It was not until 1866 that the cable was successfully relaid.

The rate of sending was slow at first. The first transatlantic cable could only send fifteen letters per minute. It was also very expensive, at a minimum of 100 dollars per message. Improving technology allowed the rate of sending to speed up, so that by 1950 it was possible to send 2,400 letters per minute.

Morse's telegraph was a revolution in communication. It made possible for the first time the fast transmission of important political and commercial news right across North America, and then more or less instant communication right round the world. It was one of those great leaps of technological change that made the world a much smaller and more connected place. It also revolutionized the business world; Western Union became a huge and powerful monopoly. It was becoming a world of powerful American monopolies too, with the parallel rise of John D. Rockefeller's Standard Oil Company. And Morse himself became very rich; he became one of the new generation of American philanthropist tycoons.

THE INVENTION OF THE BICYCLE

(1839)

THE BICYCLE IS a machine that has evolved gradually, step by step, and it is hard to put a specific date to its invention. The earliest machine to look like a bicycle was probably the *célérifère*, which was invented by the Comte Mede de Sivrac in 1790. This was a two-wheeled scooter-like contraption, but because it incorporated no means of propelling it other than the rider's feet scooting the ground past on each side, it cannot really qualify as a bicycle. An English version of this was called the hobby-horse, and it was about as useful as a means of transport as the child's toy of the same name.

In 1816, Baron Karl von Drais de Sauerbrun invented a two-wheeled device which he named the Draisienne. This was like a modern bicycle in having a steering bar attached to the front wheel. The rider sat between the two wheels, but still there were no pedals. The only way of moving the machine along was by the rider scooting it. So this machine was not a true bicycle either.

The first true bicycle, a two-wheeled vehicle powered by pedals, was invented by Kirkpatrick Macmillan (1812–78), a blacksmith in Dumfries, in around 1835. The frame and wheels of the modern bicycle had already been invented by Sauerbrun and Niepce as early as 1816. Macmillan added a pedal system and a brake, which in effect made it into a working machine, the recognizable predecessor of the modern bicycle.

The new bicycle was on the heavy side. It had iron wheels without tyres; the front wheel was thirty-three inches in diameter and the rear wheel forty-two inches. The bicycle made it possible for someone to travel under his or her own power faster than they could run, but the lack of tyres meant that the ride was rough and uncomfortable. Not for nothing were these early machines nick-named bone-shakers.

In 1861, the Parisian coach-builder Pierre Michaux invented an improved version of the Macmillan bicycle, called the velocipede. This had its pedals directly on its front wheel. Michaux built some prototype models and sold 142 of them in 1862. Further improvements were made in 1870 by the Coventry machinists James Starley and William Hillman. Their machine was lighter in weight and was also the first to have wire-spoked wheels, like most modern models; they called it the Ariel and put it on the market at a retail price of £8 – or £12 with a speed gear.

From here it was a short step to the appearance of the modern bicycle in 1885. This was rather like photography in being the invention of two people working independently on two sides of the English Channel. One of the inventors was the French engineer, G. Juzan. His bicycle had two wheels of equal size, with a chain-driven rear wheel; it used a stronger chain than the one used on the very first rear-drive bicyclette designed by Andre Guilmet in 1868. The Rover Company of Coventry introduced the safety bicycle designed by James Starley, also with wheels of equal size and solid rubber tyres and a chain-driven rear wheel. The French and English bicycles built in the 1880s converged on a design almost identical to the one drawn by Leonardo da Vinci in 1493. The new bicycle was much more suitable than the earliest versions for general use by the public and it quickly caught on as a very popular means of transport, enabling ordinary people to get to work easily and inexpensively, and it also made possible virtually-free recreational travel.

THE INVENTION OF VULCANIZED RUBBER

(1839)

VULCANIZATION IS THE chemical process by which rubber is hardened and stabilized. During the process the springy rubber molecules become to some extent cross-linked, which makes the material harder, more durable, more resistant to attack by chemicals, less likely to disintegrate. It also makes the surface of the rubber smoother, more water-repellent and prevents it from sticking to metal.

Rubber had been known about for a long time. It was a natural tree product that had been in use for hundreds of years before Columbus introduced it to Europeans. It seems the rubber tree was originally known as cahuchu ('weeping wood') by the Native Americans, and the sap or latex was harvested by cutting the tree's bark. It was its property of erasure that led to the use of the name rubber.

Rubber was known about from 1492, but in its natural uncured state it presented a range of problems. For one thing it started to break down after only a few days, a process called perishing. The proteins in rubber break down and the large rubber molecules oxidize on contact with air, which also causes disintegration. Rubber that has been poorly vulcanized may also perish, but at a slower rate. The process of perishing is accelerated by exposure to sunlight.

The process of vulcanization was a major breakthrough because it turned a curiosity substance that had virtually no value because of its instability into an extremely useful material with many applications. Rubber appeared in Europe in the later eighteenth century. In 1770, it was on sale in a shop in Cornhill, London; cubes of natural rubber were sold for three shillings per half-inch cube. The rubbers (they were being sold as pencil erasers) had a high novelty value, which explains their exceptional price. Rubber continued to have this curiosity value, apart from its use in the rubberized mackintosh raincoat, until vulcanization was invented. Rubber before vulcanization was difficult to use because of its tendency to break up and crack; it also smelt bad.

When vulcanization is done properly, the rubber is long-lasting, stable, waterproof, has high flexibility and no odour.

A thirty-nine-year-old former Philadelphia hardware merchant called Charles Goodyear was the man who pioneered the effective commercial use of rubber. In 1839 he acquired formal legal rights over a hardening process for the treatment of rubber. Whether this was the same as the process for which he acquired another patent in 1844 is still not clear. Precisely when Goodyear discovered his new process, called vulcanization, is not known. He claimed that he found it by systematic research, but he seems to have discovered the process more or less accidentally when he unintentionally overheated a mixture of rubber, sulphur and white lead. The overheating changed the nature of the rubber, making it much harder and more durable.

Charles Goodyear sold the patent on his father's invention, which was the first tined steel pitchfork, in order to pay for his experiments on raw rubber. A process for hardening rubber had already been invented in 1823, when Mackintosh invented a method for making durable waterproof sheeting for making raincoats, or mackintoshes as they became known. The usefulness of rubber as a substance had long been suspected, but its tendency to become sticky in hot weather and rigid in cold weather was a limitation on its practical use and commercial potential.

The invention of vulcanization was a matter of great controversy in the nineteenth century. Charles Goodyear acquired a patent for vulcanization in 1844 but claimed that he had invented it earlier, in 1839. The process was known about long before. We now know that a form of rubber hardening process was carried out as early as 1600 BC in Central America. Goodyear nevertheless did not know that and might even so be credited with re-inventing the process and making it available on an industrial scale. He wrote his version of his discovery of the process in an 1853 autobiography, which he gave the extraordinary title *Gum-Elastica*. The controversy arose because someone else also claimed to have invented vulcanization. The engineer and scientist

Thomas Hancock acquired a UK patent for it in November 1843, two months before Charles Goodyear applied for his UK patent.

Goodyear may or may not have been the first to invent vulcanization, but he vigorously defended his patent, taking a man called Day to court at Trenton in New Jersey in 1852 for infringement. Goodyear was brilliantly represented by Daniel Webster, who painted Goodyear as a broken and disappointed man, emaciated by privation.

> *Is Charles Goodyear the discoverer of this invention of vulcanized rubber? Is Charles Goodyear the first man upon whose mind the idea ever flashed, or to whose intelligence the fact ever was disclosed, that by carrying heat to a certain height it would cease to render plastic the India Rubber and begin to harden and metallize it? If Charles Goodyear did not make this discovery, who did make it? Why, if our learned opponent had said he should endeavour to prove that some one other than Mr Charles Goodyear had made this discovery, that would have been very fair.*
>
> *On the contrary they do not meet Charles Goodyear's claim by setting up a distinct claim of anybody else. They attempt to prove that he was not the inventor by little shreds and patches of testimony. Here a little bit of sulphur, and there a parcel of lead; here a little degree of heat a little hotter than would warm a man's hands; and yet they never seem to come to the point. I think it is because their materials did not allow them to come to the manly assertion that somebody else did make this invention, giving to that somebody a local habitation and a name. We want to know the name, and the habitation of the man who invented vulcanized rubber, if it be not he who now sits before us . . . I say that there is not in the world a human being that can stand up and say that it is his invention, except the man who is sitting at that table.*

Of course the defendant's lawyer could have named Thomas Hancock, but that would have been no help at all to Day, and Daniel Webster shrewdly guessed that Hancock's name would not be mentioned in court. The answer to each of the rhetorical questions was 'Thomas Hancock', but it was not a pub quiz and no one in that courtroom was going to own up to knowing the answer. The court found in Goodyear's favour and the decision clinched Goodyear's legal ownership of the right to be regarded, at least in law, as the inventor of vulcanization.

The consequences of the invention itself were to be far-reaching. The new hardened rubber had a huge range of applications ahead of it. It was the ideal material from which to make bicycle tyres. When the motor car was invented, it was the ideal material from which to make car tyres. Vulcanized rubber is all around us, in all sorts of unobtrusive corners. It is, for instance, used to make the seals round the edges of double-glazed units for windows, and to make air-tight seals round the edges of doors. It is used to make air-tight seals round the doors of refrigerators and freezers, and watertight seals round the doors of washing machines – a vital material for modern living.

THE INVENTION OF INCANDESCENT ELECTRIC LIGHT

(1840)

As EARLY AS the year 1801, Sir Humphry Davy demonstrated that electrical energy could be converted into light. Yet it was to take a whole century for the modern electric light bulb to develop, with many experiments by many inventors along the way. Davy experimented with a platinum filament, making platinum strips glow when he passed a current through them. The strips evaporated after a few minutes: too quickly for them to be of any use in lighting a room. A few years later, in 1809, Davy developed a carbon arc lamp in which a very bright light was created when an electrical connection was made between two charcoal rods connected to a battery.

It was not until 1835 that a constant electric light was invented. James Bowman Lindsay demonstrated this in public in Dundee, producing a light that was bright enough to enable him to read a book. Lindsay unfortunately did not spend time developing his electric light further, but switched his attention to wireless telegraphy.

The next development came from another British scientist, Warren de la Rue. In 1840, de la Rue put a platinum coil inside a vacuum tube and passed a current through it. The experiment was based on the idea that platinum's high melting point would allow it to glow at a high temperature, while the vacuum would contain fewer gas molecules to react with the platinum. This would make the platinum last longer before burning out. The design worked well, and established the concept of an incandescent electric light bulb. Even so, it was not a practical design with any commercial prospect simply because platinum was too expensive.

In the following year, 1841, another British scientist, Frederick de Moleyns, was issued a patent for an incandescent electric lamp. This used powdered charcoal heated up between two platinum wires. Like de la Rue's lamp, the device was contained inside a vacuum bulb. In spite of this patent, de la Rue should be regarded as the inventor of the first incandescent electric light bulb.

Meanwhile, in America, work on an electric light was developing along similar lines. In 1845, John Wellington Starr was given a patent for an incandescent light bulb using carbon filaments. The story of the light bulb for the next few decades seems to be one of repetition and re-invention without much progress. In 1872 Alexander Lodygin repeated the de la Rue invention, gaining a patent for it in 1874. In 1893, the German scientist Heinrich Gobel claimed that he had invented the light bulb forty years earlier, but when he tried to assert this in a patent interference suit, the judge ruled against him, saying that his claim was 'extremely improbable'.

The first major advance after de la Rue's invention of the incandescent electric light bulb came in 1878. Joseph Swan of Sunderland developed a working light bulb as early as 1860, using carbonized paper filaments inside a glass bulb vacuum, but the technology available to him then had not allowed him to create a complete enough vacuum. By the 1870s, more powerful pumps were available, and by 1878 he had perfected his light bulb, now using thin rods of carbon as filaments. Because the vacuum was almost complete, there was almost no oxygen left inside the bulb to ignite the filament. This meant that the filament was able to glow almost to white heat without catching fire. It was, at last, an electric light bulb that worked. Swan started manufacturing his light bulbs and selling them for installation in houses and public buildings.

The great American inventor, Thomas Edison, was not far behind. His first successful test of an electric light bulb with a carbon filament came on 21 October 1879. It is Thomas Alva Edison, one of the world's most prolific inventors, who is nearly always credited with inventing the electric light bulb. As we have seen, several other inventors were there before him. To be precisely accurate, Edison refined and perfected the light bulb. Electric light bulbs had been created as much as thirty years before, but they did not work well. What Edison managed to do was to devise a light bulb that was a practical, working proposition, a light bulb that could be sold and fitted in people's homes.

Edison started working on the electric light bulb in the spring of 1878. He was thirty-one and he took a vacation with George Barker, another college lecturer. Edison was already famous for inventing the phonograph, and Barker suggested that Edison should bend his mind to developing a domestic light bulb. Edison warmed to the idea and when he went back to work at his 'invention factory' in Menlo Park, New Jersey, he assembled a work team and made a rash announcement that they were going to achieve the goal in six weeks. In fact it took a lot longer.

Edison's specifications were exacting. The light bulb had to be long-lasting, run on a small electric current, and remain functioning even if other bulbs on the circuit went out. A major problem with bulbs before Edison was that the filaments quickly burned out or melted. In the open air this happened within seconds, so it was quickly realized, even before Edison applied his mind to the problem, that the filament needed to be enclosed in a glass globe with the air pumped out of it. Edison used a Sprengal pump, which was very effective in creating a near-absolute vacuum.

Then Edison worked on producing the perfect filament. A platinum filament lasted only ten minutes before it melted. It was in any case a very expensive metal. Edison experimented with a carbon filament, and succeeded in making one that gave light for as much as two hours before burning out. Finally he devised a fragile filament made out of cotton thread baked into carbon. This was the filament that was used in the successful experiment late at night on 21 October 1879.

The filament glowed with a dim red glow, perhaps the equivalent of one watt. It was unspectacular, but it did not go out. It went on and on burning. Edison tried turning the current up, which made the bulb brighter, and still it went on burning. Eventually the filament did burn out, but only after more than thirteen hours.

This marked the beginning of the age of the electric light. Within three years, Edison had set up the Edison Electric Light Company, built a power station at Pearl Street in the city of New York and run a network of wires through old gas pipes to reach the homes of his customers. By 1900, one million people had electric light in their homes. The light revolutionized people's lives, turning night into day, making it possible for them to stay up later, reading, studying, making music. But it also made it possible for people to work at night just as easily as in the day-time and so opened workers to potential abuse by their employers. The electric light was a serious threat to leisure, just as the earlier invention of gaslight had been.

THE INVENTION OF THE SUBMARINE TELEGRAPH CABLE

(1842)

THE UNDERWATER OR seabed telegraph cable was invented in 1842 by Samuel Morse, who had invented the telegraph in 1839. From that moment on the idea of a submarine telegraph cable across the Atlantic became a long-term aspiration for the future. As early as 1840 Morse publicly declared his intention to go in this direction and two years later he experimentally submerged a wire in New York harbour. The experimental cable consisted of a wire insulated with tarred hemp and encased in India rubber. Morse succeeded in sending a telegraph message through it, underwater. Charles Wheatstone carried out a similar experiment in 1843 in the bay at Swansea.

The main concern was to find reliable insulation to cover the wire and prevent any contact with water; this was essential to ensure the success of a long submarine cable. Morse used India rubber, which the Russian electrician Moritz von Jacobi had tried as early as 1811. In 1842 another option became available in the form of gutta percha. This was an adhesive derived from an Indian tree, the Isonandra Gutta tree, introduced into Europe by William Montgomerie, a Scottish medic who worked for the British East India Company. Montgomerie's interest was in the potential that gutta percha had for making surgical apparatus. Michael Faraday and Charles Wheatstone found that gutta percha had electrical insulation properties. In 1845, Wheatstone proposed that the telegraph cable to be laid on the seabed between Dover and Calais should use gutta percha for its insulating covering. It was tried out experimentally on a wire across the River Rhine at Cologne and again on the seabed off Dover.

In 1850, the first submarine telegraph cable was laid across the English Channel by John Brett's Anglo-French Telegraph Company. It consisted of a copper wire coated with gutta percha. This was a hasty and technically inadequate action, designed mainly to keep the company's concession. On 13 November 1851, Brett replaced the first Channel cable with a properly constructed cable. It was laid by a government vessel, the *Blazer*. A start had been made.

In 1852 a submarine telegraph cable was laid across the Irish Sea to join England and Ireland. In the same year a second Channel cable linked London and Paris for the first time. In 1853 a cable was laid across the North Sea bed from Orford Ness to The Hague; this was laid by a paddle steamer called the *Monarch*.

Cyrus Field was the main activist behind the first transatlantic cable, which was laid in 1858. This was a much bigger project than the initial British cables. The transatlantic cable was not only much longer: it had to be laid in very deep water. The project was beset by all sorts of technical problems. Later attempts to lay a cable in 1865 and 1866 were more successful.

The transatlantic cables had an outer protective layer of steel wire, surrounding a layer of India rubber. This in turn surrounded a gutta percha layer round the central multi-strand copper wire. The stretches closest to the land at each end were given an extra sheathing of protective wire. The gutta percha turned out to have almost perfect insulating properties for this purpose. It would not be replaced as a cable insulation until the 1930s, when polyethylene was introduced.

The discovery of a near-perfect electrical insulation was not the end of the story, though. There were still major electrical problems produced by the sheer length of the submarine cables. Nineteenth-century technology did not allow for repeater amplifiers within the cable, which were the ultimate solution to the problem. As early as 1823 it had been noticed that electric signals were slowed down when passing through an underground core, and the same effect was later seen in the underwater cables. Faraday explained it in terms of capacitance between the wire and the surrounding water. The electric charge in the wire induces an opposite charge in the water, the two charges attract each other and the charge in the wire is retarded. The result was that the speed of any signal through a submarine cable was reduced.

Faraday had understood the principle, but those responsible for engineering the cables did not. Whitehouse, the Atlantic Telegraph Company's electrician, was bullish about it, insisting that the problem could be overcome by pushing up the voltage. Because of the high voltages set by Whitehouse, Cyrus Field's first transatlantic cable did not work properly. When Whitehouse increased the voltage beyond the cable's designed capacity, the cable simply short-circuited to the ocean.

Lord Kelvin devised a solution that involved increasing the sensitivity of the receivers so that they could detect faint telegraph signals. Kelvin invented a mirror galvanometer to do this, and became rich on the royalties from it.

There was an unexpected outcome of the submarine telegraph cables. They were evidently affected by atmospheric electricity and the Earth's magnetic field. These problems provided one of the scientific motives for polar exploration.

That first generation of submarine communications cables carried telegraph messages. A second generation of submarine cables carried telephone messages, then data communications. All the most up-to-date cables use fibre-optic technology to carry digital messages, which are used to take telephone traffic as well as Internet traffic. There are now submarine communications cables linking all the continents together; Antarctica is the one continent that remains outside this global communications system.

THE REINVENTION OF ANAESTHETICS

(1846)

ANAESTHETICS WERE INVENTED, forgotten and reinvented several times in the course of history. It was Joseph Priestly who discovered and isolated the gas nitrous oxide in 1772. In 1800 Humphry Davy was experimenting with the gas when he noticed that it created a sensation of euphoria – he called it laughing gas – and made people insensitive to pain. Davy recommended surgeons to use it to alleviate pain, but they ignored his advice. Michael Faraday made similar observations about ether which, like nitrous oxide, was used for entertainment by students. People inhaled the gas at parties, enjoying its intoxicating properties. Laughing gas and ether were the first recreational drugs (after tobacco).

An American doctor, Crawford Williamson Long, experimented on himself with ether in the early 1840s, and on 30 March 1842 he tried ether out for the first time on a surgical patient. James Venable inhaled the ether from a saturated cloth. It rendered him unconscious, and while unconscious he had a tumour removed from his neck. The operation was both successful and painless. Remarkably, Long kept quiet about this major breakthrough. He went on using small doses of ether as an anaesthetic for minor surgery.

William Morton, a Boston dentist, attended a lecture by Charles Jackson, in which Jackson mentioned that inhaling sulphuric ether could induce unconsciousness. Morton tried it out successfully on himself and his dog, and then used it on a patient while extracting a tooth. The patient excitedly leaked the news of the painless extraction to the press, and the secret was out. Dr Henry Bigelow, a Boston surgeon, read about it in the newspaper and persuaded Morton to demonstrate his technique at the Massachusetts General Hospital. There, John Warren used ether on 16 October 1846 to anaesthetize a patient during surgery for a superficial tumour.

Morton took out a patent the following month, but it was unenforceable. It went into immediate use in surgery on casualties of the Mexican War. The name anaesthesia was coined by Oliver Wendell Holmes for this new phenomenon and it was soon in widespread use by surgeons all over America and Europe. In 1847 the London doctor John Snow introduced (in effect rediscovered) ether as an anaesthetic. But even that was not the end of the story. In 1884, cocaine was found to be equally effective as an anaesthetic. The New York surgeon William Halsted pioneered the use of cocaine by injection. Unfortunately, during the course of his experiments, Halsted became addicted to cocaine. It was an early example of the bad side effects of drugs used inappropriately or to excess.

General anaesthesia was a very important advance. With patients unconscious it was possible for surgeons to work more carefully, slowly and meticulously, and undertake altogether more ambitious surgery. As early as the 1890s, it was possible for surgeons to contemplate open-heart surgery for the first time.

THE INVENTION OF THE SEWING MACHINE

(1846)

SEWING IS AN ancient craft that has gone on for at least 20,000 years. The first needles were splinters of bone and antler; the first threads were animal sinew and strips of animal hide. The remains of Bronze Age and Stone Age clothing found by archaeologists clearly show that both fabric and leather have been sewn for thousands of years. Iron needles were not invented until as late as the fourteenth century AD. The first needles with eyes for carrying the thread did not appear until the fifteenth century. Throughout this long period sewing was done by hand.

The first attempt that we know of to mechanize sewing came in 1755. In that year a British patent was issued to a German inventor called Charles Weisenthal. The patent was for a needle designed as a component for a sewing machine, though the machine itself was not described; it is not even clear whether such a machine existed, but by implication it did. It was the English cabinet maker and inventor Thomas Saint who was issued with the first patent for a complete sewing machine, in 1790. The patent describes the machine as having an awl that punched a hole through leather and passed a needle through the hole. It is not known whether Saint actually built his sewing machine, but the fact that he was a cabinetmaker suggests that he had the necessary practical skills to do so. On the other hand a later attempt to build a reproduction of Saint's machine was a failure; the machine did not work.

Thomas Stone and James Henderson were issued with a French patent in 1804 for their invention of 'a machine that emulated hand sewing'. John Scott Duncan also patented an 'embroidery machine with multiple needles' in that year. Neither of these machines worked.

In 1810 a German, Balthasar Krems, invented a sewing machine specifically designed to sew caps. He did not apply for a patent and his machine did not work very well. An Austrian tailor called Josef Madersperger made several attempts to design a sewing machine and was issued with a patent in 1814, but none of his machines worked properly either. In 1818, the first attempt at an American sewing machine was made by John Adams Dodge and John Knowles. The machine was only able to sew a few stitches before malfunctioning.

There were major problems confronting would-be inventors. One was the fact that the needle and thread needed to be moved back and forth through the cloth, while the cloth needed to be moved past the needle at a controlled rate to make the stitches a regular length. The other was that a very high standard of accuracy was needed to make machine sewing anywhere as neat as hand sewing. It was not surprising that so many of the early attempts were failures.

The first sewing that worked was invented by a French tailor, Barthelemy Thimonnier, in 1830, the year of the Captain Swing Riots in England. The machine relied on a single thread carried on a hooked thread; it made an embroidery chain stitch. The implications for tailors and seamstresses were obvious. Many of them would be put out of work by the new invention. Although the new machine might help the factory owners to make a lot more money, large numbers of workers would suffer job losses. An angry crowd of French tailors attacked and burnt down Thimonnier's factory and almost killed Thimonnier himself.

In 1834 Walter Hunt invented the first successful sewing machine in America, though he lost interest in it when he saw that it would cause unemployment. Hunt did not apply for a patent. In 1846, a US patent was issued to Elias Howe for a similar machine, using 'thread from two different sources'. Howe's machine had a needle with an eye close to the point. The needle passed through the cloth and made a loop of thread on the far side. A shuttle running along a track slid a second thread through the loop, making what is known as a lockstitch. Howe had difficulties in defending his patent and marketing his machines. He began a nine-year struggle to raise interest in his machines and defend them against imitation. A major problem was that several people were now coming up with similar solutions to the technical problems.

Isaac Singer invented the up-and-down mechanism which has

become standard on sewing machines; the earlier machines had all operated on a side-to-side action. He built the first commercially as well as technically successful sewing machine in the 1850s. Another innovation introduced by Singer was the treadle. This was a great advance as it allowed the seamstress to work the needle with her feet and keep both her hands free to handle the cloth. Howe was furious, launched a suit against Singer for the infringement of his patent, and won. He made a huge fortune from his invention.

Initially, the new sewing machines were installed in garment factories and mass production of clothing became possible. It was only later, in fact not until 1899, that a sewing machine model was designed and marketed for use in the home.

THE INVENTION OF THE SAFETY ELEVATOR OR LIFT

(1853)

SIMPLE MECHANICAL LIFTS or hoists have been in existence for hundreds of years. In the lofts of some medieval cathedral towers in Europe, the large wooden treadmills still survive that were used to winch up blocks of stone for the building of the parapets and spires. These treadmills could be powered by donkeys or people. It was not until the nineteenth century that the development of modern lifts as we know them began. Some models used hydraulic power; people stood in the lift compartment, while water was run into a hollow tube until the weight of the water hoisted the cabin up.

A problem encountered in these early hydraulic lifts was the absence of control over the sometimes alarming speed of ascent and descent. Valves and regulators were fitted to control speed.

Then lifts were raised and lowered with ropes that ran through systems of pulleys and counterweights. The modern counterparts of these are the dumb waiters that are still widely used in British hotels for moving food, bed linen and other items from one floor to another.

The first power lift appeared in America in the idle of the nineteenth century. It was designed as a straightforward freight hoist operating between two floors in a building in New York. The breakthrough moment was applying this idea to the movement of people up and down inside a building. It was Elisha Otis who designed the first powered passenger lift in 1853. He fitted an important safety device that automatically stopped the passenger compartment from falling if there was any mechanical failure of the lifting machinery.

The first Otis passenger elevator was installed in a department store in New York for 300 dollars. Elisha Otis's sons went into business manufacturing his invention. By 1873, over 2,000 of them had been installed in commercial buildings, mainly office blocks and stores, all over America. Many of them are still in use, still functioning perfectly. A particular refinement, which was essential for safety, was that the Otis lifts could be set so that they automatically stopped to within a fraction of an inch at the correct floor level; people did not have to step up or down into an Otis lift. The invention was a huge success.

In 1884, Frank Sprague invented the first electric lift, which was installed in a cotton mill in Lawrence, Massachusetts. Sprague also invented the push button control that has become a standard feature of modern lifts. The first use of an electric passenger lift in a commercial building was in New York in 1889, in the Demurest Building. The early electric lifts used ropes winding round a drum, and this technique was satisfactory for low-rise buildings, though not for high-rise. As the buildings became taller, the ropes had to be made longer and longer, and the drums became impossibly large to house.

In 1903, a gearless traction engine lift was devised, and this could operate in a building 100 storeys high. Later still, multi-speed lifts were developed to shorten the ride up and down the very tallest skyscrapers. It was the development of the passenger lift that made the invention of the skyscraper a practicable proposition.

THE INVENTION OF THE BUNSEN BURNER

(1855)

THE BUNSEN BURNER was invented in 1855 by Robert Bunsen. Apart from the test tube, the Bunsen burner is the most commonly used piece of equipment in a science laboratory. It uses natural gas as a fuel supply. It has a simple design, consisting of a rubber hose from the gas supply source, a circular base and a chimney that is a metal tube. Gas enters the chimney at the base through a small aperture that produces a jet of gas. Round the base of the chimney, level with the jet, is a movable cylindrical collar. The collar has a hole in it that may be lined up with a similar hole in the chimney.

When the two holes are lined up, the maximum amount of air enters. As the collar is turned round, decreasing amounts of air enter. In this way a variety of different sorts of flame can be produced. When the hole is closed, a luminous yellow flame is produced, because not all of the fuel is being burned. By increasing the air flow, more of the fuel is burned. When the holes at the base are aligned, a transparent blue flame is produced. There is also a roaring sound. This is the hottest flame the burner can create: about 1,500 degrees Celsius.

The Bunsen burner is a technically simple device that is extremely effective and useful, and can be used in an enormous number of different situations. It is also very cheap to mass-produce and is therefore ideal for use in school and college laboratories.

Robert Bunsen (1811–99) was a German chemist and teacher. He invented his burner specifically to help him in his research in isolating chemical substances. The high-intensity transparent blue flame does not interfere with the coloured flames given off by various chemicals being tested. Bunsen worked at the University of Heidelberg, and asked a technician, Peter Desaga, to build the first Bunsen burner for him; he gave Desaga the detailed specifications.

The laboratory burner was not the only thing Robert Bunsen invented. He also invented a photometer to measure the intensity of light, a hydrojet filter pump and a chemical battery. In collaboration with the physicist Gustav Robert Kirchhoff, Bunsen invented the Bunsen-Kirchhoff spectroscope in order to undertake spectral analysis of materials. They used the spectroscope to discover two new elements, rubidium and caesium, in 1860. Bunsen was blinded in one eye by an explosion in a chemistry experiment that went wrong. As we saw with Humphry Davy, invention is not without attendant risks.

THE INVENTION OF
THE BESSEMER
CONVERTER

(1856)

HENRY BESSEMER OBTAINED a patent for his steel converter in 1856. The traditional blast furnace produced pig-iron, which contained quite a high percentage of carbon, and the carbon made the iron brittle. This had been discovered in 1750. The brittle iron was suitable for castings that required little strength, such as ships' anchors and firebacks. To make steel, the carbon needed to be removed. It could be burnt off in the furnace, but making steel, especially in large quantities, was expensive.

Bessemer's revolutionary converter removed all the carbon from melted pig-iron simply with a blast of air. The oxygen bubbled through the pig-iron combined with the carbon and dispersed it by turning it into carbon dioxide.

The Bessemer process required iron ore that was free of impurities, but otherwise guaranteed an unlimited supply of steel. At a stroke, the Bessemer process allowed the mass production of steel. In the early nineteenth century, ships were made of iron, a rather heavy metal in relation to its strength. By the end of the nineteenth century nearly every ship was built of steel. Hulls could be built thinner and lighter – and larger. With the Bessemer converter, the age of the modern supertanker became possible.

The Bessemer converter is another case of near-simultaneous invention. An American called William Kelly developed the steel-making process apparently independently, but he applied for his patent just after Bessemer in 1856, so there can be no doubt that Bessemer rightly takes the major credit.

THE INVENTION OF THE WASHING MACHINE

(1858)

WASHING CLOTHES IS an activity that has gone on since antiquity. There is evidence of ancient soap in Rome, where the fat of sacrificed animals was used as soap. Along the western wall of the classical city of Troy, there were stone washing troughs, where the townswomen went to wash their families' clothes. These troughs were supplied with water by a subterranean passage leading to a triple spring underneath the town itself. They were in use 2,000 years ago, and they probably replaced similar washing facilities at the same spot in King Priam's Troy in the Bronze Age 1,200 years before that.

In 1797 the scrubbing board was invented. In the nineteenth century, many people washed their clothes in an open wooden or tin tub. They dropped their clothes into the tub with hot water and soap flakes, and then agitated the clothes with a dolly. This was a wooden device that looked rather like a three-legged stool on the end of a handle; there was a cross-piece at the top, so that the dolly could be firmly gripped with both hands. Using a dolly was very hard work. Some of the later automatic washing machines tried to imitate the rotating backwards and forwards motion of the dolly. The first device to look like a modern washing machine was invented in 1851 by the American James King. This machine contained a rotating drum, but it was still hand-powered, like the mangle for squeezing wet garments dry.

In 1858 Hamilton Smith was given a patent for his invention of a rotary washing machine. The first electrically powered washing machine, called the Thor, was invented in 1908 by Alva J. Fisher and manufactured by the Hurley Machine Company of Chicago. This was another machine with a drum, but this time with an electric motor.

THE INVENTION OF THE OIL DERRICK

(1859)

IN THE NINETEENTH-century oilfields, the most conspicuous feature of the landscape was the forest of wooden derricks. These were obelisks of timber scaffolding designed to hold the weight of the drilling equipment. They were named after a once-famous, though now long-forgotten, seventeenth-century English hangman by the name of Derrick, because they looked rather like gibbets.

The invention of the derrick followed the discovery of oilfields in America and the need to extract the oil from underground. The properties of mineral oil had been known about since ancient times, but until the nineteenth century it had only been collected at sites where it naturally seeped out at the surface. For a long time people had realized that drilling down to the oil reservoirs, the underground rock layers where the oil was concentrated, was the only way to extract large quantities, but the technology to achieve this was slow to develop.

An American called Edward Drake was the man who made the significant change. In 1859, Drake devised and built a derrick and a steam-powered drill and began using this new equipment to drill for oil at Titusville in Pennsylvania. Augering by hand was only possible with light-weight drill bits, and only down to shallow depths. The derrick allowed the use of much heavier drill bits that would penetrate greater distances underground. The drilling process was too slow for Drake's financial backers and at one point they wrote to him, telling him to give up, but the mail service was also slow and unreliable, so Drake carried on.

He had drilled down sixty-nine feet into the ground and stopped work for the day just as the drill bit entered an underground cavity. The following day, one of Drake's workmen went out to check the gear,

341

looked down the pipe and saw that there was oil in it. During the night, oil had seeped from the reservoir under pressure and risen some way up the pipe. It was the beginning of a gigantic new industry that would transform not only certain states but the entire American economy. America was to become exceptionally rich and a world superpower on the back of oil.

Within a very short time after Drake's discovery, the landscape became spattered with oil derricks. Deep underground, reservoirs were found where the oil was at enormous pressure and drilling them produced black fountains of oil at the surface. One of these gushers in 1910 released more than nine million barrels of oil in a year and a half.

The principle of the derrick was used in the twentieth century, but translated into steel. The wooden derricks stood up to eighty feet high; the steel rigs could be higher. Often, once the oil is tapped, the oil well is fitted with a walking beam or 'nodding donkey', an electrically powered steel pump with a horizontal arm that rocks up and down. Drake's invention was not only a breakthrough that kick-started an entire industry; it turned out to be capable of development and extension. As oilfields have been discovered on the seabed of the continental shelves, the steel rigs have been mounted on drilling platforms that stand in water that is sometimes 300 feet deep. These are very tall structures indeed, as tall as cathedrals. Drake was drilling down to shallow depths of only seventy feet. Today there are rigs that drill down almost one mile.

It is hard to exaggerate the importance of Drake's innovation. The oil industry that he allowed to expand came to dominate the world economy of the twentieth century, and looks as if it will dominate the twenty-first century too. The biggest consumers of oil have been America and the economically developed states of Western Europe. The existence of huge reserves of oil in the Middle East has meant the growth of a powerful trade in oil between these regions. To cater for it, huge technological leaps have taken place in the transport of oil. Pipeline technology has developed. Special freighters have evolved to carry large volumes of oil by sea, leading to the construction of colossal supertankers, many of them over 1,000 feet long.

The fact that the advancing technology has outstripped the resource and it is a finite resource means that huge political tensions have built up, especially in parts of the world where further oil reserves are thought to exist. Britain's struggle to retain control of the Falkland Islands may be explained by the presence of oil reserves in the shallow waters round them. The continuing military intervention by Britain and America in Afghanistan and Iraq may also be explained by the presence of rich oilfields. The oil industry seems likely to go on having enormous geopolitical significance – on a global scale.

THE INVENTION OF THE IRONCLAD

(1859)

THE FIRST IRONCLAD ship, *La Gloire*, was designed by the great French naval architect Dupuy de Lôme in 1859.

The idea of the ironclad warship had its origins a few years before, in an incident in the Crimean War. The Battle of Sinope in 1853 was an encounter between the Russian Black Sea Fleet and a Turkish flotilla. It was a very unequal match. The Russians were equipped with brand-new Paixhans guns that fired explosive shells at the Turks in their wooden ships. The Turkish flotilla was completely destroyed.

What the battle showed was that weapons technology had reached a point where there needed to be a responding development in the technology of armour. Wooden ships were suddenly very vulnerable. While the Crimean War was still continuing, the French and British navies collaborated to design ironclad floating batteries which could be used to weaken the Russian defences. So far the Russian defences had withstood bombardment by wooden-hulled battleships. The French used the newly designed floating batteries in 1855. The British were unable to get theirs into position before the war came to an end.

The principle was, even so, established. Wooden hulls needed to be sheathed in metal sheets to give them more resistance to the new naval guns. *La Gloire*, the first ocean-going ironclad warship, was launched in 1859. The new warship was built of timber in the traditional way, but with her hull covered in nailed on iron plates. Dupuy de Lôme would probably have designed a hull entirely of iron, but for the fact that the French iron-making industry had not the capacity to produce enough iron. The new ship was, however, a significant advance technically over existing warships, and her launching was a great political coup as far as the balance of power between France and Britain was concerned. The French now had the edge over the British with regard to naval technology, but it was not to last. The British iron and shipbuilding industries were more than equal to this weapons race. In 1860, the Royal Navy launched HMS *Warrior* (classified at the time as an armoured frigate) and the following year HMS *Black Prince*, both of which were iron-hulled.

The French went on to build sixteen ironclad warships, including two more single-deck ships like *La Gloire* and the only two-decker ironclads to be built, the *Magenta* and the *Solferino*. These two were the first warships to be fitted with the spur ram; the bow projected forwards at the waterline, so that in the event of a collision with an enemy warship the enemy ship would be fatally holed below the waterline. The French also built two ships with iron hulls.

The first time steam-powered ironclads were put to the test in combat was during the American Civil War. The first of the ironclads to become involved in action, in 1861, was the CSS *Manassas*, which had formerly been known less glamorously as the steam tug *Enoch Train*. The first engagement between two ironclads was the Battle of Hampton Roads. This was a duel between the USS *Monitor* of the United States Navy and the CSS *Virginia* of the Confederate States Navy. It is said that the *Monitor* was the first ironclad warship to be commissioned by the US Navy, but in fact the first was the USS *Baron de Kalb*, designed by James Eads and commissioned late in 1861, a few months before the *Monitor*.

The *Monitor* was designed by the Swedish engineer John Ericsson and launched in 1862. She was a strange-looking vessel, described as 'a cheesebox on a raft'. The hull was no more than a floating platform, its deck only just above the waterline, carrying a heavy round iron turret that housed two huge cannon. The *Virginia* was a conventional timber ship with her hull sheathed in iron plates.

The two ironclads fought each other for four hours without doing any serious damage to one another. The battle was a tactical draw, though the *Monitor* succeeded in preventing the *Virginia* from breaking the Union blockade. On the other hand, the *Virginia* remained in position while the *Monitor* was towed away after her captain had been hit in the eye. The two ships never fought one another again, though the *Virginia* occasionally steamed out to square up to the *Monitor*.

The *Monitor* was considered successful enough for more *Monitor*-class warships to be built, some with two- or three-gun turrets. The last was decommissioned as late as 1937. The *Monitor* herself was extremely unseaworthy, floating far too low in the water. She sank in a storm on 31 December 1862 while under tow off Cape Hatteras with the loss of sixteen crew. The wreck was located in 1973 and many of her parts were brought up to the surface; they are on display at the Mariners' Museum at Newport News, Virginia. In 2003 the remains of two of her crew were found, still trapped inside the gun turret.

THE INVENTION OF LINOLEUM

(1860)

LINOLEUM IS A synthetic floor covering made from solidified linseed oil combined with wood flour or cork dust on a canvas backing. It created a very smooth and waterproof surface for a floor, and with a certain amount of practice and skill it could be cut so that it exactly fitted a room from skirting board to skirting board. Once laid, it was easy to clean. In poorly built houses, it was a way of covering unsightly, low-quality floorboards.

Another characteristic of linoleum was that it could be made in any colour. The finest linoleum floors (still made) are made by cutting and inlaying pieces of differently coloured linoleum; this is a craft as advanced as marquetry. These high-quality floors were and are very durable, because the colours penetrate well below the surface. The best quality linoleum was called 'battleship', literally because it was used as a flooring in battleships. This was an unfortunate application, as one of linoleum's negative characteristics was its flammability.

By the 1950s, linoleum was beginning to acquire an old-fashioned feeling, and its reputation was not helped by the availability of cheaper, lower-quality lines which had patterns printed on them and were therefore prone to visible wear quite quickly. But for 100 years from its invention in 1860, linoleum supplied hundreds of thousands of ordinary homes with cheap, smart, clean and hygienic floor coverings. In the late nineteenth century, it was a favourite floor covering for high-use areas like hallways, passages, and as a filler to cover the floorboards round the edges of carpets. In the twentieth century, because it was waterproof and easy to wipe clean, it was a favourite covering for kitchen floors. As well as being hygienic, it reduced the breakages of crockery because of its resilience.

But from early on people experimented with linoleum. At the beginning of the twentieth century, the writer of the *Mapp and Lucia* novels, E. F. Benson, cut sheets of black and white linoleum into squares and laid them on his floor in a chessboard pattern, to create a rich effect of black and white Italian marble.

Linoleum was invented by Frederick Walton in 1860. He had the idea when he saw the tough, flexible, lustrous skin that formed on the surface of a tin of linseed oil-based paint. In 1864 he set up his Linoleum Manufacturing Company. His invention was an instant success and in demand from home-owners everywhere. It was in demand from tenants, too; you could roll up your linoleum and take it with you when you moved. By 1869, Walton was selling linoleum not only in Britain, but in many countries on the European mainland. He was exporting to America too. By 1877, Kirkcaldy had become the biggest producer of linoleum in the world; there were six linoleum factories in the town.

In collaboration with Frederick Palmer, Frederick Walton went on to invent Anaglypta and Lincrusta, varieties of embossed wall coverings that were very popular in Victorian homes. Lincrusta was made of a linseed oil mixture. Anaglypta was made of cotton pulp. Anaglypta enjoyed a long popularity as a covering to disguise unevenness in walls.

Today, linoleum has been almost completely replaced by vinyl. This has similar properties of durability, toughness and flexibility, but also offers brighter and more translucent colours – marble effects are possible. Vinyl is also less flammable, though in a fire vinyl gives off some toxic combustion products. There are still some reasons for preferring linoleum. Because it is composed of organic materials, not synthetic, it offers fewer problems with certain allergies. The old associations of linoleum with poverty have disappeared, and there has been a recent revival of interest, in particular in the upper end of the range, inlaid linoleum, which can be crafted by experts to reflect the architecture of a particular room, and create an incredibly luxurious effect.

THE INVENTION OF THE GATLING GUN

(1861)

IN 1863, THE American engineer Richard Jordan Gatling invented the Gatling gun. This was the first machine gun, and it was capable of firing hundreds of rounds of ammunition per minute. The Gatling gun first saw serious use in warfare in the American Civil War, starting in 1864.

In 1883, the British engineer Hiram Stevens Maxim designed and built an improved version which was fully automatic. His Maxim-Vickers gun was adopted by the British Army in 1889 and after that by every other major army. Maxim's gun was a significant improvement on the Gatling gun in that the energy from the recoil of each bullet fired was used to eject the spent cartridge, insert the next round and fire it.

In 1893, when the Matabele tribe rebelled against Cecil Rhodes's British South Africa Company, Jameson was able to cut the rebels down with rapid machine-gun fire, suppress the revolt and force the Matabele king Lobengula to give up his capital Bulawayo and go into exile. The new machine gun played a decisive role in this operation. It was used again in the Battle of Omdurman in 1898, a battle that gave Kitchener a decisive victory over the Khalifa of the Sudan. The British used Maxim-Vickers machine guns to kill 11,000 dervishes and wound 16,000 more, while sustaining only forty-eight casualties on their own side. When the Boer War broke out in 1899, as a matter of course the British troops were equipped with Maxim-Vickers guns. In a very short space of time the machine gun had become standard equipment in modern warfare.

The victory at the Battle of Omdurman was a vivid demonstration of the huge military superiority the machine gun could give; it could be the decisive factor in winning battles. It was also a horrific demonstration of the escalation of bloodshed that resulted from the industrialization of war. The Industrial Revolution had produced many inventions that led either directly or indirectly to significant improvements in standards of living, to the mass production of useful materials and to the saving of human labour. But it was also producing machines for mass destruction. The invention of the machine gun was an indication of the direction in which technology was taking the human race, and in particular the rich countries of the West, able to afford the latest in weapons technology. It was preparing the way for the bloodbath of the twentieth century, when more people would die violently than ever before in the history of the human race.

THE INVENTION OF THE CYLINDER LOCK

(1862)

LOCKS SEEM TO be almost as old as houses. Certainly they are as old as wealth. The earliest known lock was found by archaeologists in the ruins of a Mesopotamian palace near Nineveh. It was made in about 2000 BC. It was an ancestor of the pin tumbler type of lock, a type that would become common in contemporary Egypt. The lock consisted of a large wooden bolt that secured the door. It had a slot in its upper surface including several holes. When the bolt was in place, wooden pegs were pushed down into the holes to prevent the bolt from being slid back. Even more elaborate wooden locks were made by the Minoans in Bronze Age Crete between 2000 and 1400 BC.

For a long time in prehistoric and later Europe the only real security that people had for their homes was a substantial bar fitting into iron brackets on the door jambs. This method of securing a door was almost universal, from the drystone neolithic houses of Skara Brae to the doors of great medieval castles. The double doors of the great Lion Gate defending the citadel of Agamemnon's Mycenae in 1250 BC were secured by a heavy baulk of timber in exactly the same way.

Metal keys and locks were made in the Middle Ages, but the problem was that very often one key opened many locks. The idea of a unique pairing of key and lock had either not been thought of or it had not been possible for locksmiths to devise enough different shapes. Even if one did not have the key, it was often quite easy to pick a lock.

In 1778 Robert Barron made the first serious attempt to improve security. He invented a double-acting tumbler lock. In 1784 Joseph Bramah invented the safety lock, which was considered to be impossible to pick; it was a thief-proof lock.

In 1857 James Sargent invented the first working key-changeable combination lock. This was seen as so secure that it was the lock of choice for manufacturers of safes. The US Treasury Department also used it. Sargent went on to invent a time lock mechanism in 1873; this became the forerunner of those commonly used today in bank vaults.

The first jemmy-proof locks were invented in 1916 by Samuel Segal, who was a retired New York policeman. In 1921 Harry Soref set up the Master Lock Company and invented an improved padlock. It was both strong and inexpensive. He built it out of layers of metal – the same concept as the metal-layered doors of a bank vault.

The pin-tumbler lock was invented in 1848 by Linus Yale, Senior. His son made improvements to the Yale lock by making the key smaller and flatter, with serrated edges, in fact just like the keys still used today on Yale locks. Linus Yale, Junior (1821–68) patented his improved pin-tumbler lock in 1862. The great advance that this new generation of locks offered was that no two locks were the same. The five pin-tumblers acted in effect like a combination lock, and there was no way of knowing from the external appearance of the lock what that combination might be. Only one key, or an exact copy of that key, would open each Yale lock. It marked a huge advance in security, both for homes and for businesses. Linus Yale died in 1868 before he could gain any real benefit from his great invention. It has been said that the Yale lock was 'the first product to be mass-produced in non-identical form'.

THE INVENTION OF THE STAPLER

(1866)

THE STAPLER WAS invented twice in 1866, by the Patent Novelty Manufacturing Company and George McGill. A stapler is a useful device for fixing together sheets of paper by driving a thin metal staple through the sheets and bending the ends over at the back.

There were various devices for fastening pieces of paper together as early as the eighteenth century. It is said that Louis XIV used such a device. As industry and commerce expanded in Western Europe in the nineteenth century there was an increasing need to find ways of organizing paperwork. What has sometimes been described as the first modern paper fastener was invented in 1841 by Samuel Slocum, but the patent description shows that it was little more than a machine for sticking a number of pins into a piece of paper for the purpose of marketing the pins Slocum was manufacturing.

In 1866 the Patent Novelty Manufacturing Company took out a patent on a device that was much closer to the modern stapler. It could be loaded with a single staple and was used to bind papers together, but it could also be used to staple carpets, furniture or cardboard boxes. The company had the initiative to make and sell staples of several sizes; the smallest was three-sixteenths of an inch, the largest half an inch. This 1866 stapler was very like today's staplers, except that it could only be loaded one staple at a time.

Other stapler designs followed. In 1866 George McGill invented a small bendable brass fastener; the following year he invented a device for pushing the brass fastener into the paper. McGill went on developing his idea for a stapler through the 1870s. In 1879, he patented the McGill Single-Stroke Staple Press. This weighty piece of office equipment fired one staple at a time through several sheets of paper. The Clipless Stand Machine, which was sold in America from the 1880s through to the 1920s, was another development of the stapler, though this one worked without staples. Instead it cut a tongue into the paper and folded it back to hold the sheets together.

The stapler has been gradually improved to the point where it can be loaded with twenty or more staples and punch them through several sheets. Heavy-duty staplers are available that can staple many sheets of paper together. Today, there are few commercial or administrative offices without staplers. Indeed, there are few home offices without them.

THE INVENTION OF DYNAMITE

(1866)

FOR A VERY long time, gunpowder was the only explosive available for industrial blasting in mines and quarries. The problem with gunpowder was that both making and storing it were hazardous; it was also not all that easy to detonate.

In 1866 the thirty-three-year-old Swedish engineer Alfred Nobel perfected his invention of dynamite, an explosive that was a great improvement on gunpowder in terms of reliability. It was a very important step forward. Ascanio Sobrero discovered the power of nitroglycerine in 1847, and what Nobel did was to find a way of harnessing that power safely. Nobel, who studied mechanical engineering in the USA, mixed nitroglycerine with absorbent earth, wood meal or charcoal. Adding the nitroglycerine to an inert solid material reduced its sensitivity to shock, so that it was far less likely to explode by accident while being handled. This created a safe blasting powder which replaced black powder. Dynamite was a great invention, aiding the mining and quarrying industry enormously. It would also be harnessed for military use in due course. In the USA this became known as straight dynamite.

In time, the use of pure nitroglycerine was discontinued because it tended to freeze at fairly high temperatures and did not store well in cold climates. Instead, nitrated mixtures of glycerine or sugars were used. The standard dynamite used in American mines is forty per cent nitroglycerine, forty-four per cent sodium nitrate, fifteen per cent wood pulp and one per cent calcium carbonate.

In 1875, Nobel made another discovery, blasting gelatine. He found that seven per cent of collodion nitrocellulose dissolved in nitroglycerine converts the nitroglycerine into a stiff jelly that could be detonated with a blasting cap. This jelly was not widely used as it was very sensitive to ignition by flame.

Man-made explosives are dangerous, but the main danger in the mining industry has always been the explosion of natural gases. Many disastrous explosions in coal mines, for instance, have been caused by the ignition of fire damp (methane mixed with air) and of coal dust. In the mining and quarrying industry, Nobel's dynamite was nothing but beneficial. It was just unfortunate that dynamite could equally well be used for munitions.

THE INVENTION OF THE DEPTH-SOUNDING MACHINE

(1866)

THE DEPTH SOUNDING machine was invented in 1866 by Thomas Walker. It was a weighty, heart-shaped object made of brass, and it was designed to be lowered straight down to the seabed from a ship, and record the depth of the water at that point. The Walker sounding machine was manufactured by Thomas Walker & Son of Birmingham, a firm that was well known for making logs, which were devices that were trailed in the water behind a vessel for recording the distance travelled.

The Walker sounding machine had a self-contained unit that housed a distance measuring calculator. It had a large impeller (or what today we would call a propeller) which drove a calibrated disc round. The disc was calibrated in fathoms, the traditional nautical depth-measuring unit of six feet. The device measured to depths of 900 feet. As the weight descended to the seabed, the impeller drove the disc round and the disc recorded the depth. There was also a flap which acted as a lock, so that the maximum depth recorded on the disc was held while the machine was wound back to the surface.

In 1872, Professor William Thomson invented an improved depth-sounding machine. This used steel piano wire instead of the hemp rope that had been used previously. Wire was much faster to operate, soundings could be taken much faster and it was therefore possible to survey larger areas in detail. As a result of Thomson's invention, or his improvement to Thomas Walker's invention, huge strides were made in the exploration of the oceans. Thomson's machine was used aboard the

USS *Tuscarora* in 1873–74, for a survey of the seabed of the North Pacific Ocean, and it revealed the existence of previously unknown sea mounts, trenches and ridges. In 1875, the crew of the USS *Gettysburg* used Thomson's depth-sounding machine in the North Atlantic, where they discovered an underwater mountain 130 miles off the coast of Portugal.

The sounding machine was a simple and serviceable device for measuring water depth. Ships used sounding machines to compile accurate maps of the seabed. It was in this way that in the 1870s and 1880s major discoveries were made about the nature of the seabed. Huge tracts of submerged terrain suddenly came into view on detailed maps. The continents turned out to be surrounded by continental shelves, areas of shallow water less than 600 feet deep and generally with very smooth surfaces. At their outer edges, there was a steep drop (the continental slope) down into the ocean basins. These were regions of very deep water, where the seabed was spattered with thousands of extinct volcanoes (sea mounts), in some areas engulfed in deep layers of mud.

The most important discovery made during the earliest days of machine-sounding was the Mid-Atlantic Ridge. This was a colossal submerged mountain range passing right down the centre of the Atlantic Ocean. The sounding machine opened up a whole new phase of exploration and discovery that was every bit as important scientifically as the so-called great age of discovery had been commercially. It would lead on, in the middle of the twentieth century, to major new insights into the nature of the planet and its regenerative processes, and the whole new science of plate tectonics.

THE INVENTION OF
THE TORPEDO

(1866)

THE MODERN SELF-propelled torpedo was invented by Giovanni Luppis and Robert Whitehead in 1866. Before that date there were several attempts at explosive devices with a similar intention. The *Turtle*, an early submarine, was supposed to attack British ships by diving under them and attaching a bomb by means of an auger; then the bomb would be detonated with a clockwork time fuse. On the occasion of the only known attack, the *Turtle*'s crew failed to penetrate the hull of HMS *Eagle*. No allowance had been made for the *Eagle*'s copper sheathing.

The first use of the word torpedo was by Robert Fulton, who used it to describe the gunpowder charge towed by his submarine, the *Nautilus*, in 1800–05. This primitive type of torpedo was still in use during the American Civil War in the middle of the nineteenth century. The Confederate submarine *H. L. Hunley* was specifically designed to wield this type of weapon.

In the American Civil War the word torpedo really meant a floating mine of the same type used in the First and Second World Wars. Some floated freely on the surface of the water. Others were tethered to the seabed by a line and weight, holding the contact mine just below the surface. It was exactly this type of tethered mine that was responsible for sinking HMS *Hampshire*, with Lord Kitchener on board, off the west coast of Orkney in the First World War. The early torpedoes were indistinguishable from what are now called mines.

As the first self-propelled devices, the real torpedoes, were introduced the tethered devices were at first referred to as stationary torpedoes and later as contact mines. The first torpedoes to be directed towards specific targets were spar torpedoes. The spar torpedo had a bomb fitted to the end of a spar forty feet long, projecting forwards below the water line from the bow of the attacking vessel. The attacker simply steamed at the target, detonating the bomb and holing the enemy ship below the water line. Spar torpedoes were used by the *H. L. Hunley* and also the David class of torpedo boats.

The first self-propelled (i.e. true) torpedoes were created by Giovanni Luppis, an Austrian naval officer from Fiume, a port of the Austrian Empire. In 1860, he devised a floating weapon which he called the salvacoste, or coast saver, which was powered by ropes. The Austrian Navy did not take up his invention. Luppis was acquainted with Robert Whitehead, the English manager of a factory in Fiume. Whitehead was born in Bolton in 1823, educated at the local grammar school and left at the age of fourteen to become an apprentice engineer. He attended the Mechanics Institute at Manchester. In 1844 Whitehead left England to work in France. In 1847 he set up his own engineering business in Milan. Within a few years he was working for the Austrian government as an engineer in Fiume.

In 1864, after the torpedo design had been turned down, Luppis entered into a contract with Whitehead to refine his invention. The result was a submarine weapon which they called the *Minenschiff,* and this was the first real self-propelled torpedo. It was powered by compressed air and it carried an explosive charge of gun cotton. The new weapon was officially presented to the Austrian Imperial Naval Commission in December 1866.

This time there was a government decision to go ahead and invest in the torpedo. Robert Whitehead and his son set up the first torpedo manufacturing plant in Fiume. By 1870 Whitehead had improved the torpedo so that it could travel 1,000 yards at up to six miles per hour. The torpedo quickly caught on and the demand for it was huge. By 1881, Whitehead's factory was supplying torpedoes to ten countries outside the Austrian Empire: Britain, Russia, France, Germany, Denmark, Italy, Greece, Portugal, Argentina and Belgium. Whitehead went on developing the torpedo. He progressively raised its speed, improving its effectiveness in two ways: the impact on the target hull was stronger and more likely to cause a detonation, and it was harder for the target to take avoiding action. By 1876 he had trebled the maximum speed to eighteen miles per hour; by 1890 he had raised it to thirty. Whitehead's torpedoes were nicknamed the Devil's device.

The British Navy naturally wanted torpedoes. In 1871, the Admiralty paid Whitehead £15,000 for access to his new weapons technology, and Whitehead opened a torpedo factory close to Portland harbour in 1891.

The biggest Whitehead torpedo was eighteen inches in diameter and nineteen feet long, made of polished steel or phosphor-bronze, with a 200-pound gun cotton warhead. Whitehead went to a lot of trouble to make his torpedoes regulate their own courses and water depths. To help with this he bought the rights to the gyroscope.

The first ship to be sunk by torpedoes was the Turkish steamer *Intibah*. She was sunk in the Russo-Turkish War of 1877–78, on 16 January 1877. The torpedoes were launched from torpedo boats working from the tender *Velikiy Knyaz Konstantin*. By the time the *Blanco Encalada* was sunk by torpedoes from the gunboat *Almirante Lynch* in the Chilean Civil War in 1891, the torpedo boat had won itself a reputation as a useful addition to any navy. Torpedo boat destroyers were designed to combat the torpedo boats; one innovation followed another.

During the First World War, torpedoes were regularly fired from warships and submarines. It was torpedoes fired from German U-boats that sank the battleship HMS *Royal Oak* and the liner *Lusitania*.

In the two decades between the World Wars, economic depression meant that most navies were on tight budgets. There was consequently little in the way of testing or innovation in the field of torpedoes. It was only the Japanese who had fully tested torpedoes when the Second World War broke out. By then the torpedo was a standard part of the armoury of all classes of warship, from battleships to destroyers. The bigger warships were by now heavily armoured, often with double-skinned hulls (at least at the sides) to make them torpedo-resistant. The technology of the torpedoes was modified in response. A magnetic exploder was invented. The torpedo was sent underneath the enemy ship and the magnetic exploder caused it to detonate directly under the hull, breaking the ship's back.

Ethical questions surrounded torpedoes from the start. Warships were clearly legitimate targets for torpedo attack once war had been declared. Regarding enemy merchant ships as targets was prohibited by the rules of war. Nevertheless, in both World Wars, German U-boats systematically attacked and sank Allied merchant vessels, and this was regarded by the Allies as a major breach of convention.

THE INVENTION OF ANTISEPTIC SURGERY

(1867)

AT THE BEGINNING of his career, the British surgeon Joseph Lister made some important observations on the coagulation of blood and inflammation. When he started work at Glasgow, he busied himself by writing articles on amputation and anaesthetics for a textbook on surgery. His ideas on inflammation took a new direction when his attention was drawn, in 1865, to the work of Louis Pasteur. What Pasteur was discovering, Lister could see, had enormous implications for medicine. Pasteur had shown that putrefaction, like other fermentations, was caused by airborne microbes. Lister saw straight away that if wounds were infected by microbes from the air there must be some way of cutting the seat of the wound off from the air and preventing the putrefaction from setting in.

Lister decided to experiment with chemical agents and his first experiment was to apply undiluted carbolic acid to a wound. The acid formed a crust, together with the blood, and the result was a conspicuous success. Lister wrote up his result in the medical journal *The Lancet* in 1867. The article's title was *On a New Method of Treating Compound Fracture, Abscess, etc.* Carbolic acid was unsuitable in many situations, so Lister worked to find a better alternative. He first settled on carbolic acid and shellac spread on calico, afterwards painted with a solution of gutta percha in benzene. Later he experimented more successfully with impregnated dressings of absorbent gauze.

Lister's next step was to try to develop a germ-free environment for surgery, to reduce the danger of infecting wounds still further. This involved creating a very clean environment for antiseptic surgery by ensuring that the air in the operating theatre was free of microbes. He achieved this by ensuring that the room was kept scrupulously clean.

All over Western Europe, medics immediately realized the importance of the modern antiseptic approach to surgery. It was an obvious breakthrough. It was certainly one of the greatest breakthroughs in human perception of the nineteenth century, and one for which Lister will be for ever gratefully remembered. This one change ensured that from then on far more people would survive surgery and the aftermath of surgery than ever before, and this in turn raised life expectancy.

THE INVENTION OF CELLULOID

(1868)

CELLULOID WAS INVENTED in 1868 by John Wesley Hyatt. Celluloid was made out of cellulose, and Hyatt was trying to make an artificial substance suitable for making billiard balls instead of ivory, which was becoming scarce and very expensive. He first attempted to use cellulose nitrate and found that it was not tough enough until camphor and alcohol were added. The new substance was known as xylonite or artificial ivory: celluloid was a trade name. Hyatt was not working entirely blind; there had been earlier experiments on the same lines in England, by A. Parkes and D. Spill.

The new material that Hyatt invented could be moulded with heat and pressure into almost any desired shape. It had a range of highly desirable properties: resilience, high tensile strength, toughness, high density, high lustre, uniformity, resistance to water, oil and dilute acids and workability. At normal room temperatures, celluloid could be sawn, drilled, lathe-turned, planed, buffed and polished. It could also be made in a virtually unlimited range of colours. Besides billiard balls, the other major use for celluloid was as the first flexible photographic film. It was used for both still photography and motion pictures. Once the motion picture industry was under way in 1900, there was an ever-expanding market for celluloid.

The invention of celluloid was highly significant in other ways too. Celluloid was the first synthetic plastic material to be manufactured. It marked the beginning of a new industrial era, the age of plastics.

THE INVENTION OF
THE TYPEWRITER

(1868)

CHRISTOPHER SHOLES, CARLOS Glidden and Samuel Soule were granted a patent for their invention of a typewriter in 1868. It was a major invention, one that would revolutionize office work and have the negative effect of undermining the craft of handwriting. James Densmore and George Washington Yost, Sholes' backers, encouraged Sholes to lay his keyboard out in such a way that the most commonly used keys were widely separated on it. This was for purely mechanical reasons, so that there would be less likelihood of the metal arms bearing the type letters jamming. Sholes came up with the QWERTYUIOP layout that has been used ever since. It is still used on computer keyboards, which work electronically and there is no possibility of jamming.

Densmore and Yost bought the patent from the inventors and negotiated with the Remington Fire Arms Company to manufacture the new machine. Philo Remington was confident that the typewriter would sell well and acquired the sole rights to the Sholes typewriter. The Remington typewriter, as it came to be called, went into factory production in 1874. The retail price was set too high and only eight machines were sold. The Remington typewriter was displayed at the Philadelphia Centennial Exhibition. In 1878, Remington brought out an improved model with a shift key system, making it possible to select upper or lower case letters using the same key. It was a major step forward.

In 1895 the New York ribbon and carbon merchant John Underwood set up his Underwood Typewriter Company to develop a new design of typewriter. This was a major advance on the Remington, enabling the typist to see what had just been typed.

The machine became more and more popular. Perhaps it was

inevitable that the typewriter was slower to catch on in Europe than in America. The Olivetti Company was founded near Milan in 1911; Camillo Olivetti built the first Italian typewriter. It was a daring venture as few Italian firms had been prepared to allow their staff to use steel-nibbed pens and they were unlikely to be impressed by the newfangled typewriter. But by the end of the first year Olivetti had landed an order for 100 machines from the Italian Navy. By 1933, his factory would be producing 24,000 machines a year and exporting to twenty-two different countries.

In 1933 a new era in typewriters began, when IBM entered the typewriter production business, acquiring a firm that had been trying for ten years to perfect an electric office typewriter. The IBM Selectric typewriter designed by Eliot Noyes was introduced in 1961. This machine worked on a different principle, a moving golf-ball cluster of type. Three years later this was linked with a magnetic tape recorder that permitted automated letters to be typed but with individual addresses. This led to an age of irritating, pseudo-personalized junk mail; not every invention increases the sum of human happiness.

By 1975, the IBM Selectric accounted for seventy-five per cent of the electric typewriter market, though there were alternative models available, many of which were simply the old-style typewriters but with power-assisted keys; the keys had only to be touched – they did not have to be pressed any more. This made the work of professional typists much easier. From there it was but a short step to the personal computer, and this operates, like the one in front of me now, from something that looks remarkably like the Sholes typewriter keyboard of 1868.

THE INVENTION OF MARGARINE

(1870)

IN THE RUN-UP to the Franco-Prussian War, butter was scarce and expensive in France. A cheap but tasty substitute for butter was needed. The prospect of war also meant that the authorities were looking for a substitute that would store better than butter. The Emperor Napoleon III sponsored a competition; a prize was offered for the best butter substitute.

A French research chemist by the name of Hippolyte Mège-Mouriès mixed suet fat in water, heated it and added milk. He named the end result oleomargarine, because he thought beef fat contained margaric acid. It doesn't. In spite of the inappropriate name of his product, Mège-Mouriès won the prize. His butter substitute was nutritious and cheap, and production on a commercial scale was taken on by the Dutch firm Jurgens.

Margarine became popular in France, then in other European countries. During the First World War, when there was a shortage of butter in Canada and America, margarine became a popular butter substitute there as well.

Ever since its invention, margarine has played a major part in the diet of the Western countries. It was recognized as a useful food product, though only after its recipe was seriously modified. Today, margarine is usually made out of a variety of vegetable oils. Depending on the use for which they are designed, margarines and spreads can be made to different recipes. UK margarines commonly use rapeseed, sunflower, soya bean, palm and palm kernel oils, all thoroughly refined and blended. Flavouring, salt, milk and vitamins A, D and E are added. The product is emulsified, pasteurized and chilled. One reason why margarines have become so popular is that they can be spread on bread straight from the refrigerator, which butter usually cannot.

THE INVENTION OF LAWN TENNIS

(1874)

LAWN TENNIS WAS invented in North Wales by Major Walter Wingfield. Like many other inventions, Wingfield's game had antecedents. It is not clear exactly where tennis came from, but there were several games of the tennis family in the Middle Ages. In medieval England, aristocrats played real (royal) tennis inn special courts, not unlike modern squash courts, with sloping roofs round the walls. These were derived from the pentise or lean-to roofs that were commonly built round courtyards; the cloister garth of an abbey had a pentise roof round all four sides. Real tennis must have had its origins in a game played with racquets and balls in such a courtyard. The sloping roofs would have made the ricochet of the ball harder to predict. Real tennis was, above all, a game played on hard surfaces, indoors, and it was played as early as 1200.

In France in the Middle Ages there was a tennis-like game called the *jeu de paume*, the palm game, and it too was probably played in the cloisters of monasteries and cathedrals. The word 'deuce', which has survived in modern lawn tennis, probably came from the French term *à deux*. It is also likely that the French demand for 'attention', *tenez*, was the origin of the name of the game itself. In 1292, there were at least a dozen manufacturers of tennis balls in Paris. In the sixteenth century, Shakespeare referred to the insulting gift of tennis balls sent by Henry V of England to the French Dauphin. In fact several French kings in the sixteenth century were keen tennis players, and Shakespeare would certainly have known this. Francis I had a real tennis court at the Louvre Palace in 1530; by this time a hand-held racquet had replaced the palm of the hand for hitting the ball. In England, both Henry VII and his son

Henry VIII were tennis players, and Henry VIII's court at Hampton Court is still in use.

The game of tennis seems to have remained one that was played indoors or in stone courtyards because it was impossible to get a ball to bounce up from grass – until the arrival of rubber. Vulcanization, and the invention of improved rubber balls with plenty of bounce, suddenly made an outdoor version of tennis possible. It was J. M. Heathcote, a real tennis enthusiast, who invented the distinctive rubber ball covered with white flannel that became the lawn tennis ball.

In Wales, there was (allegedly, and there is one anecdotal source for this) an outdoor game apparently called Cerrig y Drudion. Major Wingfield saw the possibility of adapting this traditional Welsh game (though no one seems to know anything about it) and combining it with real tennis to make it more sophisticated and give it more general appeal. He developed it on the well-kept lawns of his house at Nant Clwyd, and took out a patent on it in 1874 and published a book of rules. He called his new game 'sphairistike or lawn tennis'. Others immediately claimed that they had invented it, including the Marylebone Cricket Club, which set about establishing the modern rules of the game.

In 1875, the All-England Croquet Club at Wimbledon allocated one of its croquet lawns for the new game of lawn tennis, and added 'Tennis Club' to its name the following year. A committee at Wimbledon made some adjustments to the rules and reshaped the tennis court. Wingfield had used a traditional hourglass-shaped badminton court, whereas the Wimbledon committee decided on a rectangular court twenty-six yards long and nine yards wide (for singles). The real tennis scoring method, counting in fifteens, was adapted to lawn tennis and the server was allowed one fault. These were decisions that have remained integral elements of the game ever since.

In 1877, Wimbledon Tennis Club held its first championship under the new rules. Two hundred spectators paid one shilling each for tickets to watch the final, which was won by Spencer Gore, the first winner of the Wimbledon Lawn Tennis Championship.

Wingfield's new game was immensely popular, both in Britain and in the USA. Because it could be played on any lawn, and therefore involved less expense, it was far less exclusive than real tennis, and open to people of all socio-economic classes. The name 'sphairistike' was never going to be popular, and the game has been known as lawn tennis ever since.

THE INVENTION OF BARBED WIRE

(1874)

NON-BARBED WIRE was made as early as AD 400. Hot iron was pulled through dies to make short lengths of smooth wire of various thicknesses. By the second half of the nineteenth century the technology of wire manufacture had developed to the point where it was possible to buy high-quality wire of many different gauges and almost any length. By 1870 wire was often used for farm fencing, with two or three strands stapled to wooden posts at intervals. This type of fencing served as a useful boundary marker, but was no deterrent to straying animals.

On the Prairies, farmers who wanted to fence off plots for arable were in difficulties because of the huge herds of half-wild cattle that roamed across the landscape looking for pasture. Where there was fencing, the cowboys were able to stampede the cattle and make them trample down both fences and crops. These were the decades of the cattle kingdoms.

The solution came from a New York teacher who bought a farm in Dekalb in Illinois in 1843. His name was Joseph Farwell Glidden. At a local fair he saw an attempt at a barbed cattle deterrent. It was a wooden rail with sharp nails sticking out of it at intervals, and it was hung inside a conventional wire fence. Glidden started thinking of ways of improving on this. He wanted to find something that would give a cow a jab that was nasty enough to make it turn back, but without severely injuring the animal. He finally devised a twist of metal wire with a point at each end, tied at intervals into a double-stranded fence wire, also twisted.

Glidden patented his brilliant invention in 1874. The trail drivers were bitterly opposed to the new invention, seeing that it could mean the end of the unrestricted cattle drives. But farmers, large and small, all over the Prairies started using barbed wire to defend their crops – and it was very effective. One hundred years later an even nastier form of barbed wire was invented, called 'razor wire', and this is commonly used in endless rolls round the tops of prison walls.

THE INVENTION OF THE TELEPHONE

(1876)

ALEXANDER GRAHAM BELL was Scottish by birth, American by choice. He opened a training school in Boston for teachers of the deaf, and also gave lectures on the mechanics of speech. He devoted himself to the teaching of deaf mutes and the promotion of his father's system of 'visible speech'. Bell's father devised an alphabet system in which the alphabet symbols were diagrams of the positions of the organs of speech needed to make the sounds.

The younger Bell experimented with various acoustic devices, finally building a primitive telephone. He patented his invention in March 1876. Bell had to defend his patent against Elisha Gray, who invented the telephone independently, applying for a patent only hours after Bell. Bell's device, conversely, was not developed entirely independently of Gray's. When Bell applied for his patent, he had not built a fully working machine and he used Gray's *Notice of Invention* to help him do it. The Bell telephone, or perhaps we should call it the Bell–Gray telephone, consisted of two identical microphones and receivers, each made up of a solenoid mounted next to a metal membrane. Vibration set up in the transmitting membrane induced a current in the solenoid that travelled down the wire and caused the membrane in the receiving telephone to vibrate in the same way. It was on 10 March 1877 that Alexander Graham Bell gave his historic demonstration of the newly invented telephone. The first publicly witnessed phone call was to his assistant: 'Mr Watson, come here. I want you.' A few weeks later, Bell made his first long-distance telephone call, from New York to Boston, again to Thomas Watson.

The idea of the telephone had been around for a long time. In his 1627 book *New Utopia*, Francis Bacon described a long speaking tube to enable people to talk to each other at a distance, but it remained no more than an idea for 250 years. Really, the groundwork for Bell's invention was done by a string of inventions and discoveries. In 1729 Stephen Gray succeeded in sending electricity down a length of wire; in 1820 Christian Oersted discovered that electricity creates a magnetic field; in 1821, Michael Faraday discovered electrical induction. Thirty years later it was suggested that electricity could be used to transmit speech. These were necessary steps on the way to the invention of the telephone.

Bell was skilful in exploiting his invention, and the first switchboard was installed as early as May 1877 at the Boston offices of Edwin Holmes, the proprietor of Holmes Burglar Alarm Service. In 1879 Leroy Firman invented the multiple switchboard, and it was this that made the telephone a runaway commercial success. It enabled the number of callers to be increased, and the number of subscribers in America went up from 5,000 in 1880 to 25,000 in 1890.

American businessmen found the telephone invaluable for making business deals and pressed for the initial daytime service to be extended. Soon, instead of starting at 8 a.m. it was available from 5 a.m.

Bell's invention was to transform the twentieth century. Along with the motor car, the telephone was responsible for a major acceleration in the pace of life. It transformed business. It transformed politics, making it possible for political leaders to discuss matters from country to country, from continent to continent. Ironically it was to do nothing to improve international relations or understanding, and the twentieth century proved to be the bloodiest in human history.

THE INVENTION OF THE PHONOGRAPH

(1877)

THE ORIGINS OF recorded sound go back to the Egyptian Bronze Age. A colossal statue of Memnon was built at Thebes in 1500 BC. It was built in such a way as to reproduce the sound of a harp string being plucked every day, to greet the Goddess of the Dawn, who was Memnon's mother. How this harp sound was reproduced is not known, and it cannot be discovered by archaeologists because the statue was destroyed by an earthquake in AD 27. The secret of sound reproduction was lost.

For a long time the wheel was used to record and re-create sound. Pegs could be positioned at intervals round the rim to strike chimes as the wheel was rotated by hand. In the Middle Ages, music could be reproduced by hand-operated rotating cylinders with metal pins attached; the pins were made to strike specific bells or keys as the drums rotated. Automated carillons were made in the fourteenth century. The earliest surviving barrel organ was made in 1502. The musical boxes that were popular in the eighteenth and nineteenth centuries and the later pianolas were based on similar technology.

The phonograph was invented in the midst of a flurry of inventions of electrical appliances, at a time when almost anything seemed possible. Some people refer to this phase as the second Industrial Revolution. The invention of a machine that could reproduce sound came at about the same time that experiments were going on to make machines that could reproduce images – moving images. The two, when successfully invented, would together make the talking picture possible, the cinema.

The idea of a sound reproduction machine began as a means of

375

preserving events, of creating historical archives. But once the machine was in existence it was used to a great extent for music.

The earliest successful sound recording machine was called the phonautograph, and it was invented in 1855 by Leon Scott de Martinville. This had a horn that functioned as a mouthpiece, and a membrane attached to a stylus that recorded the sound waves on smoked paper wrapped round a rotating cylinder. But this was a long way from the later gramophone, as it recorded sound visually, as silent wave patterns, rather in the way that a seismograph would record earthquakes. The phonautograph could not play back the sound, so it was in effect only half of an invention.

For all its unsatisfactoriness, the phonautograph stimulated several inventors in various countries to work on a solution to the problem of playing back. It would take a further twenty years before Thomas Edison produced his phonograph in 1877. This was a machine that could both record and play back. Edison used essentially the same structure and mechanism as Scott, but replaced the smoked paper with tin foil. The stylus was directly attached to the membrane and it traced the recording as a furrow of variable depth (hill and dale).

Throughout 1877 and 1878, Edison did public demonstrations of his phonograph. He made a wide range of sounds, including talking and coughing, while recording, and then astounded his audiences by playing the recording back. Scarcely able to believe what they were hearing, some members of the audience made outlandish noises, such as imitations of animals, to see if the machine would reproduce them. The phonograph needed development, but unfortunately Edison's attention moved quickly on to other challenges. He was preoccupied with inventing a working light bulb at that time, and for the time being abandoned the phonograph.

In 1885, Alexander Graham Bell and Charles Tainter took out a patent on their invention, the graphophone. This was an adaptation of Edison's phonograph, using wax cylinders instead of tin foil. It was a huge improvement. The graphophone recordings were clearer and longer, and the wax cylinders were rather more durable than tin foil. The graphophone advance rekindled Edison's interest in the project and, recognizing the superiority of the wax cylinders, he adopted them for his own machine. Wax cylinders were used for many years, until they finally went out of use in 1928. Wax was better than tin foil, but it brought with it a range of problems. It softened and melted when heated, and it went mouldy. A better substitute was found in celluloid, an early plastic made from dissolved cellulose; it was lighter, tougher and resistant to mould and fungi.

The first celluloid cylinders were made by Thomas Lambert, they were bright pink and they were known as Pink Lamberts. Edison eventually acquired the patent rights to the celluloid cylinder, and replaced his wax cylinders with Blue Amberols in 1912. These were a bright azure blue and backed with plaster, which caused some problems by swelling and warping the cylinders.

Other inventors of sound reproduction machines were experimenting in other directions. In 1893 Emile Berliner invented the gramophone, which used discs instead of cylinders. Edison had tried discs in 1878, but abandoned them in favour of cylinders. Berliner's gramophone discs were pressed by a zinc master and made of hard rubber. They turned out to be far easier and cheaper to produce, and mass production became a distinct possibility. Any number of discs could be pressed from the master, whereas the cylinders had to be engraved individually.

The major problem with the Berliner discs was the poor sound quality, which deteriorated progressively as the relatively soft rubber became worn. This led on to the development of Duranoid, which was a tough plastic, though brittle. The brittleness of the 'Seventy-Eights' as the gramophone records were often known, after their running speed of seventy-eight revolutions per minute, was overcome in 1948, when vinyl was introduced. Vinyl is short for polyvinyl chloride, or PVC.

From the start of the twentieth century a major worldwide industry developed, and it became possible to buy wind-up gramophones relatively cheaply along with recordings, eventually, of every kind of music. The shortness of the playing time of both cylinders and discs meant that light music was favoured. Short music hall songs like *One of the Ruins that Cromwell Knocked About a Bit*, *Clementine* and *Don't Dilly-Dally on the Way* lent themselves to this kind of recording, but it was not long before classical music, often in edited-down form, was also available. The record industry transformed the musical life of the world. It became possible for people to acquire almost any piece of music and listen to it any number of times, rather than the single hearing they would get if they went to a concert. Instead of hearing Beethoven's Seventh Symphony perhaps once a year, it has become possible to listen to the piece daily, to hear the secondary melodies and inner parts, and really get to know it inside out. This was a great aid to people who wanted to gain a better understanding of the classical repertoire. For many people, it became their main access to serious music.

For composers it meant that their work would no longer be written off after a possibly inadequate or unsympathetic first performance; a recording allowed people to get to know not only a piece but an interpretation of a piece intimately over a number of years. Audiences became better informed and more discerning. And composers became more adventurous. It is possible that the increased diversity and adventurousness of musical styles across the whole range has come about partly because of the record industry. The recording had to be definitive, and of the highest quality, to stand up to repeated hearings and this too has had a profound effect on the music industry, pushing performing standards ever higher.

THE INVENTION OF
THE CASH REGISTER

(1879)

JAMES RITTY INVENTED the first working mechanical cash register in 1879. He called it the Incorruptible Cashier. He incorporated the bell device which became the distinctive sound of a cash sale at shop counters all round the world. Ritty patented his register in January 1883.

When John Patterson read a description of Ritty's cash register as sold by the National Manufacturing Company, he decided to buy both the patent and the company. In 1884 Patterson renamed it the National Cash Register Company. He also improved the cash register significantly by adding a paper roll to record sales. This was a major step forward, enabling shop owners and managers to check sales by the day and therefore develop a clearer view of their businesses. It also meant that it was far more difficult for shop assistants to mischarge customers or pocket change. The amount the customer was being charged showed on metal flags that sprang up from the top of the machine, so the customer knew what was being recorded on the paper roll. Businesses both large and small benefited enormously from the arrival of the cash register; and customers everywhere received fairer treatment.

While working at the National Cash Register Company, Charles Kettering designed an electric cash register in 1906. This eventually replaced the mechanical, hand-powered registers, but in small shops all over the world the original registers continued in use for fifty years.

THE INVENTION OF LAVATORY PAPER

(1880)

LAVATORY PAPER IS one of those commodities that we take for granted, and cannot imagine needing to be invented. Huge quantities of it are used. It is said that the average American uses more than 100 rolls of lavatory paper a year.

There was a pre-paper age, when people used a variety of materials. The ancient Romans used a specially manufactured L-shaped stick, like a hockey stick. At public lavatories, people used sponges on sticks that were cleaned in brine after use. Roman soldiers carried their own sponges as part of their routine gear. In other cultures, sand, earth, stones, mussel shells and even corn cobs have been used.

By the late fifteenth century in Europe, paper was becoming more widely available and it began to find a regular use as lavatory paper. Old letters, pages from discarded books, old paper bags were used. In many of the poorer homes in the nineteenth and twentieth centuries, sheets from torn-up newspapers were commonly used, perforated and hung on a string loop in the privy.

Deliberately manufactured lavatory paper was invented in 1880. It was developed and marketed first by the British Perforated Paper Company. The first lavatory paper was smooth, hard and sold in squares packed in boxes. Some lavatory paper is still sold in this form. The first rolls of lavatory paper were manufactured by the Scott Paper Company in America in 1890. These early rolls were not perforated, and lavatory paper dispensers were fitted with serrated edges to tear the paper off.

For a long time people were too embarrassed to ask for lavatory paper in shops. In the days before supermarkets, customers had to ask an assistant for the products they wanted, so when it came to rolls of lavatory paper, customers were reduced to asking for, 'Three, please', expecting the assistant to understand. The items appeared on the counter in plain brown wrappers.

The manufacture and sale of lavatory paper, paper towels, paper table napkins, paper tissues and kitchen rolls – all disposable, all made of paper – have became one large industry. The raw material is bleached pulp with relatively little refining, which leaves the paper soft, bulky and absorbent. Paper for sanitary use is also crêped; the paper is dried on a cylinder, then scraped off with a metal blade, and this slightly crimps it. The crêping makes the paper softer to the touch, and also allows it to break up in water. Given that the paper is to be flushed into the sewerage system, it is important that lavatory paper does disintegrate. It would be a problem if it did not.

Lavatory paper is offered to customers in a variety of forms. It can be one-ply or two-ply; it can be embossed or quilted; it can be coloured; it can be scented. The most important choice is between lavatory paper that is virgin paper, made directly from freshly chipped wood, and paper that is recycled. Many people are conscious of the need to conserve forests and prefer to reserve virgin paper for writing on, using recycled paper for everything else. The cardboard cylinders round which the lavatory paper is wound is almost always recycled material.

THE INVENTION OF
THE SKYSCRAPER

(1884)

TALL STRUCTURES WERE attempted by various civilizations long before the first modern skyscrapers were built. The ancient Egyptians built pyramids, and they were huge structures. The Great Pyramid of Khufu at Gizeh rose to a height of 481 feet, about 150 feet higher than St Paul's Cathedral in London. Ancient Mesopotamia had its own distinctive tall structures: the ziggurats. Pyramids and ziggurats were designed on the same basic principle. The pyramid began as a solid rectangular single-storey funerary building called a mastaba, and the Egyptian architects found that they could make it look more imposing by building another, slightly smaller, mastaba on top of it, and then another, smaller again, on top of that. In this way the Step Pyramid at Saqqara was constructed. The ziggurats were similarly built as a series of solid square plinths raised on top of one another. In the Middle Ages, more ambitious towers were built for cathedrals. These were constructed in a very different way; they were hollow, had near-vertical sides, and stood on very small bases.

All of these early structures were sacred buildings of one sort or another – temples, mausolea or churches – but for ordinary dwellings or commercial buildings, much lower buildings of one, two or three storeys were the norm.

In the late nineteenth century several changes took place to force the invention of the skyscraper. Vigorous urbanization was under way in Britain and America and, as the towns and cities grew, land in the city centres became more and more valuable and in shorter and shorter supply. Building upwards was a way of getting the maximum rent out of a plot of land, and also of getting the maximum use out of it.

Another change was technological. Iron- or steel-framed buildings had much greater strength. The walls might still be made of brick, but were no longer load-bearing. The metal girders used to build a frame for the building, rather like the old timber-frame buildings of medieval Europe, helped to distribute load and other stresses through the building. The steel frame was the first serious advance in building construction technique since the invention of the early English arch and the flying buttress. The word skyscraper itself was not new; it was already in existence in the Middle Ages, to describe the tall towers raised in thirteenth-century Italy, and some of those were almost 300 feet high. The steel-frame construction method changed the shape and style of buildings enormously. In the earliest designs the method was cunningly concealed, but gradually as the decades passed it came out more and more into the open.

A key invention was the Otis elevator. The average fit and healthy adult can easily walk up stairs to the top of a four-storey building, but ten storeys would become a problem. The invention of a reliable lift or elevator that could stop predictably at each floor within half an inch of the correct floor level was crucial to the development of the skyscraper.

Another was the great fire in Chicago in 1871. It destroyed a large area of the city, which had to be rebuilt quickly and economically. This was why Chicago was the birthplace of the skyscraper. The building that is pointed out as the very first skyscraper is the Home Insurance Building, which was designed by the architect William LeBaron Jenney and built at the intersection of La Salle and Adams Streets in 1884–85. It was ten storeys high, steel-framed, with marble-faced outer walls and polished granite columns covering the steel girders. The Home Insurance Building was demolished in 1931.

LaBaron Jenney's lead was eagerly followed by other architects, and a whole new wave of multi-storey buildings in the same style was created in Chicago in succeeding years. The Tacoma Building in Chicago was the first to be built using a load-bearing steel skeleton throughout. William Holabird was the Chicago-based architect of the Tacoma Building and it was his firm, Holabird & Roche, that became the major developer of Jenney's concept into a landscape of modern office buildings. It was Holabird who transformed the Chicago skyline, adding the Caxton Building in 1890, the Pontiac Building in 1891, the

Marquette Building in 1894 and the Tribune Building in 1901.

The most famous early skyscraper is probably the Fuller or Flatiron Building, which was built in 1902 in New York. It got its name from its distinctive triangular plan, made necessary by the shape of the building plot at the diagonal junction of two roads. It is decorated in carved limestone, once again concealing the structural engineering. The Flatiron Building was the first of the multi-storey buildings to go up in New York.

Then came a wave of much taller buildings, the second generation of skyscrapers. The first of these was the Woolworth Building, an elaborate Gothic-style tower built in 1913 and which soared 800 feet above Broadway. Then Walter Chrysler, the car manufacturing magnate, wanted a building to bear his name and he wanted it to be the tallest in the world. The element of commercial competition had entered the story of skyscrapers. Companies aiming for a high profile wanted a high building to reflect their status. Skyscrapers became company billboards. The buildings became taller and taller. Chrysler hired William van Alen, an architect trained in Europe, to design his building. The initial design was to be 925 feet high. Then van Alen introduced a vertex spire that was assembled inside the building and then lifted into place; that made the finished structure 1,046 feet high, and the tallest in the world.

The Chrysler Building was spectacular when it was new, and it is a great tribute to its architect that it is still breathtaking. It is one of the finest examples of Art Deco architecture in existence. But it was doomed to be overtopped relatively quickly, like a tree in the canopy of a rainforest. By 1929, building wor   k had started on the Empire State Building. Within three years it was finished. At 1,250 feet high and with 102 floors, the Empire State Building was for a long time the tallest building in the world. Again, lift technology made the great height manageable; it had sixty-seven high-speed lifts.

The steel-frame technology became steadily more visible in later buildings, and the latest generation of towers is conspicuously glass-and-steel. The most famous of these were the twin towers of the World Trade Center in New York. These were two identical 110-storey towers standing very close together and dominating the Lower Manhattan skyline. Whether the tragic collapse of the twin towers under determined terrorist attack in 2001 will slow down the development of skyscrapers remains to be seen, but the American cities have already been surpassed by Kuala Lumpur. The Sears Tower in Chicago, built in 1974, is 1,450 feet high. The Petronas Towers completed in Kuala Lumpur in 1998 are 1,483 feet tall.

THE INVENTION OF THE FOUNTAIN PEN

(1884)

THE FOUNTAIN PEN was invented in 1884 by Lewis E. Waterman, a New York insurance salesman. There had been earlier attempts at pens with ink reservoirs, some as early as the tenth century, but Waterman's pen was the first that really worked. It became the main handwriting instrument for the next sixty years and is still the preferred pen type for many people.

The German Daniel Schwenter invented a type of reservoir pen in 1636. He made his pen out of two quills. The inner quill acted as the ink reservoir and it was sealed at each end with corks. The ink was squeezed through a tiny hole towards the outer quill point. Progress towards a working pen was slow, and an added problem was that the composition of the ink made it corrosive and therefore destructive of any delicate mechanisms to control its flow. The Romanian Petrache Poenaru invented a fountain pen while he was a student in Paris in 1827. Other attempts followed. From 1850, there were more frequent attempts at a reservoir pen. It really only became viable once three other innovations were in place: free-flowing, sediment-free ink, iridium-tipped gold nibs and vulcanized rubber.

The modern fountain pen contains a reservoir of water-based ink that is fed to the nib by a combination of gravity and capillary action. There have been various methods for refilling. One is by using disposable and replaceable cylindrical ink cartridges. Another is filling the reservoir with an eyedropper. Another is an internal mechanism that squeezes a rubber reservoir and allows it to reflate and suck ink up through the nib. The problem with the rubber reservoirs is that they eventually perish, giving the pen a relatively short life. Fountain pens were a great advance on the steel-nibbed pens that preceded them, because they removed the need to dip the nib in an inkwell after writing every fourth word.

The first pens were made in antiquity. The ancient Egyptians used reeds to write on papyrus. Over the centuries, pens were developed into sharpened instruments with slits cut into their ends. Quills made out of feathers came into use in the sixteenth century, partly because they were pliable and broke less often under pressure.

By the middle of the nineteenth century, pens were regularly made with metal nibs mounted on wooden handles. Lewis Waterman was one of many thousands of people who became irritated with the inconvenience of returning to the inkwell after every fourth word, and he found a solution. In 1884 he patented the fountain pen. The idea of a pen with an ink reservoir had been thought of before, but the problem had been to find a way of regulating the flow of ink onto the nib. Waterman overcame the problem by adding additional ducts to allow air into the pen, to replace the ink as it flowed down onto the nib. But at least the reservoir meant that writers did not have to return to the inkwell to refill more than once a page.

The age of mass-produced fountain pens began in the 1880s. The great market leaders at that time were Waterman, based in New York, and Wirt, based at Bloomsburg in Pennsylvania. Waterman outstripped Wirt, remaining the clear market leader until the 1920s. Fountain pens became prestige objects in their own right and manufacturers created celluloid barrels in a wide range of colours, trims and finishes, including some with exotic marbled effects, to encourage this process of individuation.

Later developments included the roller ball and the ballpoint pen. In a ballpoint pen, the ink flow is controlled by gravity. The ink dries immediately, which is for many uses a great advance on the fountain pen. For fast use, the ballpoint is superior as the writer is less likely to smudge the writing; this makes the ballpoint pen a firm favourite with students in examinations.

THE INVENTION OF THE STEAM TURBINE

(1884)

THE STEAM ENGINE was not a new invention. Hero of Alexandria had made a simple form of steam engine in 130 BC. In the eighteenth century the first generation of modern steam engines was invented, the most effective of them by James Watt. The steam engine Charles Parsons invented in 1884 was the first of the second generation. It was a smaller, lighter and more efficient engine than the old steam piston engine.

Parsons' steam turbine was the forerunner of the powerful turbo-generators that produce most of our electricity today.

Charles Parsons, described as one of the greatest British engineers of all time, was an aristocrat, a son of the Earl of Rosse, who was himself a distinguished astronomer. Charles learned a lot about engineering and science from his father. He went to Cambridge University to study mathematics and then, rather surprisingly, went to work in Sir William Armstrong's factory as an apprentice; he was determined to learn engineering.

The turbine engine Parsons invented was far more effective than earlier engines. The expanding steam was made to work continuously across many turbine stages, passing from one to the next. This made maximum use of the steam power. The action was a continuously rotating movement, which gave maximum efficiency. Its smoothness of action gave, overall, an absence of vibration. Together with high speeds, the new turbine was ideal for generating electricity. The existing engines for generating electricity were so big and noisy that one power station in Manchester had to be closed down because of complaints about the high noise level. Parsons and his partners saw that there was huge potential for the steam turbine, but suitable electrical generators needed to be designed to accompany the turbines. Parsons duly invented a turbo-generator. Three of his four-ton generators were installed at the Cambridge Power Station in 1895, and were used to supply electricity to the first electric street lighting system in Cambridge.

This new ability to convert coal quietly and efficiently into electricity paved the way for a whole raft of new industries in the twentieth century, industries that depended on electricity. The industrial estates fringing most towns and cities in the Western world house a bewildering range of light industries which nearly all depend on electricity to power them. This transformation would not have been possible without the invention of the Parsons steam turbine.

Charles Parsons pioneered the installation of steam turbine engines in ships. In a spectacular publicity stunt, Parsons gate-crashed the 1897 Naval Review at Spithead with his vessel the steam turbine-powered *Turbinia*. In the *Turbinia*, Parsons was able to steam past the lines of anchored warships at a speed of thirty miles per hour, easily outstripping the naval patrol vessels policing the event. Not long after this high-profile display of speed, Parsons steam turbine engines were fitted as standard to all large ships and they remained standard for fifty years, until the advent of the large diesel engines.

THE INVENTION OF THE FINGERPRINTING IDENTIFICATION SYSTEM

(1885)

IN 1885, THE British scientist Francis Galton invented a system for identifying people from their fingerprints. Galton was sixty-three and he had earlier in his career founded the pseudo-science of eugenics with a book that he provocatively entitled *Hereditary Genius*. With this new idea, however, he had come up with something really useful – that no two people have the same fingerprints.

Galton's idea was not new. The knowledge that everyone has different fingerprints was available in antiquity; the impression of an Eastern monarch's thumb was his signature. The permanent (fixed for life) character of the fingerprint was first put forward scientifically by J. E. Purkinje, a professor of physiology who read a paper about it in Breslau in 1823. He identified nine basic patterns and suggested a system of fingerprint classification. Nobody took any notice of Professor Purkinje.

Francis Galton worked on the very specific idea that a person's fingerprints might be used to trace his or her criminal activities. From there it was but a short step to the idea of taking the fingerprints of people suspected of committing crimes, by inking their fingertips and dabbing them onto a piece of paper or card, and then comparing them with the fingerprints found at the scene of a crime. It was found that a light dusting of white powder could make fingerprints visible on a great variety of smooth surfaces, from glass window panes to metal door handles and polished wooden table tops.

To be sure of the match, Galton recommended taking the prints of all ten fingers of a suspect. They could be arranged on an identification card in a set order, left hand on the left, right hand on the right. It was the Bengal Police who adopted the system first, under the administration of E. R. Henry, who afterwards became the Chief Commissioner of the Metropolitan Police.

One great advantage of fingerprints was that they did not change during life. Once the police had someone's fingerprints they could be used at any time after that for identification. The patterns lend themselves to easy classification into arches, loops, whorls and composites. Each of these types has subclasses.

The idea quickly caught on and became the standard means of identifying the presence of individual people at crime scenes for much of the twentieth century. The identification division of the American FBI was set up in 1924, and at that time the several previously separate collections of fingerprints were put together to make a single database. By 1952, the FBI had amassed a collection of more than 125 million fingerprint cards – not all of them relating to criminals. The ultimate problem with this kind of collection is that it turns into the collection of information about citizens for the sake of it. Now, with computerization, it has become very easy to amass huge quantities of information about whole populations, and politicians and administrators find it almost impossible to resist the temptation to go on collecting, which inevitably means encroaching further and further on people's privacy.

The fingerprint technique was only superseded by DNA in recent years, and in the course of a century, identifying people by fingerprints has led to the solution of hundreds of thousands of crimes worldwide.

THE INVENTION OF THE MOTORCYCLE

(1885)

GOTTLIEB DAIMLER INVENTED the first gas-engine motorcycle in 1885. It was really a fairly crude grafting together of two separate inventions: the gas-powered engine and the bicycle, in this case made of wood. Daimler used a newly invented engine crafted by Nicolaus August Otto. Otto had invented his four-stroke internal combustion engine in 1876. Daimler simply mounted the Otto cycle engine onto the wooden bicycle and created the motorcycle. Probably the machine had a spray-type carburettor, as Daimler's assistant, Wilhelm Maybach, was working on this invention at the time.

Daimler's motorcycle had the normal two wheels, front and back, but it also had a smaller spring-loaded wheel on each side, acting as outriggers. The main structure was wooden, with iron-banded wooden-spoked wheels. The machine was nicknamed the bone-crusher.

It is sometimes said that Daimler is only credited with inventing the motorcycle, as if he didn't really invent it. The reason is that there were some earlier attempts. In 1869, Sylvester Howard Roper of Roxbury in Massachusetts had invented a steam cycle, which was a bicycle with a small steam engine, a charcoal-powered two-cylinder engine, mounted between the wheels. The engine had connecting rods that drove a crank on the back wheel. The Roper steam cycle was shown off at fairs and circuses in America.

The motorcycle was not of enormous interest at the time; there was far more interest in vehicles with three or four wheels. During this phase of experimentation, it was difficult enough to get machines running, and keep them running, without having the extra problem of having them falling over. But there were some low-profile attempts at better two-

392

wheeled motorcycles. After Daimler came a more sophisticated-looking machine, the five-cylinder Millet of 1892. This had its engine mounted at the hub of the rear wheel; in fact its crankshaft formed the rear axle.

The first truly successful two-wheeled powered vehicle was the 1894 Hildebrand & Wolfmueller. This had a step-through frame and a fuel tank mounted on the down tube. The engine was mounted low down on the frame, with connecting rods linked directly to a crank on the back axle. To store energy between the firing of the cylinders, it used a pair of heavy-duty elastic bands. Then in 1895 the French firm DeDion-Buton designed an engine that made mass production and the everyday use of motorcycles possible. This was powered by a small, lightweight, high-revving four-stroke engine, which was widely imitated by other manufacturers including Harley-Davidson in America.

THE INVENTION OF THE ELECTRIC TRANSFORMER

(1885)

A TRANSFORMER IS A piece of electrical equipment for transferring energy from one circuit to another by way of a magnetic coupling. It is a fairly simple device, yet the details of its design have been improved many times over for a hundred years.

Transformers can be very small, such as the thumbnail coupling transformer in a stage microphone, or very large, such as the units that connect large segments of national electricity grids.

Michael Faraday made a transformer in 1832, though unwittingly. He used the device to demonstrate electromagnetic induction and had no idea of the use that would be made of it at the end of the century. In London in 1881 Lucien Gaulard and John Dixon Gibbs exhibited a device they called a secondary generator, and this seems to have been the first practical transformer. They sold their device to the American company Westinghouse. They exhibited a similar device in Turin in 1884, and there it was taken up for use with an electric lighting system. Meanwhile one of the engineers at Westinghouse, William Stanley, succeeded in building the first practical transformer after Westinghouse had bought Gaulard's and Gibbs' patents; it was first used commercially in 1886.

In 1891, Nikola Tesla, who worked for but later quarrelled with Thomas Edison, invented what became known as the Tesla coil. This was a transformer used for generating very high voltage at high frequency.

Transformers functioned like electrical gearboxes; they could be used for stepping voltages up or down. They turned out to be essential for transmitting electricity at high voltage, and high voltages were necessary for transmitting electricity long distances. In other words, transformers made it possible for power stations to supply consumers living a hundred or more miles away. It was only because of the invention of transformers that national electricity supply grids became possible.

THE INVENTION OF ESPERANTO

(1887)

IN 1887, AN idealistic twenty-eight-year-old Polish philologist called Lazarus Zamenhof invented a new language. He called it Esperanto and he hoped it would achieve world peace. The principle of this new and artificial language was that it would extract ideas and roots of words from a range of existing languages and compile a new one that would not be tainted with any particular country, culture or events.

The idea was a good one. Some years ago, there was a suggestion that the cause of European unity would be helped by having not just a common currency but a common language. The French vetoed this idea, because they knew that the common language would be English. Another idea was that the common currency unit should be called the ECU, an acronym standing for European Currency Unit. I liked that proposal because it was going to be easy to pronounce and, as it happened, there was at one time a French currency unit called an écu. It was a word with a history, a word with background. But the Germans vetoed this proposal – just because it seemed to be favouring the French. And this is how we ended up with a sterile, cultureless and bureaucratic word like euro, which is actually quite difficult to say in some languages and sounds unpleasant in all of them.

Esperanto was not the only universal language to have been devised. In 1880, J. M. Schleyer published a language called Volapuk, which was popular for a decade and then was heard of no more. There was Zamenhof's Esperanto in 1887. Then in 1907 Louis Couturat and Louis de Beaufront published an edited version of Esperanto that became known as Ido. An international committee set up in France in 1907 to choose the best language for international use came to the conclusion

396

that Esperanto edited in the direction of Ido would be the best. The supporters of pure Esperanto rejected this proposal. Already, ironically, new language loyalties and patriotisms were developing. Between 1907 and 1950 no fewer than fifty schemes for a new language were proposed, most of them wanting a reformed Esperanto or an edited version of Ido. For a time the Esperantists thought they were gaining ground. Their language was the only constructed language that had a worldwide group of living speakers – there are thought to be a million of them.

Esperanto should have been a huge success in the world of the European Union, but it has not been. One problem with it was that it operated on rather complicated rules and was therefore difficult to learn. Another was that there was no literature written in Esperanto.

Esperanto was one of those ideas that should have been big, but was not. When Esperanto was failing, it looked as if French was going to be the common language, not only of Europe but of the world. That was why everyone in my generation was made to learn French at school. Now it looks as if English is fast becoming the world language; it is certainly the second language of vast numbers of people. Now there can be no hope for Esperanto.

THE INVENTION OF THE ELECTRO-CARDIOGRAM (ECG)

(1887)

THE CENTRAL KEY to human well-being is a regular heartbeat. This regular beat is maintained by a group of pacemaker cells that generate pulses of electricity, causing the heart muscles to contract. The heart consists of several chambers, the right and left atria and right and left ventricles, and they all need to contract and relax in harmony. What the electrocardiogram machine does is to record, in graph form, the pattern of electrical activity within the heart, in other words the pattern of signals that the various muscles are receiving.

The ECG picks up three distinct elements in the heartbeat. One is the P wave, which is the electrical activity spreading over the atria. The second is the QRS complex, which is the electrical activity spreading over the ventricles. The third is the T wave, which is the recovery phase of the ventricles.

The ECG machine has evolved over a long period. Its ancestor is the galvanometer that was devised in 1794, which was able to sense, though not actually measure, electrical activity in the heart. This was enhanced in 1849 by Emil du Bois-Reymond, who made a device called a rheotome. A refinement of the rheotome made in 1868 by Julius Bernstein was used to take electrocardiograms of frogs' hearts. For these experiments, the electrodes were placed directly on the heart itself.

The first successful measurements of the electrical activity of the human heart were carried out by Augustus Waller in 1887. In his initial report, he referred to the technique as a cardiogram. Later he called it an electrogram. Later still, the two terms were joined together to make the term that has been used ever since, electrocardiogram.

Meanwhile, Willem Einthoven started making his own version of the galvanometer in 1900, independently of Waller's work, and based more on developing du Bois-Reymond's galvanometer of decades earlier. Einthoven's machine was initially manufactured by Edelmann and Sons in Munich. The first ECG machine to appear in America was an Edelmann machine imported by Alfred Cohen in 1909.

During the next few years, the ECG proved its worth. There was a dramatic demonstration of its value on 20 May 1915, when a patient had a coronary while undergoing an ECG, and the ECG itself gave the diagnosis. The machine itself underwent several major improvements. By 1928, the machines had been reduced to one-twentieth of their original weight, and the size of the electrodes placed on the patient had also been reduced. Another improvement has been the development of direct-writing barbed instruments. These are able to translate the electrical impulses into ink marks on a page, to give a continuous graph read-out of the heart's activity. Recent developments include wiring the electrodes to a tape recorder, so that a patient can have a twenty-four-hour ECG. This enables doctors to pick up any occasional heartbeat irregularities.

THE INVENTION OF THE PNEUMATIC TYRE

(1888)

THE INVENTION OF the pneumatic tyre came about in a way that could never have been predicted. The inventor was a veterinary surgeon from Edinburgh, so one might have expected him to invent some new technique in animal surgery, but his invention was nothing to do with his work. His little boy had a tricycle and complained that riding it was bumpy. John Boyd Dunlop's response to this juvenile complaint was unexpected. He carefully fitted inflatable rubber hoses to the wheels of his son's tricycle to give him a smoother ride.

Unknown to Dunlop, the principle of inflatable rubber tyres had already been patented forty years earlier, by Robert Thomson in 1845. It was a case of something invented, forgotten, and then re-invented. Dunlop was impressed with the tricycle tyres and decided to take the idea and turn it into a practical reality. He took out his own patent in December 1888. In 1889, Dunlop set up a business to manufacture inflatable tyres on a commercial scale. This business, trading as the Pneumatic Tyre and Booth Cycle Agency, was run in partnership with William Harvey Du Cros. Dunlop sold the pneumatic tyre patent to Du Cros for a moderate sum and took 1,500 shares in the company. Difficulties then arose because it emerged that the pneumatic tyre idea had been patented forty years earlier, but the company managed to hold its position because it held patents on various accessory processes.

Dunlop successfully produced practical pneumatic tyres for bicycles and went on to produce them for cars as well. He did not make a vast fortune out of his important invention, as he took no part in the great development that followed the sale of the company to E. T. Hooley in 1896. By then the pneumatic tyre business was worth five million pounds.

The effect of Dunlop's innovation was to make riding bicycles far more comfortable, and so make long-distance cycling more practicable. Whether Dunlop should be credited with inventing the pneumatic tyre or not, he certainly made regular cycle journeys to work possible, as well as cycling for leisure. Dunlop tyres made cycling into the major recreational activity it has become today. Similarly, it is impossible to imagine the modern motor car without pneumatic tyres. Solid rubber tyres on a car would inevitably have meant short journeys at low speeds. Dunlop's invention – if that is what it was – made long-distance driving at speed possible. Dunlop made his contribution to the great car economies of the West.

THE INVENTION OF GESTETNER TYPEWRITER STENCIL

(1888)

A HUNGARIAN IMMIGRANT living in London, David Gestetner, invented the first typewriter stencil in 1888. It was seven years earlier that Thomas Edison invented the first stencil duplicating machine. But the duplicating machine Gestetner designed for his typewriter stencils was a significant improvement on Edison's. It worked on the principle of a waxed sheet fixed to a cranked rotating drum. Handwriting, a drawing or a map could be drawn onto the waxed sheet with a metal stylus. The lines where the stylus scraped the wax off allowed ink to pass through a membrane onto the sheets of duplicating paper as they were fed through the copier. In this way it was possible to write a list of instructions for employees, run them off on the copier and give them to the employees.

The fact that these stencils could be typed opened up the possibility of copying whole documents, and each turn of the crank handle produced another copy. It was very easy – and cheap – to produce hundreds of copies from one stencil.

This innovation made mass communication in the factory, the office, the school and the university significantly easier. It also made it easier for teachers and lecturers to produce their own teaching materials, encouraging creativity instead of slavish dependence on textbooks. It also helped to generate more administrative documents and increased the amount of paperwork that everyone at work would have to deal with in the twentieth century. The Gestetner duplicator is one of those great, but greatly underestimated and undervalued, inventions that have changed the texture of modern culture. It has to a great extent been superseded by spirit duplicators, which allowed the use of several colours, and the photocopier. But it is still used – and it is cheap.

THE INVENTION OF THE ZIP FASTENER

(1892)

ELIAS HOWE, ONE of the inventors of the sewing machine, also invented the zip fastener. He patented 'an automatic, continuous clothing closure' in 1851. It sounds from his description very much like a zip fastener, but it seems that Howe never actually made it. He may have been so preoccupied with the sewing machine that he was unable to pursue the zip.

As it was, a period of forty-four years went by before the zip was re-invented. Whitcomb Judson patented his zip fastener in 1893, though he called it a clasp locker. Because Judson actually made, produced and sold his fastener, he must take credit as the inventor of the zip fastener. The clasp locker was on show at the 1893 Chicago World's Fair, but it had conspicuously little success.

It was Gideon Sundback, an electrical engineer, who developed the Judson fastener to the point where it became recognizable as the modern zip fastener. Sundback was born in Sweden and emigrated to Canada. He worked for the Universal Fastener Company. He was a skilled designer, and he also made a successful career-move in marrying the plant-manager's daughter; he became Universal's head designer. In that position, he worked on improvements to Judson's C-curity Fastener, perfecting it in St Catherines, Ontario in 1913.

Sundback increased the density of fastening elements from four per inch to ten. The fastener had two facing rows of metal teeth that were pulled together into a single strip by a slider. Sundback was issued with a patent for his hookless fastener or separable slider in 1917. He also designed a machine for manufacturing the new fastener; the intention was to mass-produce it. Within no time, Sundback's zip machine was producing several hundred feet of zip fastener a day. But the device was still not called a zip or zipper. That happened in a roundabout way in 1923. The Goodrich Company manufactured a line of rubber overshoes or galoshes, for which it coined the name Zipper. It was the shoes that were the Zippers, but they were fitted with the revolutionary new fastening device. Gradually it was the fastener itself that became known as the zipper.

THE INVENTION OF THE DIESEL ENGINE

(1892)

RUDOLF DIESEL'S DREAM was to create an engine that would free ordinary people from the laborious drudgery of operating the machines that existed in many mid-nineteenth-century factories. He was motivated by the desire to help, rather than the need to make money out of his invention. Unusually, the diesel engine arose out of a philanthropic rather than an economic impulse.

Diesel was born in Paris of German parents, and when the Franco-Prussian War broke out he and his family were deported to London as undesirable aliens. He went to Germany for a technical education, then returned to Paris. He worked on designing an engine that was more powerful and efficient than the gas engines that were then in use. One feature that he saw as inefficient in the existing gas engine design was the external ignition of the gas. He improved on this by igniting the air-fuel mixture internally, in the driving cylinder. Under compression, the mixture's temperature would increase to the point where it would spontaneously ignite; there was no need to have an external mechanism to ignite it.

The first experimental test was disastrous. The engine exploded and almost killed Diesel.

A significant improvement in Diesel's engine was that it would run on cheap fuel. He had initial ideas that it might run on coal dust or animal fat, but later settled on cheap crude oil, which became known as diesel fuel. The engine also cost less to repair and needed a shorter warm-up period. It also functioned without a water supply, unlike the steam engine that had powered the first Industrial Revolution.

Diesel patented his first successful engine in 1892, though it was scarcely ready to go into widespread use at that stage. But it was not long before the diesel engine found uses where a heavy-duty power source was required. Later it would be installed in tractors, ships, lorries, buses, locomotives and submarines.

The diesel engine is what Rudolf Diesel intended it to be, a relatively cheap and economical machine, more efficient than a petrol engine. It does carry with it some serious problems, though. It is inclined to be noisy, and it gives off serious pollution.

Diesel was an odd mixture of brilliant inventor and impracticality. He made enormous amounts of money from his invention, yet was constantly in financial difficulty because he could not manage money. He also became increasingly depressed by the prospect of war in Europe, war that his engines would power. He was crossing the English Channel on a night ferry in 1913, the year before war broke out, when he vanished over the side.

THE INVENTION OF THE TRACTOR

(1892)

THE TRACTOR WAS invented by John Froelich in 1892. This was a development from the invention of the motor vehicle; the tractor is a specialized motor vehicle designed for farm work. It was designed to do the work of a horse, and it has in effect replaced the horse in the rural landscapes of the Western world, which many people think is regrettable.

The first tractor equivalents on farms were steam tractors. They depended on steam engines, were dangerous and liable to explode. They also used exposed belts to drive attachments, and these were responsible for a string of horrific accidents in which farmhands were injured. John Froelich built the first working petrol-powered tractor in Clayton County, Iowa. Only two of Froelich's tractors were sold. It was only after the Waterloo Gasoline Traction Engine Company worked on and improved Froelich's design in 1911 that the tractor began to sell.

In Britain the earliest tractors seem to have been oil-powered. The Hornsby-Ackroyd Patent Safety Oil Traction Engine was on sale for the first time in 1897. The first commercially successful design was the three-wheeler Ivel tractor built in 1902 by Dan Albone. Then Saundersons of Bedford developed a four-wheel design in 1908. That was very successful commercially and Saundersons became the biggest manufacturers of tractors outside America. It is hard now to see why tractors were so unpopular at first, but they certainly began to gain in popularity during the First World War. In Europe, the huge loss of horses during the war itself may help to account for the switch to tractors. Certainly by the 1920s, it was normal for farms to have a tractor with a petrol-powered internal combustion engine as basic equipment.

The tractor is a versatile machine that can be used for pulling or pushing a variety of farm trailers and machinery, and can be used for ploughing, harrowing, planting, spraying. The classic tractor design has two very large driving wheels on an axle below and behind a single central driving seat, and two small steerable wheels under the engine. This design has remained virtually unchanged in 100 years, with the single exception of the driving seat. For the first few decades, the seat was unenclosed. Farm workers driving tractors along slopes were very vulnerable if the tractor turned over. The commonest fatal accident on farms was caused by tractors toppling over and crushing their drivers. A protective cab, which also gives protection against the weather, is now standard on tractors.

THE INVENTION OF HENRY FORD'S CAR

(1893)

BY THE AGE of thirty, Henry Ford had worked his way to the top of the Detroit Edison Company. He was responsible for Detroit's electricity supply, but had a great deal of free time. He used it to experiment with a design for an automobile. By 1893 he had built himself and road-tested a gasoline buggy, a motor car powered by a petrol engine. Three years later, Ford produced an improved version, a tiller-steered Quadricycle or horseless carriage which caused a stir when he drove it through the streets of Detroit early one morning. Henry Ford was not only a first-rate mechanic: he had a flair for publicity and self-promotion.

In 1899, Ford joined the new Detroit Automobile Company as its chief engineer, but this promising beginning proved to be a false dawn. Only two years later, the company went bankrupt after selling a total of only five cars. Ford was then hired as an experimental engineer by the businessmen who had bought the company's assets. A car designed by Ford won a high-profile race, and then some of the old Detroit Automobile stockholders rallied round to form the Henry Ford Company, giving Ford one-sixth of the stock in the new business. This venture too proved to be a false dawn, and he dropped out after less than two years.

In 1903, the Ford Motor Company was founded, based on what Ford thought was a marketable car. Ford was allocated 225 shares for his car design and held the seventeen patents on its mechanism, and then production began in a converted wagon factory on Mack Avenue, Detroit. From this emerged Ford's first truly commercial vehicle, the Model A Ford. This had a two-cylinder, eight-horsepower, chain-driven engine. It weighed half a ton, was ninety-nine inches long, and it cost just 750 dollars. Ford was pounced on by the Licensed Auto Manufacturers, who claimed he was in breach of the 1895 patent on the petrol engine. Ford defended himself robustly, claiming his engine was different. In court he lost, but won on appeal in 1911.

Henry Ford's big success came with the Model T Ford in 1908. This car, which was a neat-looking convertible, soon outsold all its competitors. Nicknamed the flivver, it cost only 850 dollars, so that most middle-income Americans could afford to buy one. It had a wooden body on a steel frame. Customers could order a Model T Ford in 'any colour you want, so long as it's black.'

Ford's genius was in inventing something that worked, was fit for purpose, reliable and above all affordable. He knew his market and knew how to give value for money. There were lots of Americans who wanted cars. Already in 1908 car production in the USA was up at 63,500 a year. What the Ford car did was to put millions of people on the road and to mobilize the twentieth century, transforming people's lives in a thousand ways. It became easier to travel to work, easier to go on holiday, easier to keep in touch with friends and relations, easier to live out of town – and easier for towns to spread.

THE INVENTION OF WIRELESS

(1895)

THE INVENTION OF radio was dependent on two earlier inventions: the telegraph and the telephone. It is often the case that inventions feed upon one another and that more than one inventor is involved. With radio there is one name that stands out – the name Marconi – though Alexander Popov and Oliver Lodge were also radio pioneers. At the same time as Marconi, Oliver Lodge sent a signal similar to Morse code over a distance of half a mile. The difference between Lodge and Marconi was that Lodge saw the wireless signal as a scientific curiosity, whereas Marconi saw that the phenomenon of radio was potentially enormously useful, persevered with it, and went on refining it until it acquired commercial value.

Guglielmo Marconi studied physics at the Technical Institute of Livorno, where he became fascinated with the electromagnetic waves discovered by Heinrich Hertz. In 1894 he started experimenting with ways of converting them into electricity on his father's estate just outside Bologna. He used fairly crude equipment: an induction coil with a spark discharger and a simple filings coherer at the receiver. By the following year he was transmitting over short distances, successfully sending a radio message to his brother who was out of sight on the other side of a hill. At that moment, Marconi knew he had made a breakthrough.

After that, he tried to find ways to extend the range of transmission. He discovered that a vertical aerial increased the distance to over one mile, a major improvement. With reflectors fitted round the aerial to concentrate the signal into a beam, he was able to increase the distance even further.

Marconi was convinced he had stumbled on something of enormous importance, yet he received little encouragement in Italy. His mother's relatives persuaded him to take his invention to England, where it was more likely to be appreciated, and that proved to be the case. In February 1896 Marconi arrived in London, where he had a meeting with William Preece, the engineer-in-chief to the Post Office. Preece gave him all the encouragement he needed, and Marconi filed his first patent for wireless telegraphy in London in 1896.

Marconi gave several demonstrations of his wireless telegraph system, trying out kites and balloons to give his aerials greater height in order to extend the transmission distance. He succeeded in this way in sending radio signals four miles on Salisbury Plain and twice that distance across the Bristol Channel. The demonstrations and Preece's lectures promoting the technology attracted a lot of public interest in England and abroad.

At this point the Italians became interested. Marconi was able to erect a wireless station at La Spezia and a communication link was established with warships up to twelve miles away. A major development came in 1898 when he set up a station on the South Foreland in Kent to communicate with Wimereux in France, thirty miles away; he used this to send radio signals across the Channel. Two years later, with money from his cousin, Jameson Davis, Marconi was able to set up the Marconi Wireless Telegraph Company. The great advance of Marconi's spectacular invention was that it allowed practical communication between widely separated places without using any connecting wires – hence the name wireless.

Marconi's greatest breakthrough was still to come. Physicists thought the curvature of the Earth would put a natural limit on the distance radio waves could be transmitted. Probably 100 miles would be the limit. In December 1901 Marconi proved the physicists wrong. By bouncing radio waves off the upper layers of the atmosphere, he succeeded in sending signals in Morse code the full width of the Atlantic, from Poldhu in Cornwall to St Johns in Newfoundland. This success caused a sensation. The principle of international and even intercontinental radio transmission was firmly established.

By 1918, Marconi had developed his system to the point where he could send radio signals right round the world, from England to Australia. For his epoch-making invention, which was one of the handful of inventions that transformed the twentieth century, Marconi was awarded the Nobel Prize for physics, together with Karl Braun. Marconi's invention made instant worldwide communication possible. First it was used merely for messages, but later it was used for entertainment and eventually it spawned television. Without in any way intending to, Marconi launched the age of mass entertainment.

THE INVENTION OF MOTION PICTURES

(1895)

THE INVENTION OF still photographs in the early nineteenth century was followed just a few decades later by the invention of cinematography – motion pictures. The first known experiments with motion pictures were carried out by Eadweard Muybridge. He used a series of cameras to record people and animals running. To show the photographs in rapid succession and show the movement at natural speed was a problem. Muybridge invented the zoopraxiscope or wheel of life. The photographs were mounted round a wheel and viewed through a slit while the wheel was turned with a handle. One of the most surprising things to emerge from Muybridge's studies of locomotion was the fact that when a horse gallops there is usually at least one hoof on the ground; the eighteenth-century hunting prints that invariably show the 'flying gallop' with all four legs stretched out in the air fore and aft were inaccurate. It was the first but not the last time that film would enable us to clarify and explain the world around us.

Muybridge's films were little more than the equivalent of film clips, but they established the principle that many still images shown in rapid succession could create an illusion of continuous movement. It depended on a particular phenomenon of human vision, which is called persistence of vision. An image is retained in the visual memory for a fraction of a second; we go on seeing something for a moment after it has gone. Because of this momentary persistence we do not see the breaks between the frames of a cine film.

At the same time as Muybridge, Thomas Alva Edison was working on a system for producing moving pictures. With W. K. Laurie Dickson, Edison developed one of the first working systems, which was patented as the kinetoscopic camera in 1891. The camera took a series of photographs on a continuous band of celluloid. They could be viewed on a peepshow, just like Muybridge's, but Edison had also invented the light bulb which enabled him to project the image onto a screen. Dickson supervised the building of the first movie studio in the world, the Black Maria. The first Western was made there in 1899, *Cripple Creek Bar-Room*. The first really famous film was *The Great Train Robbery*, which was filmed in New Jersey in 1903; it was famous for its exciting fast-action effects.

Simultaneously in France, Louis Lumière was also working on a system for motion pictures. He invented a portable cine camera, film processing unit and projector, all in one machine which he named the cinematograph. Louis Lumière and his brother Auguste were the first to present moving photographic images to a paying audience, and many believe that they were really the first in the race to produce moving pictures. The first commercial presentation of an Edison motion picture in America took place on 20 May 1895 in New York; an audience in a store watched a four-minute film of a boxing match. The first theatre showing of a Lumière motion picture took place two months earlier, on 22 March 1895 at 44 Rue de Rennes in Paris. The Lumière brothers' presentations were sensational. Audiences ducked in terror as railway trains appeared to career (silently) towards them from the screen.

If Edison was just beaten by the Lumière brothers in producing silent motion pictures, he had sound to offer as well. Edison had independently invented the phonograph in 1877, and was well placed to turn the silent movie into talking pictures. Sound and moving pictures together made a very powerful new medium, one that would transform the twentieth century, and its view of itself.

From the beginning, motion pictures were an American phenomenon, and from the beginning they were extremely influential in shaping and reinforcing American beliefs, attitudes and fashions. As a result, motion pictures can be seen as a kind of unconscious propaganda. They continue not only to reflect but parade and reinforce the American way of life. Some of the early films failed to do this. Edwin S. Porter's 1903 five-minute film *Life of an American Fireman* tried to reflect life as it was really lived and showed the potential of film as a medium for information and edification, but it failed to inspire either America or Porter. Porter switched back to making mainly one-scene comedies and trick films – light entertainment. Later attempts at real life, such as the biographical films or biopics, were only able to make their way at the box office by fictionalizing and glamorizing the past, by imposing modern American values, and by having well-known stars in the leading roles: Yul Brynner as the Pharaoh Ramesses, George Sanders as King Solomon, Dirk Bogarde as Franz Liszt.

The release of the *Great Train Robbery*, also made by Edwin S. Porter in 1903, was a huge contrast to his *Life of an American Fireman*. It was profoundly influential. It marked the beginning of a century-long obsession with the formulaic Western and the start of a tradition of film making that gave America a distinctive and highly coloured view of its own past. The history of the motion picture has been in large measure an expensive and high-powered flight from reality.

THE INVENTION OF THE X-RAY MACHINE

(1895)

THE GERMAN PHYSICIST Wilhelm Roentgen discovered x-rays. It was on 8 November 1895 that he noticed an unusual phenomenon while experimenting with the cathode-ray ultra-vacuum tube recently invented by an English physicist, William Crookes. When a current was passed through the tube, a nearby piece of paper painted with barium platinocyanide fluoresced brightly. This phenomenon occurred even when Roentgen covered the tube with black cardboard and other materials. Roentgen realized he had discovered an invisible ray that could pass through some substances, though not others. 'Behind a book of 1,000 pages,' Roentgen said, 'I saw the fluorescent screen light up brightly.'

Roentgen did not know what the invisible rays were, so he called them x-rays. Undoubtedly the most significant application of Roentgen's discovery (or invention) was that the x-rays could be used to take photographs of the interiors of people's bodies. In December 1895, he took an x-ray photograph of the hand of one of his colleagues, a photograph that became one of the great icons of medical history. The bone structure showed up with exceptional clarity, as the rays could pass through skin and muscle very easily, bone not at all, so it was possible to look inside someone's body. Roentgen's photograph was a great landmark, and the application of William Crookes's cathode tube immediately provided doctors with a revolutionary diagnostic tool. It changed the quality and effectiveness of medical care throughout the twentieth century. It also improved security at airports by enabling officials to scan baggage for guns or bombs.

THE INVENTION OF THE ESCALATOR

(1897)

THE ESCALATOR IS a moving staircase to make it easier for people to move between the floors of a department store, or between the different levels in an underground railway station. The basic principle is a simple conveyor belt, a continuous loop of jointed metal steps moving on tracks. Escalators are typically installed in pairs, one going up, the other going down, though some European stores have only up escalators. Most escalators have handrails that move at the same speed as the steps; a problem with the earliest escalators was that they had fixed handrails, which made it difficult for people to steady themselves.

In 1892, Charles A. Wheeler patented his version of a moving staircase, though it was never built. Jesse W. Reno invented the escalator (again) in 1897 building it as an amusement ride at Coney Island in New York. Then he sold it to the Otis Elevator Company. Charles Seeberger developed his own version of the escalator before joining the Otis Elevator Company. In collaboration with his new employers Seeberger built the first commercial working escalator in 1899. For a time, Otis was selling both Reno and Seeberger escalators. It was the Otis Company who named the new moving staircase escalator (derived from the Latin word for 'ladder'). It won a first prize at the 1900 Paris Exposition. In about 1920 the best features of the Reno and Seeberger designs were amalgamated and the result was the modern escalator. The name 'escalator' originated as an Otis trademark, but it so successfully entered the language as a popular everyday word that it lost its (theoretically protected) trademark status in 1949.

The early Seeberger escalator had no comb effect to guide the rider's feet off the conveyor at the exit. Instead, the rider had to step off sideways. To ensure that riders did indeed step off sideways, the escalator disappeared underneath a triangular traffic island, which guided passengers to one side or the other. The first escalator to be installed on the London Underground, at Earls Court, was a Seeberger model like this. The Seeberger escalator is shown in action, though unfortunately without the triangular islands, in the 1916 Charlie Chaplin film *The Floorwalker*.

The spiral escalator invented in 1906 takes up less space, though the first model, built at the Holloway Road tube station in London in 1906, was dismantled almost immediately. In the 1980s, the spiral escalator was reinvented with far greater success.

The escalators installed in the London Underground system originally had wooden steps. After the catastrophic fire at King's Cross St Pancras tube station in 1987, the dangerous wooden steps were all replaced with metal. Some old escalators with wooden steps do still exist, such as the those in the Tyne Cyclist and Pedestrian Tunnel in Tyne & Wear in England and in Macy's department store in New York.

The longest freestanding escalator in the world stands in an atrium at the CNN Center at Atlanta in Georgia. It is 205 feet long and rises through eight storeys.

THE INVENTION OF
THE SUBMARINE

(1898)

IT IS SAID that the ancient Athenians used divers to clear the entrance to the harbour during the siege of Syracuse. Alexander used divers to destroy any submarine defences the defenders of the city of Tyre might have put in place there. These early attempts at underwater activity were undertaken in diving bells. It was not until 1580 that there was any attempt to build a craft that could move about or navigate under water. In that year, a British naval officer called William Bourne designed a fully enclosed boat that could be submerged and rowed under water. Bourne never actually built this boat, but a similar boat was built in 1605 by someone calling himself Magnus Pelagius. The first true working submarine was designed and built by a Dutch doctor, Cornelius van Drebel, who successfully navigated his craft in the River Thames up to fifteen feet below the surface.

By the early eighteenth century over a dozen different designs for submarines had been patented in England alone.

In 1775, a thirty-four-year-old Yale graduate, David Bushnell, designed and built the first military submarine. It was a one-man craft seven feet long. It was pear-shaped and made of oak staves held together with pitch and iron hoops, just like a big wine barrel. Bushnell's sub, which was called the *Connecticut Turtle*, had ballast tanks that were operated with foot pumps, as well as a conning tower with windows level with the operator's head. Fresh air was supplied through one hose and stale air was extracted through another, with automatic valves to close them during a dive. The vessel was propelled both vertically and horizontally by hand-cranked propellers and steered with a rudder.

The *Connecticut Turtle*'s purpose was to approach enemy ships while they were at anchor. It was to sidle up to them to plant explosives on their hulls, and then steal away unobserved. The submarine carried a powder magazine and a clock timer.

The *Connecticut Turtle* saw action, of a sort, too. She went into battle for the first time on the night of 6 September 1775 in New York Harbour, in a scene that might have come from the script of an Ealing comedy. The *Turtle* had an auger mounted on it, so that the operator could bore a hole into the hull of the enemy vessel and plant its powder magazine. A crucial point had been overlooked. Many of the British ships had their hulls sheathed in copper cladding to protect them against worms. Several attempts to bore through the copper sheathing proved fruitless.

This first experiment in submarine warfare failed, but it was an important first step. In 1864, in the American Civil War, another small hand-propelled submarine, the *Hunley*, carried six men into battle. The *Hunley* succeeded in sinking the *Husatonic*, a Federal corvette that had attempted to blockade the harbour at Charleston. The *Hunley* carried a torpedo suspended ahead of her as she rammed the corvette.

The modern submarine was invented by an Irish-American teacher, John P. Holland. Holland was born at Liscannor in Ireland in 1840, the son of a coastguard. He longed to go to sea, but his eyesight was too poor and he became a novice. Eventually he was released from his vows and emigrated, with his family, to America. Holland had had no formal education, but he taught himself engineering and technical drawing.

Holland was drawn to the idea of the submarine because of the history of conflict between Ireland and England. Holland knew that British sea power was too great for the Irish ever to equal it, but with submarines maybe the Irish could subvert it.

In 1875, Holland attempted to interest the US Navy in his submarine idea. One problem he had in being taken seriously was that submarines had been tried before. The US Navy remembered that the *Hunley* had been sunk in the encounter with the *Husatonic* in the Civil War. They remembered that the *Connecticut Turtle* had been a failure as a weapon in the American War of Independence. Submarines did not have a good track record, and in 1875 the US Navy dismissed Holland's idea as preposterous. As so often with rejected ideas, the idea had come from the wrong person. If it had come from a high-ranking naval officer it might have been taken seriously, but Holland was not a naval man at all. He was an Irish immigrant – and a teacher.

On the other hand, some Irish rebels called the Fenians were interested in Holland's submarine when he presented it to them. They were impressed enough to want to invest £60,000 from their skirmishing fund to enable him to build a prototype. When it was built, Holland and his Fenian backers gathered on the banks of the Passaic River in New Jersey for the launch.

The fourteen-foot vessel quickly filled with water and sank. It looked like a very unpropitious beginning for Holland's vessel. But when it was raised, Holland found that a workman had forgotten a couple of screws; this left an opening through which the water had poured. The fault was put right, the submarine was dried out and Holland took her out for her first controlled dive. She successfully dived and surfaced again.

As he worked on refining and improving his submarine, Holland also worked out the tactics for deploying it. His idea was to carry the submarine aboard a conventional ship until it was within striking distance of the British ships, then it would be released into the water through a door in the side of the ship. These plans never came to fruition. The Fenian group began to disintegrate. In 1883, one breakaway group went off with the submarine, intent on a joyride at New Haven, Connecticut, but they were unable to launch it successfully and abandoned it. When Holland heard what had happened to his submarine he was very angry and the collaboration with the Fenians came to an abrupt end.

Holland went on developing his design over a twenty-year period. In 1895, the US Navy finally commissioned one of Holland's submarines, the *Plunger*. This used electric motors under water and internal combustion engines while on the surface. Outstandingly, it used water ballast to make it submerge, and this turned out to be the most efficient way of making a submarine dive and surface. In 1897 another serious experiment was carried out. Simon Lake designed the *Argonaut*, and successfully tested her in open waters. In 1898, he successfully navigated her through November storms between Norfolk and New York – a spectacular vindication of submarine technology. A few years later, Lake sold a submarine called the *Protector* to the Russians, who liked it and ordered several more from him.

Holland's submarine design in particular was way ahead of its time. His submarine's cigar shape was sleek, elegant and far more efficient than the design of the first generation of military submarines that saw action in the First World War. In fact the cigar shape was only really rediscovered when the nuclear submarines were built, more than seventy years later.

The submarine took on a distinctive role in warfare at sea in the twentieth century, pioneered by Germany and making a real impact on the nature of warfare in both world wars. It had a major effect on the conduct of the First World War. The controversial sinking of the Cunard passenger liner *Lusitania* by a German submarine, with the consequent loss of life of American civilians, was partly responsible for bringing America into the First World War.

V

THE MODERN WORLD

THE INVENTION OF
THE SAFETY RAZOR

(1901)

THE MODERN SAFETY razor has two components: disposable blades and a guard to stop the blade from cutting the skin. As is so often the case with inventions, half of the device was already there. Jean Jacques Perret mounted a guard on a standard barber's razor as early as 1762. In appearance it was rather like a small rake mounted on the razor blade.

King Camp Gillette started his life in adversity. He was born in a small town in Wisconsin in 1855. His parents were innovators, always looking for ways of doing things better. His father was an inventor, while his mother devised many new recipes. They moved to Chicago, where his father opened a hardware business. In 1871, when he was sixteen, the Great Chicago Fire destroyed the family business and King Gillette left school to make a living as a travelling salesman. In 1887, when he was thirty-two, his mother published her *White House Cookbook*, which was a huge success and still in print a century later.

He met William Painter, the inventor of the Crown Cork bottle cap, and Painter advised him to invent a popular but disposable product – then people would not only buy it but keep on buying it over and over again. Gillette thought seriously about this good advice and turned over in his mind several possibilities. King Gillette felt that, by comparison with his parents, he was a failure. He had acquired four patents, but there was little commercial interest in any of them. By 1894 he was thirty-nine and becoming bitter. He wrote an anti-capitalist book called *The Human Drift*, which criticized the rich and their business practices. Competition was the root of all evil. He proposed a utopian, egalitarian, pollution-free hive society to replace the society created by the Industrial Revolution.

It was, ironically, immediately after publishing this anti-capitalist diatribe that Gillette made his great invention. He was shaving one morning in 1895, when he became irritated that his razor was too blunt to work properly. He called to his wife to tell her he had just thought of an invention that would make them very rich and change the way that men shaved all over the world. Having had the idea for the revolutionary razor in 1895, he then had to work on the blades. He went to the Massachusetts Institute of Technology to ask the metallurgists there about the possibility of manufacturing small thin pieces of steel to take a sharp edge. They said it was impossible. Eventually, after six years, and with the help of William Nickerson, who by chance was an MIT graduate, Gillette devised wafer-thin disposable razor blades of stamped steel.

In 1903 Gillette and his new partner Nickerson produced fifty-one razors and 168 blades. In 1904 they produced 90,000 razors and more than twelve million blades. To begin with people were wary of the new razor, but its popularity grew and grew. By 1910, when Gillette was fifty-five, he could regard himself as a success. But he had become the capitalist that he said in his book he despised; he dominated the razor business and he was a millionaire. He was also something of a celebrity as his face was printed on every packet of razor blades.

During the First World War, the US government gave the entire armed forces Gillette safety razors as standard issue. This had the effect of converting the entire country to Gillette's razors. The huge mobility that was brought about by the war also had the effect of introducing the Gillette safety razor to Europeans who might otherwise not have seen it. Within a short time, the safety razor was in use all over the world.

When asked about the secret of his huge success, Gillette said that he had refused to believe the experts who told him such a thin blade was impossible. He said, 'If I had been technically trained, I would have quit.'

THE INVENTION OF
THE AEROPLANE

(1903)

THE FANTASY OF flying through the air like birds had been in people's imaginations for hundreds of years before it became a reality. Many early attempts to fly ended in failure and death as people fixed feathers to their arms and legs and jumped off cliffs and high buildings. In the nineteenth century the theoretical groundwork for powered flight was laid by Sir George Cayley, a British philosopher and politician. He pioneered research on wing structures and the need for a lightweight power source. Among other things, he predicted that the power source would need to be the combustion of flammable fluids. Cayley got as far as building a small glider in 1804, and a larger but still unmanned glider in 1809. He persuaded a gullible schoolboy to climb aboard a third model, which successfully glided down from a hill top. It was a significant step towards flight.

Cayley was stuck for a source of power. The only available one at the time was the steam engine, obviously unsuitable for the purpose because they were so massive and heavy in themselves and required large quantities of heavy fuel and water. In 1848, William Henson proposed an aerial steam carriage with propellers. He built a scaled-down model of it which successfully left the ground, but had insufficient power to move horizontally under its own steam.

At the end of the nineteenth century Otto Lilienthal came very close to success. He built and experimented with a glider powered by a small petrol engine, but was unfortunately killed in 1896 when it crashed. He had lost control of the balance of the glider and the Wright brothers believed that his attempt to maintain equilibrium just by shifting his body weight was inadequate. Two years earlier, Hiram Maxim built a steam-powered biplane powered by two engines and two propellers. This left the ground but remained tethered. It is not clear why Maxim gave up this promising experiment. It was left to the Wright brothers to make the first recognized powered flight.

The Wright brothers developed the theory that the air pressure exerted on different parts of the machine could be altered by making the wings adjustable, and that this would maintain equilibrium. This system, now known as aileron control, is in use today on all modern aircraft. The Wright brothers took out a patent on it. They conducted workshop experiments using a wind tunnel, to test their aileron principle.

Starting in 1902, the brothers developed a full-sized, power-driven heavier-than-air machine. Though quite large, the plane weighed only 750 pounds. It was powered by a four-cylinder petrol-fuelled motorcycle engine; it had an engine block made of cast aluminium to give it a high strength-to-weight ratio. It was piloted by Orville Wright on its maiden flight on 17 December 1903 at Kitty Hawk in North Carolina. On that historic day, the first aeroplane made four sustained free flights, the longest lasting fifty-nine seconds at an altitude of fifteen feet and a speed of 30 mph. Several newspaper men witnessed the flights, but for some reason it was not considered very newsworthy. Only three newspapers reported it.

Other people in France and Germany were working on the problem of powered flight at the same time. The German aviation pioneer Karl Jatho, claimed that he had made a flight on 5 August, over four months before the Wright brothers' flight. Jatho flew a petrol-engined biplane he had built in 1899, and went on to set up an aircraft factory at Hanover in 1913. Inventions are quite frequently made independently in two places at the same time, and the invention of the aeroplane is one of them.

The Wright brothers carried on with their experiments. In 1905, they learned how to prevent the tail-spin that had made short turns a problem. Then their flights became longer and more ambitious, and in the September of 1905 Wilbur flew the plane in a circle over a distance of twenty-four miles. The continuing improvements and tests convinced the US government in 1909 that this was a practical and reliable aircraft. The aeroplane had won official acceptance. In 1908 and 1909, the Wright brothers flew their plane at many demonstrations in Europe.

The flights, including one at Rome, attracted huge crowds.

The demonstrations reinforced the validity of their claim to have made the first powered flight in 1903. Powered flight was a reality. Within ten years there were planes that could be used for reconnaissance in a war zone, and even as fighting platforms. The German air ace Baron Manfred von Richthofen led the 'flying circus' that brought down hundreds of Allied warplanes in the First World War. The Red Baron was himself shot down in April 1918.

Already by 1910 aviators were taking planes much higher in the sky. The French aviator Louis Paullan flew his plane to over 3,700 feet over Los Angeles. In 1919 the first flying boat was built. This was the NC-4, designed by Jerome Hunsaker, and it was used to make the first transatlantic crossing by air, leaving Newfoundland on 16 May 1919 and arriving at Lisbon eleven days later; the pilot was Albert Read, leading a five-man crew. The first non-stop transatlantic flight was achieved by John Alcock and Arthur Whitten-Brown, who flew from west to east in sixteen hours in 1919.

Aviation developed incredibly fast, with longer and longer flights and larger and larger planes. By 1924 the first flight round the world had been achieved; two World Cruisers made it in only fifteen flying days. Aviation shortened distances between places, made it easier for people to travel from country to country and from continent to continent. The development of very large passenger airliners eventually brought relatively cheap air travel within the reach of millions of people.

THE INVENTION OF ANIMATED CARTOON FILM

(1906)

AN ANIMATED CARTOON consists of drawn images shown in rapid succession to give an impression of continuous movement. The principle of the animated cartoon, which depends on the optical quirk of persistence of vision, was developed several decades before the advent of the motion picture. One of the earliest commercially successful devices was the phenakistoscope, invented in 1832 by the Belgian Joseph Plateau. This consisted of a sequence of images on a cardboard disc that gave an illusion of movement when spun and looked at through a viewing mirror. In 1834 William Horner invented the zoetrope, which was a drum lined with a replaceable strip of cardboard with images printed on it. Again, when the drum was spun, the viewer saw the images moving.

In 1876 this parlour entertainment was adapted by Emile Reynaud to make it into a form that could entertain an audience in a theatre. Reynaud made beautiful hand-painted strips of celluloid, and the images from these were sent by a system of mirrors to a screen. For the first time cartoons showed animated characters that had some warmth of personality, fore-shadowing the great age of Disney in the following century.

The development of cine film, with sprocket-driven strips of celluloid, made what we now recognize as animated cartoon films possible. This was a process of evolution, but there is a case for regarding J. Stuart Blackton as the inventor of the first film-based animated cartoon. He produced *Humorous Phases of Funny Faces* in 1906, which was the first of a sequence of successful animated films. Also in 1906 Blackton experimented with the stop-motion technique, in which objects are photographed, then moved and photographed again, frame by frame. He used this in his film *Haunted Hotel*. It was the same technique that was used later by Ray Harryhausen for his special effects monsters. Parallel developments were taking place in France, where Emile Cohl was developing a style of cartoon animation using stick figures, which were far less sophisticated than Blackton's more fully drawn newspaper-style cartoons.

A great pioneer in these early days of the cartoon was Winsor McCay, who produced some elegant and imaginative films, including *Little Nemo in Slumberland* and *Dream of the Rarebit Fiend*. McCay made a hand-coloured print of Little Nemo to use in his vaudeville act in 1911. His Gertie the Dinosaur, made for the same purpose in 1914, broke new ground in setting high standards of drawing and creating more fluid and realistic movement; it also gave the animated creature an illusion of personality and a life of its own. McCay was a remarkable man. Among his other ambitious early film projects was *The Sinking of the Lusitania* in 1918.

Pat Sullivan was an Australian-born American, who opened a studio in New York. There he exploited the talent of a young animator called Otto Messmer and made one of Messmer's cartoon characters, a cunning black cat called Felix, into a major star. A series of one-reel cartoons featuring Felix the Cat was produced. The young Walt Disney watched the way Messmer drew Felix, and learned from it. Disney, working at the Laugh-o-Gram Films studio in Kansas City, borrowed the construction of Felix the Cat for his own Oswald the Lucky Rabbit. Disney would have gone on with Oswald, but for a dispute over the rights to the character; he responded by changing Oswald's ears – to make him into a mouse.

Walt Disney's public life began at the age of seventeen, when he became an ambulance driver for the Red Cross with the American troops in France in 1918. After the First World War ended he worked as a commercial artist for a couple of years before becoming a cartoonist for the Kansas City Film Advertising from 1920 to 1922.

Disney experimented with a number of animated cartoon films in Hollywood, creating the *Alice* comedies. He created and produced the first *Oswald the Rabbit* cartoons in 1926. Disney created his first successful sound picture, *Steamboat Willie*, in 1928. We take for granted that films have sound, but until then all the cartoon films had been silent. Synchronized sound made a huge difference, greatly enhancing the illusion of reality. *Steamboat Willie* took America by storm. *Steamboat Willie* featured Mickey Mouse, probably the best-loved cartoon character of all time, and from then on Mickey Mouse became a firm favourite with audiences. Disney was helped enormously by his childhood friend Ub Iwerks. Iwerks followed Disney into the film industry and became his Merlin figure; it was Iwerks who created the multiplane camera and the synchronization techniques that made the *Mickey Mouse* cartoons and the *Silly Symphonies* appear so strong and three-dimensional. Disney was working towards as complete a naturalism as he could create.

In the short *Silly Symphonies* colour cartoons Disney introduced Donald Duck, Pluto and other cartoon characters that quickly became very popular. In fact these colour cartoons were the most popular entertainment on the screen.

Mickey Mouse became something more, too. Disney had a strong commercial sense, and Mickey Mouse was the springboard for a huge and lucrative merchandizing empire.

Disney broke new ground with his full-length colour cartoon films. They required enormous numbers of hand-drawn images, and this work was shared among a team of dedicated artists. Disney was a hard task-master and demanded incredibly high standards from his staff. The first of these very ambitious projects was *Snow White and the Seven Dwarfs*, which was first shown in 1937. The film was – and still is – very popular. The reception of this film encouraged Disney to go on to make more feature-length cartoon films: *Pinocchio, Dumbo, Bambi, Lady and the Tramp* and *Sleeping Beauty*. All became twentieth-century classics.

Fantasia was a different experiment Disney took an orchestral score, conducted by Leopold Stokowski, and made an animated cartoon film to synchronize with the music. Many of the effects were surreal, and as a piece of craftsmanship it is an outstanding piece of work, but *Fantasia* lacks a story line and has never had the same popular following as *Snow White*. After the Second World War, Disney resumed full-length cartoon films, with two more huge successes: *Cinderella* in 1950 and *Alice in Wonderland* in 1953.

With *Steamboat Willie*, Walt Disney launched a minor art form, the black and white short cartoon film, and developed it into a major art form that became large-scale mass entertainment. It was a revolution in popular entertainment. He introduced a range of cartoon characters that have become icons of twentieth-century popular culture, but there is no question that the irrepressibly good-natured and optimistic Mickey Mouse was the central and most popular of them all. In his various transformations, whether on film, on car stickers, or in person at Disneyland, Mickey Mouse became a celebrity.

Disney's huge success in popularizing animated cartoons spurred many other companies to produce them, generating what amounted to a cartoon culture. Several large studios tried their hand at cartoon series. MGM produced *Barney Bear*, Paramount *Popeye*, Universal *Woody Woodpecker*, Warner Bros *Looney Tunes*, and Hanna-Barbera produced *Scooby-Doo* and *The Flintstones*.

THE INVENTION OF THE VACUUM CLEANER

(1907)

A VACUUM CLEANER was patented as early as 1869 by Ives McGaffey. McGaffey invented his wood and canvas device, which he called the Whirlwind, in a basement in Chicago and described it as a sweeping machine. It was for cleaning rugs and it was the first hand-pumped vacuum cleaner to be built.

In 1901, a British engineer called Hubert Cecil Booth took out a patent on a more ambitious vacuum cleaner. This took the form of a big, horse-drawn, petrol-powered unit that had to be parked outside the building to be cleaned. Long hoses were led from the machine through the windows. Booth demonstrated his invention at a restaurant. At about the same time, an American, John Thurman, offered a similar vacuum cleaning service in St Louis. He charged four dollars a visit, and turned up at people's homes or business premises with his petrol-powered vacuum cleaner on a horse-drawn cart. He invented and patented this machine in 1899.

The domestic vacuum cleaner as we understand it today was invented by James Murray Spangler in 1907. Spangler was one of the caretakers and cleaners at a department store in Canton, Ohio. He suffered from a cough, and deduced that he was allergic to the dust from the carpet sweeper he was using to clean the store. He experimented with an old fan motor, attaching it to a soap box stapled to a broom handle. He improvised a dust collecting bag by using a pillow case. With these simple materials and an inspired vision, James Spangler created a portable electric vacuum cleaner. Satisfied that it worked, he set about improving it. He fitted a cloth filter bag and cleaning attachments.

Spangler applied for and was given a patent in 1908. Then he set up the Electric Suction Company. Among his first customers was one of his cousins. Her husband, William H. Hoover, became a major backer for Spangler's company, and later the company's president. Hoover himself made some alterations to the design. To begin with, sales of the vacuum cleaner were slow, but Hoover offered hesitant potential customers a ten-day free trial. In the course of time, there was a Hoover vacuum cleaner in nearly every home in America.

In time there were many competitors, but the basic invention, a domestic vacuum cleaner, has transformed homes, offices, shops and other commercial premises. Dust levels have been reduced, and hygiene levels have been raised.

THE INVENTION OF
THE HELICOPTER

(1907)

WHILE LEONARDO DA Vinci did not properly invent the helicopter, he certainly had a conception of it. But he was not alone in having this conception. Both in Europe and in China in the sixteenth century children's toys were made that look very much like helicopters. The idea seems to have been a collective fantasy, like the flying saucers of the mid-twentieth century.

Several inventors tried to build working helicopters. The main problem was not the principle but the practicality of finding enough power to make the blades whirl round fast enough to lift the machine off the ground. In 1907, Paul Cornu designed and built a helicopter that did indeed succeed in lifting itself off the ground. In 1923, Juan de la Cierva, a Spanish engineer, successfully flew a machine called an auto-giro. But it was not until the 1930s that a practical working helicopter was built – by Igor Sikorsky. It was his name that became synonymous with helicopters, not the name of the original inventor, Cornu.

Sikorsky had been interested in Leonardo's drawings of flying machines, and was motivated to take an aeronautical career. From early on he imagined what the design of a practical helicopter might be like. In Paris he bought a twenty-five horsepower engine to power a helicopter with a single blade, but the engine was too weak to get the craft off the ground. Sikorsky put the helicopter to one side while he designed a series of fixed-wing aircraft. His career was interrupted when, because of his association with the tsar, he had to leave Russia for France. There he was commissioned to design a bomber for deployment in the First World War, but peace was declared and the French government cancelled his contract. In 1919 he left France for the

USA. The US government was ready to invest on a grand scale in Sikorsky's helicopter design. It was finally flown in 1939, powered by a seventy-five horsepower engine. The new machine handled well and could be manoeuvred easily in any direction.

With this design, Sikorsky's timing was impeccable. As the Second World War broke out, there was a huge demand for a variety of helicopters, to use as sky cranes, troop carriers and gun ships. Sikorsky himself was pleased with the helicopter's ability to save lives. It could land in small spaces. It could hover over the sea and haul people up from sinking ships.

THE INVENTION OF CELLOPHANE

(1908)

CELLOPHANE WAS INVENTED in 1908 by Jacques Brandenburger, a Swiss textile engineer. Brandenburger first thought of the idea of a clear protective layer when he was sitting at a table in a restaurant in 1900. Another customer spilt wine on the tablecloth, which had to be replaced. As he watched this incident, Brandenburger reflected that it would be useful if a clear, flexible, waterproof membrane existed to cover cloth, to protect it from spilt liquids.

He conducted experiments with a range of materials. He tried applying liquid viscose (or rayon) to cloth, but the viscose made the cloth too stiff. The experiment failed, but he noticed that the viscose coating peeled off in a transparent film. He realized that this might have other applications. By 1908, he had developed a machine to make transparent sheets of cellulose. The manufacturing process involves extruding an alkaline solution of cellulose fibres (usually originating as wood or cotton) through a slit into a bath of acid. The acid regenerates the cellulose to form a film. Further treatments, including washing, lead to the production of finished cellulose.

In 1912, Brandenburger was making thin flexible transparent film that he could sell to fit into gas masks. He took the precaution of getting his machinery and basic ideas patented. In 1917, he assigned his patents to a company cryptically calling itself La Cellophane Societé Anonyme. In 1923, an agreement was made between La Cellophane and Du Pont Cellophane Company: Du Pont was to be licensed to manufacture and sell in North and Central America, using La Cellophane's secret cellophane-making process. La Cellophane meanwhile retained exclusive rights for the rest of the world.

Cellophane as a product was far from perfect, though. It was not completely impermeable. It was only the research work done by the Du Pont scientist William Charch and his team that led to the production of a truly moisture-proof cellophane membrane. That invention or improvement was patented in 1927. This meant that cellophane was now ready to be used in food packaging. It was ideally suited to this use. Completely transparent and flexible, it made a perfect wrapping for food on display in shops. Customers could see exactly what they were buying, while the food itself remained completely enclosed. It meant far higher standards of hygiene in food shops.

THE INVENTION OF BAKELITE

(1909)

THE BELGIAN-AMERICAN chemist Leo Baekeland created the first plastic in 1909. The synthetic shellac polymer that Baekeland invented was made out of phenol and formaldehyde. He named his new material after himself – Bakelite – and set up a company to market the moulding powder for making Bakelite products.

Bakelite found an immediate use for the fittings on the new electrical appliances of various kinds. It turned out to be a very good material for electrical insulation, so it was very useful for making collars to protect electric light fittings. It is still used in this way.

Other uses followed. Once the domestic wireless (radio) was marketed, it seemed natural to make the case in moulded Bakelite, which insulated the electrical circuitry inside; it could also be moulded into any shape, and by the 1930s these designs became more adventurous. The most popular colour for Bakelite products was dark brown, presumably so that it looked like polished mahogany, but it could be produced in almost any colour, and children's toys were made in white, red and green. Bakelite became a substance that was seen in every Western home. It found hundreds of uses.

Bakelite was important in being the first of a whole range of different plastics created for different purposes in the twentieth century. Bakelite was very useful in its way, but it was hard, rigid and brittle. New plastics were developed over the succeeding decades with different properties. The relative softness of alkathene made it suitable for making washing-up bowls; the long chains of molecules in polyethylene made it eminently suitable for making artificial fibres.

THE INVENTION OF THE GEIGER COUNTER

(1911)

HANS GEIGER WAS born in Germany in 1882. He studied physics at Munich and Erlangen before moving to Manchester to work with the great English physicist Ernest Rutherford. He worked on topics relating to radioactivity and while doing this in 1911 he invented a device for counting the number of alpha particles and other ionizing radiation being emitted by a substance. This was the prototype of the Geiger counter.

The Geiger counter used a sealed, gas-filled metal tube acting as an electrode. A wire or needle running through the middle of the tube acted as a second electrode. Voltage was applied to the device so that a current could almost pass through the gas from one electrode to the other. When the counter was brought near a radioactive substance, the gas became ionized and the ionized particles were able to complete the electrical circuit from one end to the other. When this happened, each passing particle was made, by an electronic mechanism, to produce an audible click.

In 1914, Geiger returned to Germany to serve as an artillery officer in the German army. After the First World War, he went back to research and directed research laboratories at Kiel, Tübingen and Berlin. At Kiel, Geiger made improvements to his Geiger counter in collaboration with Walther Müller. The counter was made more sensitive and durable. The improved model is usually called the Geiger-Müller counter. In the age of nuclear arsenals, nuclear bomb testing and nuclear power stations, a portable instrument that could measure radiation levels so easily and effectively (and audibly) became very useful indeed.

THE INVENTION OF THE GEOLOGICAL TIMESCALE

(1913)

IT WAS THE great British physical geographer Arthur Holmes (1890–1965) who invented the geological timescale in 1913. A geological timescale is a diagram displaying the chronology of significant events in the geological history of the Earth, including dominant life forms and major upheavals such as mountain building episodes and mass extinctions. Holmes was only twenty-three when he proposed his timescale. Not only was he the first to propose the innovative form of presentation: he was the first to realize that the Earth was billions of years old (1.6 billion, he thought) rather than millions as most scientists had thought until then.

The timescale started with the Palaeozoic or Primary Era, the first era of life when fish were the highest life form, and it contained within it a sequence of geological periods, the Cambrian, Ordovician, Silurian, Devonian, Carboniferous and Permian. Then came the Mesozoic or Secondary Era, the middle era of life when reptiles were the highest life form; it contained the Triassic, Jurassic and Cretaceous periods. Then followed the Cenozoic or Tertiary Era, the recent era of life in which mammals were the highest life form; it contained the Palaeogene and Neogene periods. The last two million years, a very short episode in relation to the rest of the timescale, was the Quaternary Era, when the human race was the highest life form. That comprised the Pleistocene period (the Ice Age) and the Holocene period (the 10,000 years since the Ice Age supposedly ended).

One of the best things about the timescale was that it lent itself to presentation in diagram form. Posters based upon it are displayed on the walls of geography classrooms and geology departments all round the world. It continues to make a very useful framework for thinking about processes that take place across unimaginably long periods of time. It is also flexible, lending itself to endless adaptation and addition. Holmes dated the beginning of the Cambrian period at 600 million years ago. Current thinking is that this should be adjusted to 590 million, which means that Holmes was very close.

As more became known about the long aeons before the opening of the Cambrian period around 600 million years ago, the timescale could easily be extended to accommodate the new knowledge. Holmes's timescale did not start with the Creation, but with the Cambrian, because it marked the appearance of abundant fossils, which have always been the most useful means of identifying the relative ages of rocks.

The timescale was included in Holmes's book *The Age of the Earth*, which marked him as the world's leading authority on the subject. There were still many geologists who resisted his ideas, and clung to the belief that the Earth was 100 million years old, some a mere twenty million. In 1922, Holmes returned from Burma to find that the scientific establishment had swung in his favour, accepting the long timescale he had proposed. A new debate was raging about continental drift. This new idea had been put forward by a German meteorologist, Alfred Wegener, in the intervening years, and upset the establishment all over again; obviously it was nonsense, and the continents could not possibly move around. There was no force powerful enough.

Holmes once again flouted the geological establishment and taught his students that continental drift was a perfectly valid view of Earth history. Eventually he came up with a mechanism to explain how continents could be shunted around. His hypothesis was that convection currents in the Earth's mantle, driven by the heat of the core, were adequate to explain the doming effect observed at mid-ocean ridges, the sideways movement of continents, and the ocean trenches. Once the theory of continental drift was accepted by most geologists in the 1960s, Holmes's theory of convection currents acquired the status of proven fact. It does still remain only a working hypothesis, but it is the only one available. At the end of his life, Holmes with typical modesty pointed out that 'mantle currents are no longer regarded as inadmissible.'

During the Second World War, Arthur Holmes wrote his famous textbook, *Principles of Physical Geology*, which was published in 1944. It was a profoundly influential book. Then Holmes completed his most important work on the age of the Earth, the geology of Africa and the refinement of the geological timescale that he had launched in 1913. Holmes had ideas that were far ahead of his time and as a result was always considered by the establishment to be a maverick. But time has shown him to be nearly right in most of his ideas, compared with any of his contemporaries. He was described in 1932 as 'one of the few English geologists with ideas on the grand scale'. Though relatively un-recognized and certainly not honoured in his own lifetime, Arthur Holmes was in fact one of the most important geoscientists of the twentieth century.

THE INVENTION OF THE CROSSWORD PUZZLE

(1913)

THE CROSSWORD PUZZLE was invented by Arthur Wynne in 1913. Wynne was born in Liverpool and became a journalist. He devised a weekly puzzle for an American newspaper called the *New York World*. The very first crossword puzzle Wynne – or anyone – devised was a diamond-shaped puzzle published in the *Sunday New York World* on 21 December 1913. The first crossword puzzle appeared in Britain nine years later, published in *Pearson's Magazine* in February 1922.

The key feature of the crossword puzzle was finding not only words of the right length, but with the aid of clues exactly the right words, to fit the spaces available in a grid. Newspapers began publishing daily puzzles which became part of the daily recreational routine of tens of thousands of people. Many commuters solved the puzzles on their way to work in the morning. The scope of the crossword puzzle was enormous. It was possible to devise easy puzzles with straightforward clues, or difficult ones with extremely cryptic clues. The grids could be large or small. Newspapers developed the practice of publishing extra large puzzles on public holidays.

The enormous and enduring popularity of the crossword led on to the invention of other word-grid puzzles. Probably the most successful of these has been Scrabble. This word game for two or more people was invented by Alfred Mosher Butts in 1948. Alfred Butts started out as an architect, but lost his job in 1931 as a result of the depression. After that he started to develop games, including Lexico and Criss-Crosswords. Scrabble was his major success. Some revisions to the layout of the board were made by James Brunot, who also simplified the rules of the game, which became and has remained very popular.

In 1979 a major new grid puzzle appeared. This was Sudoku. The origins of Sudoku are more complicated. It seems to have had its origins in Latin squares, which can be found in medieval Arabic literature. They were seen by the eighteenth-century Basel-born mathematician Leonhard Euler, who recognized the puzzle as a type of magic square; it was Euler who named these Arabic squares 'Latin squares'. Latin squares are grids that are filled with numbers, letters or symbols in such a way that no number or letter appears more than once in each row or column. A four by four Latin square has the numbers one to four in various permutations in each row and column. A puzzle of this size is very easy – too easy – with only a small number of ways of filling it in, but in a square nine by nine there are many more. Nine by nine is the grid size for a Sudoku puzzle.

Puzzles of this kind are mathematically interesting and entertaining to do, but they can have practical applications, for example in school time-tabling or organizing sports fixtures.

The Sudoku is a special type of Latin square, a nine by nine grid split into nine boxes three by three. The aim of the game is to fill every cell with a number from one to nine, so that each number occurs once in each column, row and three by three box. As a constraint, a few numbers are given.

The Sudoku was invented by Howard Garns in 1979. The very first Sudoku appeared that year in a magazine called *Dell pencil puzzles and word games*, and it made its first appearance under the name Number Place. It became popular in Japan in the 1980s, and then it was picked up as a regular puzzle by *The Times* of London in 2004. The replacement name Sudoku means 'single number' in Japanese and has become the registered trademark of a Japanese company specializing in publishing puzzles.

It is surprising that a relatively small grid can generate so many puzzles. Two mathematicians, Bertram Felgenhauer and Frazer Jarvis, have calculated that 6,670,903,752 trillion different Sudoku puzzles are possible. Sudoku is an almost addictive puzzle, and it is very good for stretching logical thought. No mathematical skill is needed, only a systematic approach.

THE INVENTION OF THE ECHO-SOUNDER

(1914)

THE FIRST FUNCTIONING echo-sounder was invented by the Reginald Aubrey Fessenden in 1914.

When the *Titanic* sank after colliding with an iceberg in 1912, Reginald Fessenden, a Canadian engineer, came up with a proposal to use sound waves to scan the sea ahead of a ship. Sound travels at about 4,500 feet per second in seawater, so it would only take one second to detect an iceberg 2,250 feet ahead. During that time, the *Titanic* would only have travelled about forty-five feet. In theory, then, a sound wave scanner could locate an iceberg early enough for avoiding action to be taken. In practice, experiments with sonic iceberg searching in 1914 revealed that there were problems. The properties of seawater changed with depth, and sound followed curved paths that were more complex than the inventor had believed.

Undeterred by these 1914 trials, which appeared to be failing, Reginald Fessenden trained his sound-making device downwards instead of forwards. He sent out short pulses of sound and heard regular echoes returning after each one. From this chance experiment, the technology of echo-sounding was born. Fessenden called his echo-sounder a fathometer. The time-lag between the fathometer sending the sound pulse and hearing the echo could be converted by calculation into a depth measurement. As a ship passed across the ocean, a continuous log of depth measurements could be acquired. By criss-crossing over a pre-arranged grid of the ocean a continuous three-dimensional map of the ocean floor could be built up.

Sonic sounders became available in the 1920s. In 1922, the USS *Stewart* was fitted with a Hayes echo sounder, designed by Harvey

Hayes of the US Navy. She made a major survey of the North Atlantic Ocean. In 1923 the US Coast and Geodetic Survey Ship Guide was also fitted with a Hayes echo sounder and began a survey of the North Pacific. Along the way, the Guide compared old-fashioned wire-line soundings with the new acoustic soundings in depths of 600 to 27,000 feet. This made it possible to determine accurate values for the velocity of sound in seawater, which made future acoustic soundings more reliable. More and more survey ships were fitted with echo-sounders.

Suddenly, with these new echo sounders, highly accurate maps of the seabed became possible. Entire submerged landscapes came into view, and in great detail. With this new information came new insights into Earth processes.

THE INVENTION OF THE TANK

(1915)

THE IDEA OF the tank had its beginnings in the Crimean War. John Edgeworth developed a steam-powered tractor with caterpillar tracks that was able to cross a muddy field without losing traction. The invention of the internal combustion engine later in the nineteenth century brought the invention of the tank closer. In 1899 Frederick Simms designed a motor-war car, which had a powerful engine, bullet-proof plating and two revolving machine guns. Simms offered his invention to the British government, who did not think very much of it. The times were not right for the tank; it took the horrific emergency of the Western Front to make the British government think again.

When the First World War broke out in 1914, it very quickly developed into a defensive war. On the Western Front, battalions of infantrymen were ordered out of their trenches to attack enemy lines, only to be massacred by machine gun fire. The casualty figures were extremely high and the front moved very little. A sense of stalemate set in.

The constant shelling turned the Flanders landscape into a pitted quagmire, impassable on foot or in vehicles, and the complex networks of trenches on both sides of the front added to the difficulty.

As early as 1914, a middle-ranking British army officer, Colonel Ernest Swinton, was sent to the Western Front to report on the war. He saw even in the early battles that German machine gunners were able to mow down any infantry attack with machine gun fire. He also noticed that the only vehicles that could negotiate the ravaged terrain were caterpillar tractors. He wrote back that 'petrol tractors on the caterpillar principle and armoured with hardened steel plates' were the best answer to the problem. Both Swinton and Colonel Maurice

Hankey came to the realization that a vehicle with caterpillar treads and an armoured superstructure was just what was needed to give personnel protection against machine gun fire while getting them across to the enemy trenches. It would be an entirely new type of vehicle, leading to an entirely new kind of warfare.

Swinton's idea was rejected by Sir John French and his technical advisers. Swinton did not give up, but tried another route. He discussed his proposal with Maurice Hankey, who in turn passed the idea to Winston Churchill, who was First Lord of the Admiralty. Churchill thought the idea was promising and initiated a committee, the Landships Committee, to look into it and develop it further.

The Landships Committee and the newly created Inventions Committee were both enthusiastic about Swinton's proposal and drew up specifications for the vehicle. It was to have a top speed of four miles per hour on level ground and the ability to turn sharply at that speed. It was to be able to reverse. The vehicle was to have features that no pre-existing vehicle had: the ability to climb an earth parapet five feet high and the ability to cross a gap eight feet wide (i.e. a trench). It was to have enough room inside to house a crew of ten men, and it was to have a two-pound gun mounted in it.

The idea had such major strategic potential that the men who worked on the project were not told what the vehicle was really for. They were told by Swinton they were building water carriers, mobile water tanks, in fact, so they started referring to the vehicles as 'tanks'. The nickname stuck and now everyone still calls them tanks.

The commission to manufacture a small prototype landship was given to Lieutenant W. G. Wilson of the Naval Air Service and William Tritton of William Foster & Co. (Lincoln). This prototype landship was demonstrated in the presence of Ernest Swinton and the Landship Committee on 11 September 1915. The vehicle was called *Little Willie*. It looked like an upturned water tank mounted on a tractor, and it looked unpromisingly primitive. It was fitted with a Daimler engine, had caterpillar track frames twelve feet long, and was able to carry its three-man crew at a speed of three miles per hour. This fell to two miles per hour when the terrain became rough, which was disappointing. More alarming still was the fact that the vehicle turned out to be unable to cross a wide trench.

The trial was disappointing, but *Little Willie* was significantly smaller than the committee had specified, and a larger machine might well cope better with crossing a trench. Swinton remained convinced that the idea would work and that with modifications the landship would enable the Allied forces to defeat Germany.

The tanks were not used in battle until the end of 1916, at the Battle of Flers. The new technology inevitably had teething problems. Many of the first batch broke down before reaching the enemy lines, though a few did get past. They were even less successful at the Battle of the Somme. They were unreliable, and they were hot and cramped inside. Their considerable weight – even *Little Willie* was heavier than an elephant – caused them to get bogged down fairly easily; then they had to be rescued by other tanks. Even so, the signs were that the tank could play an important role in warfare, and the army ordered more to be built.

On 20 November 1917 at the Battle of Cambrai, tanks were for the first time used effectively in battle. On that day, the tank corps successfully broke through enemy lines along a twelve-mile stretch of the German front. As a result of this breakthrough by the tanks, 10,000 German soldiers and 123 artillery guns were captured. Later the Germans regained their position, but the tanks' success restored confidence in the new technology. Tanks were deployed on a grand scale on 8 August 1918, when 604 Allied tanks reinforced a twenty-mile advance.

As the First World War ended, the British had 2,636 tanks, the French 3,870. For some reason the Germans remained unconvinced of the tank's usefulness and manufactured only twenty.

The tank played a significant role in breaking the terrible, destructive stalemate of trench warfare, and introducing more mobility to the Western Front. It transformed the nature of early twentieth-century warfare, much as the chariot had done back in the Bronze Age.

THE INVENTION OF CAMOUFLAGE

(1917)

THE CONCEPT OF camouflage was well known from the animal world. A great many mammals, reptiles and insects hide themselves very effectively by adopting the colours, textures and shapes of their habitats. Lions are able to merge with their savanna environment by being sandy in colour. Stick insects are unnoticeable because they look like twigs.

The camouflage technique has obvious applications in various human activities, especially in warfare. In the eighteenth century, battles were fought out in the open, by tacit or actual agreement, and conducted almost like board games. In that context, soldiers could be dressed in striking uniforms that would establish which side they were on, rather like the strips of football teams. The British Army was renowned for its distant visibility as 'the thin red line'. The advent of guerrilla warfare, with its use of snipers, ambushes and surprise attacks, meant that soldiers needed to disappear into the landscape. From that point on, battledress (regardless of the uniforms of the parade ground) developed in various shades of grey, buff and green.

The most spectacular attempt at camouflage was dazzle or razzle-dazzle camouflage. This was a colour scheme applied to ships in the First World War. It consisted of a complicated pattern of geometric shapes in different tones, often intersecting one another. Entire battleships were painted in this way from stem to stern. The most outstanding feature of dazzle is how eye-catching it was. It seems to draw attention rather than hide. But the advantage was that in a convoy or fleet, seen at a distance, it would be virtually impossible to identify how many vessels there were, or of what type. Because the shapes cut across all the features of the ship regardless – bridge, funnel, ventilators, guns – the standard recognition features became impossible to pick out.

Dazzle also made it difficult for the enemy to pick up the ship's range, and speed, and the direction in which it was steaming. That meant that it was harder to aim guns and fire with any accuracy, using the visual range-finders available to naval artillery at the time. Dazzle similarly confused the range-finders fitted on submarines. The dazzle pattern often incorporated a false bow wave, painted on the forward end of the ship's hull; by concealing the true bow wave, this too was supposed to make it harder to estimate the ship's speed.

Dazzle did not so much make the warship invisible as to confuse the eye of the enemy and make an effective response more difficult. The effect of dazzle was rather like the jazzy markings of a tiger, which just might help the predator to hide in the dappled shade of a forest, but would certainly guarantee the confusion of any prey that it ran towards.

The French military were the first to try to develop camouflage. It was the French who coined the word, and in 1915 they set up the first section de camouflage. This military unit was organized and commanded by an artist, and its members were all artists of one kind or another in civilian life. This platoon of artists set itself the task of making military installations and personnel 'disappear'. Not long after this, similar units were formed in Britain and America. The Italian, Russian and German military hierarchies were less interested in camouflage. Even so, in the twentieth century's two world wars, hundreds of artists became camouflage experts.

The science behind camouflage was explored in the 1890s by Abbott H. Thayer, an American painter with an intense interest in natural history. He discovered counter-shading. This is a common feature of birds and animals, where the upper surfaces are darker in tone and the lower surfaces are lighter, to counteract the effects of shadows. Counter-shading can help a creature to disappear, or at least make it less visible as a solid, three-dimensional object.

In 1909 Thayer collaborated with his son Gerald on a book, *Concealing Coloration in the Animal Kingdom*, which drew attention mainly because of the controversial views expressed and the lack of scientific proof for many of the assertions. But a key idea was Thayer's assertion that concealing colours were far less effective than high-contrast patterning, which he called razzle-dazzle or 'ruptive' colouring.

The idea was that the shape and surface of the animal were broken up by a contrasting pattern. Thayer was well aware that his ideas had military applications and in 1898, during the Spanish–American War, he argued that counter-shading should be used on US warships. He proposed that the US Navy should use 'the general coloring of a seagull, worked in two shades of gray and pure white, the under part of everything being painted white, the side surfaces gray, the upper surfaces a slate colour resembling the dark back of a seagull.'

Ideas of camouflaging warships were therefore available by the time the First World War broke out in 1914, but they were not put into effect straight away. Dazzle was invented (or reinvented) in 1917 by the English artist Norman Wilkinson, while on patrol duty in the English Channel. Interestingly, some of the dazzle designs closely reflect the angular, sharp-edged shapes seen in cubist art work of the day, for instance in the paintings of Braque and Picasso. The French Army displayed their first camouflaged guns, probably painted grey, in 1915. Picasso wrote to his friend Apollinaire, who was then a soldier, 'I'm going to give you a good tip for the artillery. Even when painted grey, artillery and cannons are visible to planes because they retain their shape. Instead they should be painted very bright colours, bits of red, yellow, green, blue and white like a harlequin.' A few months after this, the poet Jean Cocteau turned up at Picasso's studio wearing a harlequin's costume under a raincoat. He asked Picasso to paint him as a harlequin, but Picasso declined. Cocteau left Picasso the harlequin costume. Picasso was delighted with the gift and said that the entire French infantry should be dressed in harlequin costumes, as the diamonds would make them visually confusing. From 1915, harlequins featured over and over again in Picasso's cubist paintings. After the war was over, the painter Braque commented that he was happy, 'when I realized that the Army had used the principles of my cubist painting for camouflage.' Someone suggested that the similarity between cubism and camouflage was a coincidence. 'No, no,' said Braque. 'You are wrong. Before cubism we had impressionism and the Army wore pale blue uniforms, horizon blue – atmospheric camouflage.'

Norman Wilkinson's cubist camouflage system was first tried out on a British merchant ship, the SS *Industry*. The first Royal Navy vessel to have dazzle was HMS *Alsatian*, in August 1917. Norman Wilkinson always denied that he was in any way influenced by Thayer.

It was seen as worthwhile, even if there was no proof of its effectiveness, and the practice spread to the US Navy in 1918. Thayer's background influence on all this was probably considerable, but always indirect. Early in the First World War, he was told that the German, French and British armies had all consulted his book as they developed their field camouflage. In 1917, when America entered the war, it was not a coincidence that one of the artists who founded the American Camouflage Corps was a cousin and student of Thayer's – Barry Faulkner.

It is still not known whether dazzle painting made any difference. The British Admiralty developed the view that it had no effect at all on the effectiveness of attacks by German U-boats. It did, on the other hand, have a significant effect on the morale of the ships' crews. The American admirals took a different view of the matter. They thought dazzle did make a difference.

In the interwar period, both warships and merchant ships returned to their normal liveries, but in the Second World War dazzle was used again. The camouflage became obsolete as rangefinders became more sophisticated. Once there were spotter planes and radar to identify the bearing and range of a ship, it didn't matter any longer what colour it was painted. Frequently, the big naval guns were firing much further than their gunners could see.

THE INVENTION OF THE AUTOBAHN

(1921)

THE AUTOBAHN IS one of those rare inventions that carries a high political charge. The commonly held view is that Adolf Hitler invented the autobahn, in part to solve a major unemployment problem, and that this was his great and lasting legacy to the German people, a legacy that in some way mitigates the horrors of his appalling regime. Hitler himself certainly promoted the idea. In his own words: 'the best way to bring the German people back into work is to set German economic life once more in motion through great monumental works ... This is not merely the hour in which we begin the building of the greatest network of roads in the world, this hour is at the same time a milestone on the road towards the building up of the community of the German people.'

The autobahn has been portrayed as the realization of Hitler's dream of an interstate highway system.

The roads were designed to be fast open roads with only the gentlest curves; they were to have no speed limit, and no gradient greater than four per cent. At the junctions there were to be long acceleration and deceleration lanes. Just as Hitler wanted his buildings and other aspects of the Third Reich's infrastructure to last 1,000 years, the roads were to be built to the highest imaginable specifications.

The reality is significantly different. The network of autobahns was not originally planned by Hitler; the design already existed before his rise to power. The scheme for a countrywide network of high-speed roads was devised under the Weimar Republic in 1921 and the technical planning began in 1924. The autobahn scheme could not be implemented because of the instability of the republic and negative lobbying from the railway interests. Interestingly, in the 1920s Hitler's Nazi Party was actively opposed to the building of autobahns.

The first short section of autobahn was built in 1929–32 from Cologne to Bonn. It was built on the initiative of the then Mayor of Cologne, Konrad Adenauer. Ironically, Adenauer survived the Third Reich and its downfall to become the first Chancellor of the Federal Republic of Germany after the Second World War was over. The word autobahn seems to have been used for the first time in 1932, not to describe the revolutionary new Cologne–Bonn road but an ambitious plan for another high speed road that would one day run right across Germany from Hamburg in the north via Frankfurt to Basel on the Swiss border in the south.

Hitler was a great propagandist and he saw the propaganda potential – for himself – of this embryonic autobahn system. The designs already existed. All that was needed was an investment of cash, labour and publicity. In propaganda terms the autobahn was a huge success. Visitors to Germany in the 1930s returned home full of admiration for the new roads. More than 1,500 miles of autobahn were completed by 1938. Now there are about 7,000 miles of autobahn covering Germany. In the immediate post-war period, many aspects of the Hitler regime were reviled and rejected, but the autobahn was still greatly admired.

Other countries wanted autobahns too. Italy built its equivalent in the autostrada network. America too was envious. It was Frank Turner who was the main instigator of the US interstate highway system. Turner was a fully trained highways engineer from Dallas and a great innovator. In 1943 he worked on the Alaska Highway project as an expediter; there he led the way in using aerial reconnaissance to select the best routes for highways. In 1946 he went to the Philippines to restore war-damaged roads and bridges.

In 1950, Turner returned to America, where he became Co-ordinator for the Inter-American Highway. There was a new vision for America now, and it came from General, now President, Dwight Eisenhower. While leading the Allied invasion of mainland Europe, Eisenhower saw the autobahn system for himself. Eisenhower wanted to build a similar road system in America. This is where Frank Turner's remit ultimately came from – the top. Even in peace time, Eisenhower thought like a general and in the context of a Cold War he thought in terms of a system of high-speed roads linking all parts of the country; it would make the rapid movement of American troops and civilians possible in the event of a military emergency.

The modern interstate highway system in America was formally instigated in 1956, when President Eisenhower signed the Federal Highway Act.

The British too wanted autobahns, but called them motorways. The first motorway in Britain was the Preston Bypass, which was later incorporated into the M6 when the system developed. Today, this first section is to be found between Junction 29 on the M6 and Junction 1 on the M55. The M1 was Britain's first full-length motorway, the first section from St Albans to Rugby opening in December 1959. The M1 has functioned well, considering that it was built to 1950s specifications, and considering that it was designed to carry traffic at a volume of about 14,000 vehicles a day and today it actually carries ten times that figure.

Autobahns, highways and motorways have been extremely successful – except that they have attracted more and more traffic. To accommodate the sometimes excessive volumes of traffic, some sections of motorway have had to be adapted. The stretch of the M1 south of the M10 originally had only two lanes in each direction and this was upgraded to three in the 1970s. Massive congestion on some stretches of the M25 Orbital Motorway similarly led to the addition of extra lanes. The M25's designers apparently overlooked the attractiveness of the road as a short-cut for local traffic travelling along it from junction to junction, and therefore grossly underestimated the volume of traffic it would carry. But altering motorways, and their junctions, is very expensive and it is to be regretted that those who planned them in the first place had so little foresight regarding their future use.

For those living in the richer countries of the West it is hard to imagine life without motorways.

THE INVENTION OF THE ROBOT

(1922)

THE IDEA OF the robot was conceived in 1922 by the Czech playwright Karel Capek. He wrote a play called *Rossum's Universal Robots*. The plot involved an army of industrial robots that became so intelligent that they were able to take over the world. Karel Capek not only invented the concept of the robot: he invented the word too. The Czech word *robota* means slavery.

Robots developed a powerful presence in fiction and film in the twentieth century, long before they were created in reality. It was a case of science fiction propelling scientists and engineers forward until it became science fact.

The popular idea of a robot is a machine, preferably made of shiny metal, that acts and looks like a human being. This was the concept behind Robby the Robot in the film *The Forbidden Planet*, an American film made in 1956, and also C-3P0 in *Star Wars* (1977). The real robots that have actually been built to do a job of work on the production line in a car factory are far from humanoid. Some look like cranes; others look like anglepoise lamps. Most are the equivalent of a mechanical arm that can pick things up, lift them, extend, swivel, and so on.

The production-line robots are programmed to carry out a specific sequence of tasks, sometimes a complex sequence, but it is always a repeating one, governed by a piece of computer software. Robots are unable to think, or decide to do things differently. The robot's computer may be set up by writing all the separate movements out as a long computer program. Alternatively, it is possible to show the robot what to do. To do this, the engineer physically guides the robot's arms through the sequence of movements required, such as spraying a section of a car body, while the computer memorizes and learns it. Once the sequence of movements is learnt, the robot can go on repeating it exactly for an indefinite period. The great benefit of robotization is that it relieves factory workers of the most repetitive and tedious jobs.

Robots are also ideally suited to undertake dangerous tasks that are far too risky for people to attempt, such as detonating car bombs.

Some robots have been fitted with vision equipment that can enhance their performance. If a robot is programmed to do welding, for example, the job can only be done well if the plates to be welded are in exactly the right place. With vision, the robot can tell whether the plates are in position before it welds them. Vision equipment of this kind is still in the development stage, and it consists of a television camera linked to the robot's computer. At the moment object recognition is still far poorer than in the human eye, and many tasks that require vision are still more efficiently done by people.

The great fear people have of robots, and a very natural one, is that robots will take away their jobs. It is fundamentally the same fear that made eighteenth-century textile workers afraid of the spinning jenny. So far, it looks as if there is less to fear from robots than once thought. Even robots designed to do not very technical tasks like housework seem to be very limited. A robot can be programmed to vacuum clean the floor of a room, but it cannot switch in an instant, as a human being can, to moving a chair to one side with a view to cleaning underneath it, and then swiftly vacuum clean a complicated staircase before moving on to dusting a shelf loaded with assorted breakable objects. The replacement of people with robots seems very unlikely.

THE INVENTION OF TELEVISION

(1925)

THE EARLIEST EXPERIMENTS in television began almost immediately after the invention of the photograph, in the middle of the nineteenth century. Even before that, in the 1830s, Michael Faraday had conducted a series of experiments demonstrating the relationship between light and electricity, and from then on there was a fitful progression towards the electrical transmission of moving pictures. A means of enabling the electrical transmission of sound, which was to be essential for television, had to be found.

A major breakthrough was made in 1873, when it was found that selenium displayed varying electrical resistance according to the amount of light trained upon it. Then it became possible to convert light into an electronic signal that, at least in theory, might be sent along a cable or through the air.

In 1883 a German engineer, Paul Nipkow, invented a rotating scanning disc that had a pattern of small holes in it arranged in a spiral. The disc broke down a picture into a sequence of dots that was then trained on a photocell. The photocell sent electrical pulses to a receiver, where a second scanning disc unscrambled the sequence of dots to re-create the picture. The end result was a rather unsatisfactory and fuzzy image, but it was the first genuine television picture. Nipkow's breakthrough triggered a series of developments over the next thirty years.

By 1925, Charles Jenkins was using a device similar to Nipkow's to send signals over the air from his laboratory in Washington DC. At the same moment, the Scottish inventor John Logie Baird was developing his own mechanical television system. Baird gave his first public demonstration in 1926. By this time radio had become a reality, so the problem of transmitting sound had been solved. The cathode ray tube, another essential component of television technology, had also been invented, to produce medical x-rays; that technology was perfected by a string of ingenious engineers: Alan Swinton, Boris Rosing and Vladimir Zworykin.

Zworykin moved to the USA after the First World War. Developing his cathode ray tube into a television receiver, Zworykin found his adopted country very receptive to new ideas. It was not long before he was working with David Sarnoff at the Radio Corporation of America. Sarnoff was one of the first businessmen to see the commercial potential of television.

In 1939, NBC began regular broadcasts. The programmes were seen on about 1,000 receivers, mostly set up in hotels, pubs and shops. Television played no role whatever in the Second World War, and the war was an interruption to the development of the medium. But in 1945, broadcasting resumed in earnest, with film of the Japanese surrender; from then on television was to play an increasingly important role in political life.

Television also became the major mass entertainment medium, swiftly overtaking the cinema. Some high-minded media managers hoped that television would become a major medium for education, but those high hopes would be dashed, with ratings wars ensuring that populist and downmarket programmes pandering to the lowest tastes dominated the broadcast output. There have been moments when programming transcended this, such as the production of Kenneth Clark's *Civilization* series, but they have proved to be all too rare.

THE INVENTION OF THE LIQUID-FUEL ROCKET

(1926)

EXPERIMENTS WITH ROCKETS were going on as early as the Middle Ages. Some were designed to go up in the air, while others were designed to shoot across the surface of the sea. The Italian Joanes de Fontana invented a rocket-powered torpedo that would run across the surface of the sea to set enemy ships on fire. In 1650, the Polish artillery expert Kazimierz Siemienowicz produced a set of drawings for a staged rocket. In 1696, an Englishman called Robert Anderson wrote a book on rocketry that included the preparation of the propellants.

During this early phase, rockets were only considered as actual or potential weapons. The British fired rockets designed by Sir William Congreve against the Americans in the war of 1812. Congreve's incendiary rocket used an iron casing, black powder and a guide stick sixteen feet long. In 1846 another British inventor, William Hale, invented a stickless rocket which was used by the Americans in the war with Mexico.

But even in the nineteenth century there were some people who saw the potential that rocketry had to reach out into space. As early as 1806, an Italian living in Paris, Claude Ruggieri, shot small animals into the upper atmosphere on rockets, recovering them with parachutes.

At the end of the nineteenth century, a major figure emerged who would explore and define some of the fundamental scientific theory behind modern rocketry. He was the Russian Konstantin Tsiolkovsky. He has a claim to being the inventor of rocketry, though it would be fairer to acknowledge the combined major contributions of Tsiolkovsky, Hermann Oberth, Wernher von Braun and Robert Goddard. What these four pioneers did was to orientate rocket research towards space exploration.

Tsiolkovsky was born in 1857 and as early as about 1875 he was becoming interested in the possibility of spaceflight. In 1895 he had a strange fantasy while looking at the Eiffel Tower in Paris. He had the idea of positioning what he called a celestial castle into a geosynchronous orbit round the Earth, with a cable leading from it 25,000 miles down to the Earth. The structure would be like the beanstalk in the children's story *Jack and the Beanstalk*. Instead of people climbing up the cable, he envisaged a lift going up it to take them to the celestial castle, which in today's language we would describe as a geostationary space station.

It was an extraordinary look into the future, and so was Tsiolkovsky's reason for building the celestial castle. He reasoned that his tower would be able to launch objects into orbit round the Earth without a rocket. Any object released at the top of the tower would also have the orbital velocity to stay in geosynchronous orbit. It was a fascinating and fruitful vision, but building the tower was not a realistic undertaking.

Self-taught in all the key areas of mathematics, astronomy and physics, Tsiolkovsky went on to develop the theory of rocket propulsion. In 1898 he submitted an article to the editors of *Science Survey* entitled 'The Investigation of Outer Space by Means of Reaction Apparatus'. Published in 1903, this famous article advocated using liquid propellants in rockets to give them greater range. Specifically, he advocated the use of liquid hydrogen and oxygen. He had many ideas, not only on the technique of flying rockets but on space stations, space suits and even designs for showers for astronauts to use under conditions of weightlessness. Tsiolkovsky's writings were a curious mixture, an alternation of highly scientific technical papers about rocketry and romantic science fiction. He was clearly strongly motivated by a desire to explore outer space. Much of his writing dealt with a favoured theme, flying into deep space. In 1911, he wrote:

> *To place one's feet on the soil of asteroids, to lift a stone from the moon with your hand, to construct moving stations in ether space, to organize inhabited rings round the earth, Moon and Sun, to observe Mars at a distance of several tens of miles, to descend to its satellites*

or even its own surface – what could be more insane! However, only at such a time when reactive devices are applied will a great new era begin in astronomy: the era of more intensive study of the heavens.

By 1926, Tsiolkovsky was proposing the creation of artificial satellites round the Earth, including manned platforms, as staging posts for flights to and from other planets. In 1929, he put forward the idea that rockets might be built as nested multi-stage rockets: he described them as rocket trains. Tsiolkovsky was an extraordinary figure. He had great vision and many brilliant insights into the way that rocket science would develop, yet he was relatively disregarded in Russia, largely because of his lowly status. He was, after all, a mere teacher – and a teacher in a remote provincial school at that. The chief rocket scientist of the East German government in the late 1940s said that Tsiolkovsky was a man 'of great efforts and few rewards'. He was nevertheless the father of the Soviet space programme in the 1950s and 1960s, and by the end of his life the Soviet authorities acknowledged it. When he died in 1935, Tsiolkovsky was given a state funeral.

If Russian scientists undervalued Tsiolkovsky, they did at least take notice of Hermann Oberth. He was born in Hungary, and published an article on rocketry and space travel in Germany in 1923. Oberth explored in detail the application of rocket propulsion to spaceflight. The Soviet scientists read Oberth's work, and it made them return to Tsiolkovsky's earlier publications, which they now read with greater interest. The Soviets became interested not only in rocket science but in promoting the idea that they were ahead of the West in rocket science. In 1924, the East Germans started publishing German translations of Tsiolkovsky's writings. That year Tsiolkovsky and others founded the Society for Studying Interplanetary Communications. The space race had begun.

Working in parallel with these developments was the younger Robert Goddard, an American physicist. In 1914, Goddard gained two American patents, for a liquid-fuelled rocket and for a two- or three-staged rocket using solid fuel. He had already proved that a rocket can propel itself in a vacuum, which meant that rockets would be able to power themselves through interplanetary space and not be becalmed once they left the Earth's atmosphere.

After a series of failures and consequent public ridicule, Goddard succeeded in launching a rocket on 16 March 1926. The launch took place on a neighbour's farm at Auburn, Massachusetts. The liquid-fuelled rocket was ten feet long and flew a distance of 184 feet. In its two-second flight it achieved an altitude of just forty-one feet. It was a very long way from interplanetary space travel, but it did show that liquid propellant could drive a rocket. Once the principle had been established, Goddard could build on it.

He moved to Roswell in New Mexico, and there he worked on multi-stage rockets, rockets with fins, rockets with gyro controls and many other refinements. By 1935 he was building rockets capable of flying faster than the speed of sound. Step by step, by way of twenty-four patents, Goddard established what we now know as rocket science.

The early rockets operated on a single engine. Once it ran out of fuel, the rocket stopped ascending and its flight was over. Goddard realized, as Tsiolkovsky had also realized, that the way to overcome this problem was to place a smaller rocket on top of a big rocket and fire it as the first burns out. Today, almost every rocket uses several stages based on this principle, dropping each stage as it empties and burns out and continuing with a smaller and lighter rocket. The first artificial satellite launched by the Americans, *Explorer 1*, was launched in January 1958 using a rocket with four stages.

The invention of the liquid-fuel rocket opened the door to artificial satellites, manned flights to the Moon, and unmanned flights to every part of the solar system.

THE INVENTION OF THE JET ENGINE

(1930)

WHEN HE WAS twenty-two, a British pilot and aviation engineer called Frank Whittle first started thinking about using a gas turbine instead of piston engines and propellers to power an aircraft. A great advantage of a jet engine would be that planes could travel longer distances and faster. By 1930, Whittle had designed and patented his aircraft jet engine, though it would be another eleven years before the jet-powered aircraft became a reality.

As with many other inventions, someone else was working along the same lines at the same time. The German engineer Hans von Chaim started working on the turbojet engine idea while at university working on his doctorate. By 1935, von Chaim had built a test engine.

Whittle's first experimental jet engine was tested in April 1937. As Whittle opened the fuel supply valve to the main burner, the engine screamed into life and accelerated out of control. Everyone ran for their lives. Whittle stayed at the controls, paralysed with fright. The teething problems were sorted out and Whittle's first jet aircraft was flown in May 1941. Von Chaim managed to get his jet plane into the air first, in August 1939. To Frank Whittle goes the credit for having thought of the idea first, and patented it first.

THE INVENTION OF THE CYCLOTRON

(1931)

THE AMERICAN PHYSICIST Ernest Lawrence invented the cyclotron, a machine designed to accelerate nuclear particles. Lawrence was born in Canton, South Dakota. His education took him to the universities of South Dakota, Minnesota, Chicago and Yale. He arrived at the University of California's Berkeley campus in 1928, persuaded to move there from Yale by promises that included collaborative links with the chemistry department. At that time it was customary for departments to operate independently, with biologists, physicists and chemists working strictly within their own departments. The idea of crossover did not exist. But access to scientists and ideas from other disciplines was crucial to what Ernest Lawrence wanted to do. The idea of crossover was the basis for the (then unique) laboratory he wanted to set up.

In 1931 Lawrence founded the Lawrence Berkeley Laboratory, the first of the national laboratories. Lawrence was therefore the prime mover in launching the new era of multidiscipline national science laboratories. In 1939 he was awarded the Nobel Prize for Physics. Lawrence's great invention, the invention that would revolutionize science itself, started out as a sketch on a scrap of paper. Lawrence was sitting in the library one evening reading an article in a science journal by the Norwegian engineer Rolf Wideroe in which the idea was floated that very high energy particles needed for atomic disintegration might be produced by a succession of very small pushes. Lawrence saw that there was a relatively simple way of doing this without using any high voltage. Because it seemed so simple, Lawrence checked his theory with fellow-physicists at Yale, to make sure he had not overlooked something.

The first cyclotron was not a high-tech machine at all. It was a pie-

469

shaped contraption made of wire, glass, bronze and sealing wax; it also utilized a wire-coiled clothes horse and a kitchen chair and cost no more than twenty-five dollars to make. When Lawrence applied 2,000 volts, he found that he was able to produce 80,000 volt particles. He referred to his machine as his 'proton merry-go-round', which made it sound like a child's toy, but it was soon known by the formal name cyclotron. Far from being a toy, the cyclotron was able to whirl particles round, boosting their energy, and then throwing them like slingshot to break open the nuclei of atoms. In this simple device, Ernest Lawrence had created the means to smash atoms. This was the door opening on America's nuclear weapons programme in the 1940s. Lawrence was not to know this, but his device made nuclear bombs, and therefore Hiroshima and Nagasaki, possible. It was also to make the scale of the Cold War possible.

The accelerating chamber of the first cyclotron was five inches in diameter. To give the whirling particles even more power, Lawrence needed to build a bigger one. His assistants built one twice the size, which would reach a million volts, but Lawrence wanted to go even bigger. With collaboration from other scientists and a larger lab – an empty building called the Civil Engineering Testing Laboratory – Lawrence constructed a cyclotron twenty-seven inches in diameter. The commandeered building was renamed the Radiation Laboratory in 1931.

The Radiation Laboratory was extended to allow collaboration between physicists, chemists and engineers; the Crocker Laboratory was built next door to house a new sixty-inch cyclotron, which started working in 1939. But Ernest Lawrence's imagination went even bigger than this. He wanted to make an enormous cyclotron with a chamber 184 inches across. This would need to be housed in a building 160 feet across and almost 100 feet high. A site was found for it on a height called Charter Hill, and this became the permanent site for the big cyclotron, which was finished in 1946.

The cyclotron was invaluable in revealing the existence of the trans-uranium elements. It led to the development of the whole new science of particle physics, and to some revolutionary discoveries about the nature of the universe. The interdisciplinary department concept is also one that has proved to be very fruitful in subsequent decades. Indeed, the concept could be taken further. The current problems in engineering, appropriate research, discussion and debate about the great burning issue of global warming are largely a result of the closed nature of the research, the egregiousness of the conferences and the reports that come out of them. Ecologists and physical geographers are aware of many of the local mechanisms, such as the effects of recent warming on pack-ice and polar bears. Physical geographers and climatologists are aware of many of the interacting processes of the lower atmosphere. Palaeoclimatologists and geomorphologists are aware of the history and prehistory of climate change and environmental change. Astronomers and physicists are aware of the variations in the sun's behaviour. No one has a grasp of the whole picture, though one would not suspect this from the way the global warming issue is depicted in either the media or the scientific literature, still less from the reports of the Intergovernmental Panel on Climate Change.

Gathering together experts from all the relevant disciplines would nevertheless not resolve the problem either, because of the way in which most of them have operated up to this point; they would simply continue to argue narrowly from their own very different partisan corners. What is needed is a very different kind of school and university education, one that is not based upon a narrowing off, upon choosing to study this subject and not that, but upon inclusiveness and the interconnectedness of everything.

THE INVENTION OF THE CAT'S EYE

(1934)

CAT'S EYES ARE the reflectors embedded in the road surface which help drivers see in the fog or at night. They were invented in 1934 by Percy Shaw. Percy Shaw was born in Halifax in 1890 and worked as a labourer in a blanket mill from the age of thirteen. He studied shorthand and bookkeeping at night school, showing an active mind. He set up a repair business with his father, mending lawn rollers and this developed into a business building paths and driveways. Shaw invented a miniature motor roller to help him in his work laying paths and drives.

The area where Percy Shaw lived was prone to fog. The roads were often dangerous for motorists because of the poor visibility. Shaw saw a solution. He set about designing some studs that would reflect light from car headlights, and the studs would be set into the surface of unlit roads. He got the idea from seeing road signs brightly lit up by the light reflected from car headlights.

After patenting his reflecting studs in 1934, Shaw set up his Reflecting Roadstuds company to manufacture his new cat's eye studs. Sales were very slow to begin with. The commercial breakthrough came in 1947, when the Junior Transport Minister James Callaghan introduced cat's eyes for use on British roads. There are now tens of thousands of cat's eyes on the roads. Usually they are set in the road surface down the crown of the road, or along the lane divisions of dual carriageways. They have proved to be a major aid to road safety.

THE INVENTION OF THE BELISHA BEACON

(1934)

THE BELISHA BEACON was another road safety device invented in 1934. The beacon is a flashing orange plastic globe mounted at the top of a black and white striped pole. The name of the designer is not known, and they were named after Leslie Hore-Belisha, the Minister of Transport who introduced them in 1934.

The beacon is a very distinctive piece of British street furniture, found marking pedestrian crossings in Britain and former British colonies. The idea of the beacon is to help pedestrians in busy urban streets to spot where the nearest safe crossing place is. When the pedestrians arrive at the beacon, they find a zebra crossing, a broad pathway across the street marked by black and white stripes; on these zebra crossings, pedestrians have right of way and motorists must stop to let pedestrians cross. The beacon has a second role in alerting the drivers of motor vehicles well in advance that they are approaching a pedestrian crossing and should be aware that they might have to stop to let people cross.

The combination of zebra crossing and Belisha beacon has been one of the most successful innovations in improving road safety in towns and cities. The beacon has been adopted in countries other than Britain, but often the idea is adapted. In Australia, the city of Brisbane had a number of true Belisha beacons marking pedestrian crossings for a short time in the late 1960s and early 1970s. Now most Australian crossings are zebra crossings marked by big round yellow signs with an icon of walking legs. It is curious that as literacy increases worldwide the number of signs not requiring literacy seems to be multiplying.

In Britain in recent years, the number of zebra crossings with Belisha beacons has declined. They are being replaced by puffin crossings accompanied by pedestrian-controlled traffic signals. The principle is that the vehicle traffic has right of way. The pedestrian arrives at the kerb, presses a button and is required to wait for the pedestrian signal to change to green; the vehicle traffic signal has meanwhile changed to red. This type of crossing allows the vehicle traffic to flow more freely, as the more frequently the button is pressed the longer the pedestrian is required to wait.

THE INVENTION OF NYLON

(1935)

NYLON WAS INVENTED by Wallace Carothers in 1935. Carothers was thirty-two-years-old when he was made director of the Du Pont Corporation's research facility. He was a specialist in polymers, which are molecules composed of long chains of repeating combinations of atoms. Polymers were poorly understood at the time when Carothers started researching them, and he made a major contribution to scientific knowledge of polymer structure and the way in which it comes about.

Carothers was researching with a view to applications, particularly in the field of artificial materials. Carothers' research team started by investigating the acetylene family of chemicals. In 1931, Du Pont began manufacturing an artificial substance called neoprene, which was a synthetic rubber. It found an immediate market in the manufacture of wetsuits. This new material was created in Wallace Carothers' laboratory.

Carothers was trying to find a synthetic fibre. By 1934, he had made a significant discovery. By combining three chemicals in a polymerization process called a condensation reaction he had created a fibre. They were amine, adipic acid and hexamethylene diamine. The fibre was too weak to be usable, though. Then Carothers realized that the water produced by the condensation reaction was inhibiting the development of the fibres. He adjusted the equipment so that all the water was distilled and removed, and better quality fibres were created. Each molecule consisted of a hundred or more repeating combinations of hydrogen, carbon and oxygen atoms strung out in a long chain. The improved fibres were long, strong, and very elastic. Carothers named this new material nylon.

This was exactly what Du Pont had been hoping for. Nylon was patented in 1935. It reached the markets in 1939 and it was instantly a huge commercial success. In particular, it was a highly sought-after replacement for silk for stockings. In a very short time, the word nylons was a replacement for stockings in everyday speech. Nylon found many other uses, in making hard-wearing, easily washable, drip-dry clothing, such as shirts and blouses, and for making toothbrushes and fishing lines. It also had many specialist uses, in surgical thread and parachutes. The new material had a profound effect on many aspects of people's lives, but probably the most important was that it marked the beginning of a whole new era of synthetic fibres – and nylon itself was by no means the best of them.

THE INVENTION OF THE BALL-POINT PEN

(1935)

THE BALL-POINT PEN was invented in 1935 by a Hungarian journalist called Laszlo Biro. In fact he was not the first to have thought of the idea. In 1888 an American leather tanner, John Loud, applied for a patent for his idea for a roller-ball marking pen. Loud's pen had a reservoir of ink and a roller ball that applied the thick ink to leather hides. Loud's pen was never produced – it never got beyond an idea. There were 350 more patents for the same invention over the next thirty years. It was a great idea, in theory. The problem was that it didn't work. The major difficulty was in getting the consistency of the ink exactly right. If the ink was too thin the pens would leak. If it was too thick the pens would seize up and fail to work at all. Some of the experimental pens did both according to temperature.

Laszlo Biro was a talented man, but had never pursued any interest far enough or for long enough to lead to a good income. He had studied medicine, art and hypnotism. By 1935 he was editor of a small newspaper. He became frustrated at the amount of time he spent fiddling with fountain pens, constantly refilling them and cleaning up blots and smudges. As with many inventions, it was his irritation with a mundane but recurring problem that led him to his big idea. Laszlo Biro noticed that the thick ink used for newsprint was fast-drying, leaving the paper dry and free of smudges. He decided to design a pen that would use the same ink. The thick ink would not flow from a normal pen nib. Helped by his brother Georg, Biro devised a roller-ball tip for his pen, a tip with a tiny ball-bearing that would roll over the paper surface, picking up ink from the ink cartridge inside and applying it to the paper on the outside.

While on holiday, the Biro brothers met an amiable old man who happened to be the president of Argentina. They showed him their design for the ball-point pen. President Justo encouraged them to go to his country and set up their factory there. When the Second World War broke out, the Biro brothers fled from Hungary, stopping in Paris to patent the invention before continuing to Argentina in 1940. Laszlo took the precaution of patenting the pen again in Argentina in 1943.

When they had found financial backers, the two brothers set up the Eterpen Company in Argentina, and began producing the Biro pen on a commercial scale. There was enthusiasm in the press for the new pen which, it was said, could write for a year without refilling. The British government bought the licensing rights to the ball-point patent, seeing it as an investment that would yield cash for the war effort. The RAF also needed a type of pen than would work at high altitudes; airmen had found that taking fountain pens up to high altitudes caused them to leak. The success of the Biro pens with the RAF gave the new pens a great deal of publicity. Unfortunately, Laszlo Biro had not seen the need to take out a US patent for his revolutionary pen, so as the Second World War ended he found himself immersed in a legal battle to protect his invention.

In May 1945, Eversharp Co. joined Eberhard-Faber to acquire exclusive rights to Biro Pens of Argentina. The pen was re-branded as the Eversharp CA (for capillary action) pen. In June, just a few weeks after the deal was finalized, a Chicago businessman called Milton Reynolds visited Buenos Aires in Argentina and saw Biro pens in a shop. He immediately recognized their huge commercial potential, bought a few as samples and returned to America with them. Safely home, Milton Reynolds set up the Reynolds International Pen Company, manufacturing ball-point pens regardless of Eversharp's licensing rights. Reynolds beat Eversharp to market, getting his Reynolds' Rockets on sale at Gimbel's store in New York by the end of October 1945. He sold $100,000 dollars' worth of pens on the first day alone. The newspaper advertising that went with the product launch oversold the pens, describing them as miraculous, fantastic and 'guaranteed to write for two years without refilling!' Eversharp's potential market was similarly undercut in England; by December, ball-point pens were on sale after being manufactured by the Miles-Martin Pen Company.

Initial sales were very promising. But the new pens were unsatisfactory. Customers found that Reynold's Rockets leaked, skipped and sometimes just did not work, and that Eversharp's pens did not work well either. A lot of pens were returned by dissatisfied customers. By 1948 the ball-point craze was over, and the price of a ball-point fell from $12.50 to fifty cents.

Then, as Reynolds' firm folded, a Frenchman by the name of Marcel Bich started manufacturing ball-point pens under the trade name BIC.

The spectacular rise and fall of the first generation of ball-points was good news for the manufacturers of fountain pens. Parker Pens, one of the leading manufacturers of fountain pens, managed to iron out some of the technical problems with the ball-point, and put its own version on the market, called the Jotter, just to make it clear that their fountain pens were what people needed for proper writing. The Jotter was a huge success. Its reservoir went on five times longer that Reynolds or Eversharp pens, and it was available in a variety of point sizes. In 1957, Parker improved their ball-points by using a tungsten carbide textured ball-bearing.

By now Eversharp was in financial difficulty, sold its pen division to Parker and went into liquidation. But Parker was not to dominate the market after all. Instead, it was BIC. By the end of the 1950s, BIC was selling seventy per cent of the ball-point pens in Europe, and went on to encroach on the American market. By 1960, BIC owned the big New York pen firm of Waterman Pens, marking the end of a remarkable trade war.

In a world where so much depends on speed and convenience, it is hard to imagine life without the ball-point pen.

THE INVENTION OF RADAR

(1935)

ONE LEGEND OF the Second World War was that British pilots were successful on night raids because they ate carrots, which enabled them to see better in the dark. This harmless piece of disinformation may or may not have fooled the Germans, but it was certainly untrue. British pilots were successful in their night flying because they had radar.

Churchill paid tribute to 'The Few who saved Britain from invasion in the Battle of Britain', but the people who invented and developed the radar that helped the airmen could not be mentioned, as radar did not officially exist.

Radar was not specifically invented because of the demands or challenges of warfare. It was developed largely in peacetime, in the period between the two world wars. The leading figure in the invention of radar was Robert Watson-Watt, who began working on the concept as early as 1915. Watson-Watt started out with an interest in radio telegraphy and then went to work for the London Meteorological Office as a researcher. As more and more aeroplanes came into service, there was a perceived need to protect them against storms and atmospheric turbulence generally.

Robert Watson-Watt worked on radar (radio detection and ranging) through the 1920s, joining the British National Physical Laboratory's radio department. There he helped to develop navigational equipment and radio beacons.

The radar system used radio signals to detect the direction and distance of obstacles, in much the same way that bats emit signals to avoid collision with one another and solid objects. A radio beacon or antenna sends out radio waves. They bounce back from solid objects and measuring the time lapse between sending and receiving the signal gives the range of the objects.

The military applications of the new ranging technique was spotted. Several companies, including some in Germany, developed it at high speed. Watson-Watt was given a post at the Air Ministry and a free hand to develop radar as fast as possible. By 1935, Watson-Watt had developed radar to the point where it could detect aircraft approaching as much as forty miles away. By 1937, he had overseen the establishment of a string of radar stations along the British coastline. It was essential to have early warning of the approach of enemy bombers and fighters.

Just as the Second World War broke out in 1939, the British scientists developing radar made a major breakthrough. They created a high-power microwave transmitter. This gave the British a huge advantage. The new transmitter gave totally accurate readings, regardless of the weather conditions. It used a short beam, which could be given a sharp focus. Another major advantage was that it could be picked up on small and compact antennae; this made it easy to install on planes.

The developed state of radar by the outbreak of war gave the British a very significant advantage over the Germans. Because the enemy aircraft were identified early, before they actually came into view, it was possible to deploy British aircraft with maximum effectiveness. The Germans took to flying at night, believing that they could then not be seen, but the British planes had small microwave units installed, which enabled them to 'see' the German planes even in pitch darkness.

Radar played an important part in defeating Germany in the Second World War. Since then, many peacetime uses have been found for radar. It can locate dangerous weather phenomena such as tornadoes and hurricanes. It helps navigation by sea. It can even be used on the ground to help archaeologists locate buried walls. It is used to measure the speed of traffic and the distances between planets. Radar has turned out to have an extraordinary range of uses that could never have been foreseen by its original inventors.

THE INVENTION OF THE RADIO TELESCOPE

(1937)

IN 1931, THE American engineer Karl Kansky discovered that interference in telephone communications was being caused by radio emissions from the Milky Way. This remarkable discovery, that radio waves are reaching the Earth from outer space, opened the way to a new era in astronomy – radio astronomy.

It was a twenty-five-year-old American astronomer called Grote Reber who built the world's first radio telescope at Wheaton, Illinois in 1937. It had a parabolic reflector thirty feet across. For almost a decade, Grote Reber was to be the only radio astronomer in the world. The areas of maximum radio emissions were mapped on the celestial sphere with the aid of the new generation of radio telescopes.

In 1942, Reber made the first radio maps of the sky. In the 1950s astronomers found themselves confronted with the problem of identifying these newly discovered radio stars, or whatever the objects might be that were the sources of the radio signals. In 1955, the American astronomer B. F. Burk discovered that not all of the emissions were coming from outside the solar system; even the planet Jupiter was giving off radio waves. In 1958 radio astronomers picked up radio waves that had been reflected off the surface of Venus, and found that this 'bouncing' of radio waves could be used to measure the distances of the planets more accurately than before.

The radio telescope gave astronomers a new way of looking at and listening to the universe. Revolutionary though they were in their day, the old Galilean telescopes simply magnified and enhanced what was already visible from the Earth. The new radio telescopes made it possible to detect things that were completely invisible, beyond the reach of ordinary telescopes. They have enabled us to probe further out into the universe than ever before.

THE INVENTION OF THE PHOTOCOPIER

(1938)

THE PHOTOCOPIER WAS invented in 1938 by an American physicist, Chester Carlson, although his machine was not to be manufactured and become commercially available until 1947.

Before that time, making copies of documents or drawings could be a laborious affair. Carbon paper had been invented, and was in widespread use in offices for making copies of commercial and legal documents. There was no technique available for making good copies of images, such as building plans. The Gestetner machine made image-copying of moderate quality possible, but for many purposes that was not good enough. The only other possibility was to hand draw additional copies, which was time-consuming.

Photocopiers use static electricity and black toner powder instead of ink and they can produce black-and-white copies at a rate of two per second. They also have extremely useful additional facilities such as double-sided copying, printing darker or lighter, enlarging and reducing. By using these facilities it is often possible to improve the quality of the original.

The electrostatic copiers are the descendants of the original Carlson machine. Carlson called his technique xerography, meaning 'dry writing' to emphasize that no liquid ink was used, a distinct attraction to those office workers and teachers who were used to finding their hands and clothes black with ink from another type of copier. The photocopier has a drum charged with electrostatic electricity and coated with a thin layer of powdered silicon, which conducts electricity when light shines onto it. The blank, white parts of the original reflect light onto the drum which removes the charge. The black parts of the original do not reflect light, so the charge on the drum remains. These charged areas or lines attract the fine black powder, toner, which forms the image on the copier paper.

Photocopying has made many tasks in offices, schools, colleges, local government and national government much easier. It is easier to share documents. The negative side has been the generation of additional paperwork which fifty years ago could never have been produced. In some occupations, employees are virtually drowning in paper, all produced by the photocopier.

THE INVENTION OF THE ELECTRON MICROSCOPE

(1940)

THE ELECTRON MICROSCOPE was invented in 1931 by Ernst Ruska. A conventional microscope depends on light for viewing an object and they can only show us structures that are bigger than the wavelength of the light reflecting off them. An electron microscope depends on electrons. The electrons are speeded up in a vacuum until they have a very short wavelength, only one hundred-thousandth the wavelength of white light. This technique produces far higher magnifications than conventional microscopes, and makes it possible to see objects as small as individual atoms. Light microscopes can produce magnifications up to a maximum of ×2,000. The most recent models of electron microscope can yield magnifications approaching ×1,000,000.

There are two types of electron microscope, the transmission electron microscope (TEM) and the scanning electron microscope (SEM). The TEM has a source of electrons, lenses and a system for projecting the image onto a fluorescent screen or photographic plate. The electrons pass through the specimen, which has to be sliced extremely thin, because only electrons passing right through are recorded. The SEM scans a tight beam of electrons over a sample. Electrons bounce off the sample and scatter onto a screen. As the electron beam scans over the entire sample, a complete image of the sample appears on the screen. The SEM works at significantly lower magnification than the TEM, but still up to ×200,000, which is one hundred times the magnification available in a light microscope.

Scientists in many disciplines have found uses for electron microscopes: medicine, biology, chemistry, metallurgy, entomology, mineralogy. The instrument has revolutionized the study of microscopic structures and surfaces. Just as twentieth-century refinements to telescope technology have made it possible to probe further into outer space, the electron microscope has made an exploration of inner space possible, producing major new discoveries in nearly every branch of science.

THE INVENTION OF POLYESTER

(1941)

IN 1941 JOHN Whinfield and James Dickson invented polyester, which was sold under the trade names Terylene in Britain and Dacron in America. In the 1950s polyester fibre was only available commercially under these two labels. Carothers' invention of nylon paved the way for this extremely useful new synthetic fibre.

In Germany especially there was a postwar tendency to want to start afresh, replacing natural fibres with synthetics where possible. But hosiery made from pure polyester, which was popular for a time, was found to have poor moisture absorbence. Garments made of pure synthetics in the 1950s and 1960s turned out to be rather sweaty. By the 1970s, more and more people were demanding garments made out of blends of synthetic and natural fibres.

Polyester has been very successfully blended with natural fibres to retain some of the best properties of the natural fibres, while also giving the cloth additional properties. These included crease-resistance and crease-retention, elasticity, ease of ironing, greater durability; the inclusion of polyester meant that the cloth would stand far more washings. Polyester has been successfully combined with wool and cotton. Shirts, slacks and blouses have been made out of polyester-cotton. Men's socks have been made out of polyester-wool.

One great attraction of nylon and polyester garments was that they could be drip-dried. Busy working people with too little time to have an old-fashioned washing day followed by hours of ironing and airing found the new materials far more suitable to their way of life.

THE INVENTION OF THE AEROSOL SPRAY CAN

(1944)

WE COME NOW to two of the most destructive inventions of the twentieth century. The first is really an object lesson in the terrible responsibility that inventors have to bear, showing how an invention can be a Pandora's box, seeming harmless at the outset but leading the way to destructive results that were completely unintentional.

The aerosol can was invented in 1944 by Lyle David Goodloe and W. N. Sullivan. They were employed by the US Department of Agriculture, and their assignment was to invent a method for killing the mosquitoes US forces were encountering while serving abroad. The tropical mosquitoes were frequently carriers of malaria, and it was vital to keep troops healthy.

A relatively primitive aerosol can had been invented in 1927 by the Norwegian Erik Rotheim. He devised a can with a valve and a propellant system, the 1944 form of which was a fully working version. The modern spray valve that keeps itself free of clogging was added in 1953 by Robert H. Abplanal.

The aerosol spray can had many applications. It continued in use as an insect repellent spray. It was also possible to use it to apply paint very thinly and evenly. The first spray paint was invented in 1949 by Edward Seymour; the idea of using an aerosol can as a paint applicator was suggested to him by his wife. He tried it out with aluminium-coloured paint, then set up in business as Seymour of Sycamore Inc, which still exists. There is no doubt that aerosol cans are useful devices. The problem is that the propellants, CFCs, find their way into the stratosphere, the layer of the atmosphere that is above the height of Mount Everest, and destroy ozone molecules. The ozone layer within the stratosphere acts as an invisible shield, screening out many of the harmful ultra-violet rays from the sun before they reach ground level. Depleting the ozone layer has led to people being exposed to higher levels of ultra-violet radiation and therefore increased risks of skin cancer.

The problem was first noticed in the Antarctic by scientists working for the British Antarctic Survey in the late 1970s. By 1985, British scientists Joe Farman, Brian Gardiner and Jonathan Shanklin were sure they were seeing significant progressive damage to the ozone layer over the Antarctic, damage that was getting worse each year. An ozone hole was opening up, and alarmingly it grew larger and more intense over the next fifteen years. The ozone hole increased until it was the size of Antarctica itself.

Consumer aerosol products were voluntarily phased out in America in the late 1970s. In 1987 an international agreement was reached: the Montreal Protocol on Substances that Deplete the Ozone Layer. The use of CFCs as aerosol propellants has stopped, but the damage will take a significant amount of time to repair. CFCs have a lifetime of 50–100 years, as measured in both the laboratory and the atmosphere, so it will be 100 years before the atmosphere is free of this problem.

We can safely use aerosol cans again now, because they no longer use ozone-depleting chemicals.

THE INVENTION OF THE ATOM BOMB

(1945)

THE ATOMIC REACTOR was invented by Enrico Fermi and Leo Szilard. Fermi was born in Rome in 1901 and became an expert in mathematics, physics and engineering. He was first of all a professor of physics, later of atomic physics. His particular research interest was the creation of artificial isotopes through neutron bombardment. His achievements in this field were so significant that in 1938 he was awarded the Nobel Prize for Physics.

It was at that time that Fermi was being persecuted in Italy because he had a Jewish wife and expressed anti-fascist views. He took the opportunity to escape from Italy when he attended the Nobel Prize award ceremony in Stockholm; he simply did not return home but went to New York instead. It was at Columbia University in New York that Fermi started working with Leo Szilard and a graduate student called Walter Zinn. Together they started experimenting with nuclear fission, on the assumption that the neutrons released during fission would begin a chain reaction – and so generate energy.

The research Fermi and his co-workers were engaged in clearly had the possibility of being developed for military use. In March 1939, Fermi tried to discuss the progress of his research with the US Navy, though nothing came of the encounter. Then Szilard explained to Albert Einstein what they were doing. Einstein was an extremely influential lobbyist. It was Einstein who alerted President Roosevelt to the military implications of the project, and suddenly lavish funding and a team of workers were made available to Fermi. Fermi's project was refereed by a second team of scientists from Princeton University. The Princeton team gave its approval and the joint team, based in Chicago, got to work on the nuclear energy project.

It was in December 1942 that the first experiment in controlled nuclear fission was carried out. It took place in a squash court at Stagg Field at the University of Chicago, and it was a successful test. In August 1944 the project was moved to Los Alamos in New Mexico, where a new tailor-made laboratory had been built under the leadership of Robert Oppenheimer. Fermi became head of the Los Alamos physics department. The new lab's specific goal was the construction of a working nuclear bomb that would win the war.

A further year's work and two billion dollars went into the development of the atomic bomb. The first atom bomb was detonated early in the morning of 16 July 1945 at the Alamogordo airbase, 120 miles south-east of Albuquerque. The bomb test was an overwhelming success, and the team set about building more atomic bombs for deployment in the closing stages of the Second World War. The Americans had the Japanese in retreat, but there was a danger that the Japanese would kill all their surviving prisoners, and refuse to surrender. The Americans made the historic and controversial decision to shock the Japanese into surrender by dropping one of the new weapons on a city on the Japanese mainland.

Just three weeks after the pilot test at Alamogordo, an atomic bomb was dropped on Hiroshima on 6 August. The ten-feet long bomb was delivered by the US Army Air Force B-29 bomber *Enola Gay*, piloted by Paul Tibbets. The bomb, nicknamed Little Boy, was released at an altitude of about 35,000 feet, and it exploded in the air about 600 feet above the city. The energy released during the explosion was enough to flatten completely four square miles of the city of Hiroshima. A total of 100,000 people in Hiroshima were killed outright. Of the survivors, perhaps another 100,000 were so horribly burnt or affected by radiation sickness that they died later. The Japanese government did not respond to what was intended to be a *coup de grâce*. There was no formal surrender. The American government decided that it had no alternative but to give a second Japanese city the same treatment. The second bomb, code-named Fat Man, was dropped on Nagasaki on 9 August. This time the atomic bomb had the desired effect, and on 10 August the Japanese government asked for a truce; the negotiation of surrender terms had begun.

The invention of the atomic bomb had an immediate impact on

world history. It spectacularly, if cruelly, terminated the Second World War by killing tens of thousands of Japanese civilians. It also became a terrible warning of what a Third World War would be like: far worse than the First or Second. The appalling photographic images of the annihilated city of Hiroshima and its afflicted people haunted the postwar generation and hung like the sword of Damocles over the Cold War. The nuclear weapons that were stockpiled during the next few decades were even more powerful and frightening than the Hiroshima bomb. The knowledge of the damage they could do was enough to create a global stalemate; fear of the atomic bomb became a major political force, the nuclear deterrent. As the Cold War developed, it was the memory of what the atom bomb did to Hiroshima that put the fear of God into the superpowers and their allies. This single invention of Enrico Fermi's created half a century of global fear.

THE INVENTION OF THE IDENTIKIT PICTURE

(1945)

THE IDENTIKIT PICTURE was invented in 1945 by Hugh C. McDonald, a detective working with the Los Angeles Identification Bureau. McDonald took photographs of the faces of around 50,000 people, cut them into twelve main sections, each representing a facial feature. He had as many as 400 distinctive pairs of eyes, noses, lips, chins, hairlines, moustaches and beards. The outlines of these features were then traced onto transparent plastic sheets, so that they could be overlaid.

The idea was to change the combinations of features in line with descriptions of a suspect until a witness agreed that the overall image was a good likeness. The composite line drawing was then reproduced for circulation to police departments and the press. This technique was used in America, and in European countries too, the UK among them.

The idea of using artists' impressions was not new. In France in the 1880s, impressions were used, sometimes successfully, to catch criminals. In England, in the case of the Whitechapel murders, newspaper artists depicted with some relish a variety of villainous characters seen round Whitechapel – and of course no one was ever found. A French criminologist called Alphonse Bertillon devised a system which he called *portrait parlé*, or speaking portrait (or, more loosely, 'spitting image'). This consisted of front and side view photographs of convicted criminals, cut into sections and mounted in such a way that particular features could be identified. It was very similar to the later development of Identikit into Photo-FIT, and was similarly designed to help police officers to recognize suspects when they encountered them.

It was in 1970 that the British police force switched from the cartoon-like and often unrealistic Identikit pictures to what they called Photo-FIT (Facial Identification Technique). Photo-FIT uses real photographs of ordinary people, i.e. not necessarily criminals, cut up in the same way as the Identikit pictures, and mounted on transparent sheets. The basic five-section kit has 195 hairlines, 99 eyes and eyebrows, 89 noses, 105 mouths, 74 chins and cheeks. These give billions of possible combinations.

The basic kit was designed for white suspects. Then further kits were made to accommodate Indian, Afro-Caribbean and other ethnic types. There was general agreement that Photo-FIT was a great improvement on Identikit. In about 1975, the Americans switched to a similar photo-based system.

Today, Photo-FIT has given way to E-FIT, a computer-aided identification system that works on exactly the same system, but allows for subtler modifications, such as complexion and skin blemishes. It also has the great advantage of losing the edges between the different sections of the face, and therefore looks much more realistic. Computer technology has now advanced to a point where a computer can look through a store of photographs of arrested criminals and select those that are closest to the E-FIT picture of a suspect in a particular crime.

THE INVENTION OF THE ELECTRONIC COMPUTER

(1946)

THE FIRST LARGE-scale general purpose electronic computer was invented in 1946 by John W. Mauchly (1907–80) and J. Presper Eckert (1919–95). Their revolutionary new computer was called ENIAC, which was an acronym for Electronic Numerical Integrator and Computer. ENIAC was built at the University of Pennsylvania.

ENIAC had its origins in the Second World War, when it existed as a secret military project that was only referred to as Project PX. During the war, artillery units were forced to fall back on tables of figures to help them work out how and where to aim their shells. Calculating the variables – the angle of the gun, the terrain, the distance – was mentally challenging under what were frequently adverse conditions. Working out a single artillery trajectory with a hand calculator might take forty hours, and even electro-mechanical devices such as the differential analyser took thirty minutes to produce a result. A faster method of calculation was needed. The challenge was to make ENIAC (or its offspring) carry out this kind of calculation automatically and quickly.

ENIAC had great historic importance because it laid the foundations for the modern computing industry. It showed, more than any previous computer, that high-speed digital computing was possible using the technology then available. The transistor had not been invented yet: this was still the age of the valve.

There were other, earlier computers, such as the Z3 in Germany, the Atanasoff-Berry Computer (ABC) and the Colossus in Britain, but

ENIAC did something that may have been more important: it stimulated the imagination of scientists and industrialists. The British Colossus, which had been an enemy code-breaking machine during the Second World War, was demolished, remained classified for decades, and so was completely unavailable for display, admiration or discussion.

After ENIAC there was an expectation that computers were going to play a major role in the way people lived in the second half of the twentieth century, which certainly they did. In the 1950s and 1960s, computers appeared at universities, national government departments, local government offices and banks. A computer was made to predict the outcome of the 1952 US presidential election, which gave the computer a completely false status. In the 1960s an American scientist, Gerald Hawkins, even tried to prove that Stonehenge had been a prehistoric computer.

ENIAC was a strange machine compared with other computers. It used a ten-digit decimal system, whereas virtually all subsequent computers used binary systems of ones and zeros. Even other computers devised by Eckert and Mauchly used a binary system. ENIAC was unable to store programs. In some ways it was based on out-of-date concepts in relation to the computers that came after it. One leading American computer expert condemned ENIAC: 'It was a monstrosity. It was rapidly overtaken by general purpose machines. There wasn't anything in it that survived into modern machines, except maybe electricity.' It was certainly true that no one else ever thought of building another ENIAC.

ENIAC nevertheless worked, at least until it was struck by lightning in 1955.

It was John Mauchly who conceived of the architecture and principles of ENIAC, and Presper Eckert who possessed the engineering skills to make the design a reality. The huge electronic engine they constructed was able to deal with 5,000 addition problems per second, which was far faster than any other computer or calculating machine had ever been able to achieve. Mauchly and Eckert were fully conscious that what they were creating was something momentous, something that would change the course of history. They were nevertheless uncertain how to break the news of their invention to the world, how to make it clear that something extraordinary had happened. They painted numbers on light bulbs and screwed the resulting 'translucent spheres' into ENIAC's panels.

The light show certainly made an impression. For ever after, dynamic flashing lights were always associated in the popular imagination with computers. No wonder people have a healthy distrust of experts.

THE INVENTION OF WARFARIN

(1947)

THE RAT POISON Warfarin was invented in 1947 by scientists at the University of Wisconsin. The way in which Warfarin was invented was typical of the way the University of Wisconsin relates to the wider community. In 1933, a Wisconsin farmer appeared at the University's School of Agriculture with a milk churn full of blood that would not coagulate. He also brought with him a dead cow and some hay. He wanted to know why his cow had died.

By 1941, Karl Link had succeeded in isolating the anti-coagulant in the blood. It was the anti-coagulant that had killed the cow, which had died of internal bleeding.

The anti-coagulant, Warfarin, takes its name directly from WARF, the Wisconsin Alumni Research Foundation. Initially it was commercially produced and marketed as a rat poison. It is no longer used for poisoning rats, partly because many rats have evolved a resistance to it and there are now more toxic and powerful poisons available. Today, coumarins are used. Coumarins are a class of drug used for controlling rats and mice in any environment, agricultural, industrial or domestic. The active ingredient is Brodifacoum, which is occasionally referred to as a super-Warfarin because it is longer acting than Warfarin. It is effective when mixed with food bait, because the rodents return to the bait and feed on it over several days, until they accumulate a lethal dose. The slow action of the coumarins thus helps to make them more effective as a poison.

Warfarin itself now plays a more positive role in human medicine, in treating thrombosis.

THE INVENTION OF THE POLAROID CAMERA

(1947)

POLAROID PHOTOGRAPHY WAS invented by the American physicist Edwin Herbert Land (1909–91) in 1947. His one-step system was a revolutionary development in photography. Previously, photographers exposed their film, then sent it off to be developed and printed: a process that might take weeks. The Polaroid camera removed the wait. Immediately the film had been exposed, it emerged from the camera, developing as it did so. The photographer had the print within a minute.

Edwin Land also developed the first modern light polarizers, which eliminated glare. He set up his Polaroid Corporation to manufacture the new Polaroid cameras and the first of these went on sale in November 1948.

The benefits of this one-step system were clear. It was possible to take a photograph and check that it had indeed come out satisfactorily. If the photograph was unsatisfactory in some way, for instance it was out of focus or the camera had been tilted or a child in a family group was sticking out its tongue – then it was easy to take another straight away.

There was also the privacy issue. Film that was sent away to a laboratory for processing was viewed by strangers. Some people liked the idea of being able to take photographs that only they would see; it was a kind of empowerment.

The invention of digital cameras, which enable people to take high-quality photographs, store them as computer files, and print them as soon as they reach their home computer, seems to be a major threat to the Polaroid. On the other hand, digital cameras require computers to operate them and there are still many people who do not have access to computers, and many who do not wish to be dependent on them. The Polaroid alternative, which is cheap medium-technology, and gives photographers a great deal of independence, still has its market. The history of inventions is in any case rarely straightforward, and it would be unwise to try to forecast the future of photography.

THE INVENTION OF VELCRO

(1948)

A SWISS ENGINEER, George de Mestral, returned home one day after taking his dog for a walk, to find that both he and his dog were covered in burrs. These small seed-sacs are very difficult to pick out of an animal's fur, or from the fabric of his trousers, and it occurred to de Mestral that the burrs were perfectly designed to stick firmly to fur in order to hitch a long ride to a new location. He took one of the burrs to examine it under his microscope. What he saw were all the tiny hooks that enabled the burr to hang on so tenaciously to the equally tiny loops in the cloth of his trousers. George de Mestral saw immediately that this phenomenon had a useful human application. He would design a two-sided fastener for clothing. One strip would be covered with stiff hooks, the other with soft loops.

De Mestral made the mental jump to human clothing because there were many pre-existing clothes fastenings that worked on that principle. What he was seeing was in effect a new version of the hook and eye. Before 1830, hooks and eyes were made of copper; after 1830 they were made of brass. They worked, but they were very small, fiddly to sew on, difficult to do up and difficult to undo; hooks and eyes were all right for the leisured classes of the eighteenth and nineteenth centuries, and they ensured the perfect fit of the 1880s bodice, but in the twentieth century people wanted to dress and undress faster and more easily.

The zip fastener was much more convenient and much faster. It too was based on the hook and eye principle.

De Mestral left his job as an engineer, persuaded a bank to lend him money to develop the idea and consulted several people working in the textile industry at Lyons, which was at that time one of the world's leading textile centres. To begin with, de Mestral found that his idea was not taken seriously, but he persisted. He entered into collaboration with a French textile worker to find exactly the right materials for the fastener and make a prototype.

He successfully processed cotton to make the hook side of the fastener, but it was too costly for mass production. It took him eight years until he found what he was looking for. As is often the case with inventions, de Mestral stumbled on nylon as the solution entirely by accident. He found experimentally that when nylon was sewn under an infra-red light, it formed the tough hooks he needed for one side of the fastener. All that was needed, was for the two strips of hooks and eyes to touch, and there was a very strong fastening. To separate them, once again, all that was needed was a sharp tug.

The finished design was patented in Switzerland in 1951, and went on to collect additional patents in ten other countries, including the USA. George de Mestral decided initially to call his invention locking tape, but later settled on Velcro, a portmanteau word made out of velour and crochet. It was to become a global replacement for the metal zip fastener.

Supported by Gonet & Co., George de Mestral formed Velcro Industries and went into production. It has become a huge industry. Velcro itself is a very useful and appropriate fastener for certain types of garment, such as fleeces.

THE INVENTION OF THE TRANSISTOR

(1948)

THE TRANSISTOR WAS invented by John Bardeen, Walter Brattain and William Shockley. Silicon was discovered by a Swedish chemist, Jons Berzelius, in 1824, and germanium by a German called Clemens Winkler in 1886. These two substances are semiconductors, which is to say that they have conductive properties halfway between metal, which conducts electricity easily, and insulation, which will not conduct it at all.

Bardeen, Brattain and Shockley met at the Bell Laboratories in New Jersey in 1945. It was Shockley who first saw the potential for a transistor to replace the vacuum tube; as early as 1939 he was already proposing using semiconductors as amplifiers.

Bardeen, Brattain and Shockley studied silicon and germanium intensively, and their work culminated in the invention of the transistor. In 1956, the three inventors received the Nobel Prize for Physics. Shockley left the Bell Labs to found the Shockley Semiconductor Lab in what became known as Silicon Valley.

The new device was given its name by John Pierce because it could transmit a current across a resistor: trans-(res)istor. The great significance of the transistor was that it made valves unnecessary. It changed the insides of many electrical devices, making it possible for manufacturers to make significantly smaller and more compact versions. The most conspicuous of these at the time was the transistor radio, a small, lightweight and portable radio, which somehow symbolized the freer society of the 1950s. But more significantly, transistors were also to make possible the development, later, of desktop computers. A whole new generation of technology had been launched.

THE INVENTION OF LONG-PLAYING RECORDS

(1948)

THE LONG-PLAYING record was invented in 1948 by Columbia. The first flat disc was invented by the German–American inventor Emile Berliner as long ago as 1887, to play on the gramophone which was also his invention. Berliner had trouble finding a suitable material for his records, trying glass, zinc and then hard rubber.

By 1915, gramophone records were revolving on their turntables at a standard speed of seventy-eight revolutions per minute (78 rpm). They were made of shellac, a hard, black, shiny substance that was attractive to look at but very brittle. It was all too easy to break records made of shellac. Another major problem with them was that because they were ten inches in diameter and revolved at 78 rpm they only held four minutes of sound. This was quite adequate for a music hall song, which might last three or four minutes, but a great limitation when it came to classical music. A movement of a nineteenth-century symphony might well last five or six minutes and a whole symphony between twenty and forty minutes. The music industry tried to find ways round this. Specially edited-down versions of popular classics were performed and recorded. One such record had Grieg's *Piano Concerto* crammed onto it, the first half on one side and the second half on the other, performed at breakneck speed in order to pack as many bars in as possible. It was an unsatisfactory solution to the problem.

The solution was to reduce the speed of rotation. In 1948 Columbia Records developed the first long-playing record (LP). This played at thirty-three and a third revolutions per minute and was twelve inches in diameter. These two changes meant that much more playing time could be accommodated. The LP was made of vinyl, which was a big advance on shellac because of its flexibility; it was still possible to scratch it, but no longer possible to break it. The LP also carried new microgrooves, which could be packed more densely on the disc. The end result was that LPs could play for twenty-five minutes on each side. This meant that an LP could comfortably carry a Mozart symphony on each side, without any cutting.

In 1949, an intermediate type of record was introduced. This was the seven-inch extended play record (EP) which revolved at forty-five revolutions per minute. The EP was an experiment that might have succeeded, but for the runaway success of the LP. The single, a small vinyl disc that carried one pop number on each side, became very widely used in the pop music industry from the 1950s onwards. The LP was also popular for collections of pop music, or albums, like the Beatles' *Sgt Pepper's Lonely Hearts Club Band*, which was released in Britain and the USA in June 1967. This forty-minute collection of songs summed up the achievement, musical values and mindset of the group, and was profoundly influential. Many critics refer to it as the most influential album ever produced. Subsequently, most pop groups have tried to do something similar in the way of producing a summative album.

The LP was a major breakthrough in the propagation of classical music too. Suddenly, people who were unable to get to concerts had access to the classical repertoire. It also meant that a lot more classical music was available than was in the concert repertoire. New composers were discovered. It was no accident that, straight after the arrival of the LP, neglected and eclipsed composers such as Mahler and Bruckner experienced major revivals.

Who was the unsung hero behind these remarkable developments, the inventor of the LP? According to some accounts, it was Peter Goldmark who invented the 33⅓ rpm record to overcome the problem of interruptions caused by constant turning over. Goldmark was a salaried employee of the Columbia Broadcasting System and received no royalties for his invention; in lieu of royalties, he receives free copies of every LP Columbia produces. Goldmark went on to become president of CBS Laboratories at Stamford, Connecticut. Another view is that it was not so much Goldmark but a team of people at Columbia who developed the long-playing record. The team included Ike Rodman, James Hunter, Vin Liebler, Bill Savory, Rene Snepvangers and Bill Bachman, working with Peter Goldmark as their supervisor. Within the team, Bachman seems to have come up with several of the key components, and it was Rene Snepvangers who developed the lightweight pickup.

An individual or a team? As so often, it is hard to tell, later and from outside the situation, where the key ideas came from. This great invention meanwhile generates colossal wealth; in America in 1972 alone, sales of LPs earned over two billion dollars.

THE INVENTION OF THE BARCODE

(1949)

THE BARCODE WAS invented in 1949 by Joseph Woodland. He was a graduate trying to devise a way of capturing information about a product. He initially thought of a system based on the Morse code dots and dashes. While toying with this idea, he drew lines on beach sand and stumbled on the idea of parallel lines. Woodland and his partner patented the idea in 1952, but it was not a success commercially. It took fifteen years before its usefulness was generally realized. It was first used to identify railway freight cars. Barcodes were attached to the sides of the cars, and as they rolled past a scanner by the railway track they could be identified. The barcode also gave information about the freight and the destination. The system did not allow for the jolting of the train as it passed the scanner, so the readings were not very accurate.

In spite of the slow and unpromising start, barcodes have now become central to the wholesale and retail trade. The usefulness of barcodes in managing retailing was first recognized in the 1970s, in the grocery business in America. Barcodes have really grown from that beginning. The Universal Product Code was introduced in 1974. The US grocery industry optimistically hoped that it would save more than forty million dollars a year, but that proved to be over-optimistic. A snag was that a fairly large initial investment was required, in for example fitting the checkouts with scanners to read the barcodes, so the huge savings were not going to be short term. But in the longer term they have made a huge difference. A recent analysis claims that the use of barcodes and the UPC saves the grocery industry seventeen billion dollars a year.

Every shopkeeper needs to know which goods are selling well and which are selling slowly, in order to restock appropriately. In large stores with many sales assistants it would be virtually impossible for managers to acquire an overview of the flow of merchandise without systematic recording. Barcodes allow retailers to record this information automatically.

A barcode is a rectangular patch of black and white parallel stripes, variable in width and spacing, and printed somewhere on the product. The stripes encode a number, which is unique to that product. One commonly used barcode is the European Article Numbers code EAN, which accords each product a thirteen-digit number. The number can be read by a laser scanner. Usually in a supermarket, where the products are small and light, the barcode is shown to a fixed scanner that sits under a glass screen in the counter. In a garden centre or DIY store, where the products are bulkier, a hand-held scanner is more likely to be used. The sales assistant has only to point the scanner at the barcode, and the computerized cash register is able to read and record the product number, and feed the product's retail price to the register, which prints it directly onto the receipt and adds it to the customer's bill.

As a precaution, the product's code number is also printed on in numerals. Occasionally the laser scanner is unable to read the barcode. If the numerals are there, the sales assistant can type them into the cash register manually. The barcode has made possible a fully automated sales system and also made the job of both sales assistants and shop managers much easier. Assistants no longer have to key prices into a cash register, a process that is always open to human error. The whole sales process is faster and more accurate than any previous method.

Barcodes have many different applications. The ISBN numbers that have for many years been allocated to books on publication can easily be presented as barcodes, so that books can be processed just like any other product. Car hire firms keep track of their vehicles by using barcodes attached to the car bumpers. Airlines use barcodes stuck to passengers' luggage to keep track of it during transit. In Japan in the 1990s there was even a fashion for girls to have barcode tattoos. Biology researchers have stuck tiny barcodes to bees in order to track their movements and mating habits.

The possibility of using barcodes with tiny transmitters is perhaps the most sinister aspect of this invention. It may be harmlessly useful in a warehouse, for instance, to be able to locate a product by using a scanner to pick up a barcode that has a transmitter embedded in it. But it would be possible for those in government in a totalitarian society or indeed an over-zealous paternalistic society to oblige every citizen to wear a personal barcode, complete with microtransmitter. This would make tracking the movements of any individual possible. No doubt police chiefs and government ministers would argue persuasively that it would ensure the safety of children, enable the tracing of illegal immigrants and be a major move towards the eradication of all crime.

THE INVENTION OF THE NUCLEAR POWER REACTOR

(1951)

THE NUCLEAR REACTOR was invented in 1942 by Enrico Fermi. His team, the Manhattan Project at the University of Chicago, produced the first atomic pile and the first nuclear chain reaction.

A nuclear power reactor is a system for controlling a nuclear chain reaction to liberate energy. The first nuclear power reactor was completed in 1951. It was the experimental breeder reactor (EBR) in Idaho and in December 1951 it generated the first usable electricity from the atom. It lit four light bulbs. Nuclear physicists already knew that nuclear power could generate electricity. The point of the EBR was to demonstrate that a breeder reactor would produce more energy than it used, a vital first step.

The first nuclear power station in the world was built in 1954 in the Soviet Union, at Obninsk, Kaluga Oblast. It was a small plant, still experimental in scale, producing enough electricity to supply 2,000 homes.

The first full-sized commercial scale nuclear power station in the world was Calder Hall, in Cumbria in the UK, and it was opened in October 1956. The weighty significance of the event was well understood by those present. The British Lord Privy Seal, Richard Butler, described it as epoch-making: 'It may be that after 1965 every new power station will be an atomic power station.' The Queen commented that the new power had proved capable of becoming a terrifying weapon of destruction but now it was for the first time harnessed for the common good. Her comment was pointed, as behind her was Windscale, the plant where the devices were made for Britain's first atomic bombs. It was certainly, as she said, the threshold of a new age.

The nearest town to Calder Hall was Workington, fifteen miles away, and it was Workington that became the first place in the world to receive its electricity, heat and light from nuclear energy. Fours hours after opening, Calder Hall electricity was reaching London.

Calder Hall was a gas-cooled, graphite-moderated reactor using the nuclear reaction in uranium rods to generate electricity.

The point of building the nuclear power station was to create an alternative power source to coal, partly because the coal reserves in British coalfields would eventually run out, partly because as the mines went deeper and deeper into the concealed coalfields the cost of coal rose higher and higher. There was an expectation that electricity from nuclear power would be significantly cheaper than electricity from coal. In the 1950s, the British government expected to save forty million tons of coal by investing in nuclear energy, and planned to generate ten per cent of the UK's energy needs from nuclear energy within ten years.

More nuclear power stations would be built, not only in Britain but in other countries too.

Then came a downturn in the prospects for nuclear energy, with a series of major setbacks. One was the Three Mile Island accident in 1979. No new orders for nuclear power stations were forthcoming in the United States, though this seems to have been a developing situation connected with dissatisfaction over greatly extended building times.

The next setback was the Chernobyl accident in 1986. Shortly after the explosion and radiation leak, over thirty power station workers and firefighters died as a result of exposure to high levels of radiation. Over the next few years, somewhere between 7,000 and 45,000 people are thought to have died in the Chernobyl area from cancer as a direct result of the radiation to which they were exposed. Chernobyl had to be abandoned and a large area around it will be uninhabitable and impossible to farm for thousands of years. The south-east wind blowing at the time of the radiation leak took high levels of radiation across eastern Europe, Scandinavia and northern Britain.

In spite of the severity of the disaster, there was no further tightening of safety regulations or building standards in the West, simply because the design of the Chernobyl reactors without containment buildings was known to be unsafe. There was a noticeable decrease in confidence in nuclear energy following Chernobyl, though. Chernobyl raised latent fears about the possible effects of accidents at nuclear power stations. In Italy, a referendum was held the year after Chernobyl and the result of that referendum was that Italy's four nuclear power plants were closed. In Britain, nuclear plants had been carefully located in relatively isolated places like Dungeness, but Britain is a small and densely populated country and nowhere in Britain is very far from towns and cities. What the Chernobyl disaster showed was that the risk is not only greater and more real than had been thought, but much farther-reaching geographically. The insidious radiation pollution could reach right across countries, right across whole continents. Whatever was said by experts about the poor design of the Chernobyl plant, the disaster showed that nuclear energy was not safe enough.

When the first nuclear power stations were built there was great excitement. Some scientists even suggested that once the power stations were built the electricity would be virtually free, as with hydro-electric stations, and certainly much cheaper than electricity produced in coal-fired power stations. This enthusiasm led successive British governments to wind down the coal industry: Britain would have no further need for coal. But the danger of radiation and the costly safety measures that had to be incorporated into power station design and management raised the cost of the electricity to the same as that derived from coal. Nuclear energy turned out to be no cheaper, and there was still the problem, the very expensive and so far insoluble problem, of disposing of the nuclear waste.

Nuclear energy was a wrong turning, and Chernobyl proved it. In a rational world, programmes would be under way to close nuclear power stations and replace them with safer ways of generating power. So it comes as something of a surprise to find a British government and its official advisers advocating the building of more nuclear power stations. What has changed the situation is the mounting panic about global warming. If temperatures are rising globally because of the release of carbon dioxide by the burning of fossil fuels, it is argued, we must reduce the rate at which we consume fossil fuels. The argument seems irresistible, but the flaw is that there is still no proof that carbon dioxide is behind the warming. If, as some scientists believe, variations in the sun are responsible for the temperature variations on the Earth, Britain will – once again, fifty years on from Calder Hall – be building nuclear power stations for the wrong reason.

THE INVENTION OF
THE HOVERCRAFT

(1955)

THE HOVERCRAFT WAS invented in 1955 by the British engineer Christopher Cockerell. The revolutionary new form of transport came into being after a very straightforward experiment that Cockerell carried out with a pair of tin cans connected to a vacuum cleaner. This satisfied him that a sufficient pressure of air squirted under a vehicle could lift it off the ground. Once off the ground, and therefore completely free of friction with it, the vehicle could be powered forwards, backwards or sideways with great economy and ease.

The first hovercraft was launched in the English Channel in 1959. It was an unqualified success, and it has been widely used ever since for both commercial and military transport.

Two jets of air are forced downwards underneath the vessel. The cushion of air that is created is held in by a flexible skirt that surrounds the base of the vessel. The skirt does not hang vertically, but is drawn in slightly round the bottom, so that air reaching the ground gets pulled back up towards the base of the craft rather than escaping out under the bottom of the skirt. The compressed air under the vessel swirls and eddies around creating invisible rollers of air that help to hold it up off the ground.

One of the most outstanding characteristics of the hovercraft is that it can move with equal ease across a land surface or a water surface, or anything in between, such as a swamp. It can also move effortlessly from one surface to another. It is possible, therefore, for passengers to board a hovercraft on land, and then be conveyed out across water. No special dock facility is needed, just a gentle ramp. When moving over land, the hovercraft needs no specially prepared surface, except that it cannot surmount obstacles that are higher than fifteen inches, which is the height of the air cushion. It is an ideal form of amphibious transport in wilderness areas, or areas where there has been little reconnaissance.

THE INVENTION OF THE CONTRACEPTIVE PILL

(1955)

THE CONTRACEPTIVE PILL was invented in 1955 by Gregory Pincus, a consultant in experimental biology at Shrewsbury in Massachusetts. In 1951, Pincus was persuaded by Margaret Sanger, a campaigner in favour of birth control, to focus his research work on reproductive biology. Pincus collaborated with M. C. Change and John Rock. The three of them studied the negative effects of steroid hormones on the fertility of mammals. Synthetic hormones became available in the 1950s, and Pincus organized some field trials of the effect on human fertility of doses of these synthetic hormones.

The trials, which were carried out in Puerto Rico and Haiti in 1954, were overwhelmingly successful.

Since then, oral contraceptives – generally referred to as 'the pill' – have gone into general use in all the more economically developed countries of the world. There have naturally been some concerns about the possible detrimental side effects on women taking the pill, especially for many years at a time. Even so, the popularity of oral contraception as the major birth control technique, particularly within marriage and other long-term stable partnerships, remains undiminished.

The success of oral contraception is almost a pharmaceutical freak. It is extremely uncommon for synthetic chemical agents to have a 100 per cent effectiveness in controlling human physiology. It goes almost without saying that if the pill was only seventy per cent reliable its practical usefulness as a contraceptive would plummet to zero.

The sociological effects of this astonishing invention have been phenomenal. The main birth control technique until the arrival of the pill was the condom. It was worn by men, so whether it was worn or not was a decision made by men. The oral contraceptive, taken discreetly by women, was something about which women could make their own decisions. For the first time in human history, women could decide when, or even whether, they would become pregnant. As a result, women have been freed to work, to have professional careers, and families have become smaller, enabling parents to give more care and attention to each child. The effects of this social, economic and demographic revolution have spread out first through the richer countries, and now through the poorer countries too.

Many human geographers see this process of reducing the birth rate as a major way of improving the quality of life for both women and children throughout the world. The invention of the contraceptive pill has changed the world on a very large scale, by staving off the population explosion that pessimists were predicting in the 1960s.

THE INVENTION OF
THE SMOKELESS ZONE

(1956)

THE CLEAN AIR Act was passed in the UK in 1956, mainly as a response to the Great London Smog of 1952. The Act was designed mainly by officials of the Ministry of Housing and Local Government and the Department of Health for Scotland. One of the provisions of the Act was that smokeless zones would be established, areas within which it would become illegal to produce smoke and to release particles of grit into the air. The ban on smoke was applied to domestic and industrial premises alike. Only smokeless fuels could be burnt within the smokeless zones. The use of cleaner coals and the increased use of electricity and gas helped to reduce sulphur dioxide levels in the city. New power stations were built away from the cities. Old coal-fired power stations like Battersea were closed down. Although domestic coal fires had contributed to the atmospheric pollution, recent research suggests that commercial and industrial sources were responsible for most of it. De-industrialization was the major factor in making London's air cleaner.

London had become a very smoky city in the nineteenth century; it was one of the many negative aspects of the Industrial Revolution. There were not only lots of factories burning coal: every household in the capital burned coal. The smoke and the soot falling from it were major pollutants. By the early 1950s all the public buildings in London were blackened with soot and the Portland stone of which many of them were built had become seriously corroded. The particles in the air formed condensation nuclei, encouraging the formation of fog. This led to the development of what were colloquially called 'pea-soupers', smoke-aggravated fogs that reduced visibility almost to zero. These smogs were a problem for drivers and pedestrians alike, bringing traffic to a near standstill. The strongly polluted air created other problems. People already suffering from respiratory problems were in serious danger from the smogs, and the death rate always increased during smogs.

As a result of the Clean Air Act of 1956 both smoke and sulphur dioxide levels in London fell. The fact that the downward trend has continued is probably due to other factors. Shifts in the economy, in particular the steep decline of heavy industry in Britain generally from 1950 onwards, go a long way to explaining the improving air quality. It cannot really be down to the Clean Air Act of 1956 alone, or its follow-up Act in 1968.

The timing of the first Clean Air Act, 1956, related to a series of severe smogs, of which the worst was the Great London Smog of 1952. There was another serious smog in 1956, when nearly the whole of England and Wales was enveloped in fog. It happened just after the passing of the Clean Air Act, but before its provisions could come into operation; it would take time for residents and industrialists to make the switch to smokeless fuels. At least six people died in road accidents caused by the poor visibility, which in East Anglia was reduced to five yards. Poor visibility caused some rail accidents; a train from Lichfield to Birmingham ran into the back of a stationary tank engine outside New Street Station, Birmingham, injuring the driver. The fog spread out over the Channel too. Three liners steaming up the Channel bringing troops home from Egypt after the Suez fiasco made very slow speeds through the fog. Flights were delayed. In the five days following the smog in December 1956, 4,000 people died prematurely.

Later legislation consisted of refinements with the same general intention. The 1968 Act introduced the use of taller chimneys for industries burning coal, oil or gas. This was in recognition of the fact that it was impossible to control or reduce the emission of sulphur dioxide, so the best that could be done was to release it as far above ground level as possible.

The cleaning up of London and other British cities – notably Glasgow – was watched with keen interest by the governments of other countries with similar problems. The United States Congress passed its own Clean Air Act in 1963 and similarly followed it up with a series of amendment Acts in 1966, 1970, 1977 and 1990. These were federal Acts and many US state governments and local governments decided to bring in their own supporting legislation too. The result has been a major improvement to the smog problem in Los Angeles.

Today, the greatest problems of atmospheric pollution are encountered in the fast developing and fast expanding cities of the less economically developed countries. Mexico City has suffered severely from the effects of rapid industrialization.

THE INVENTION OF THE VIDEOTAPE RECORDER

(1956)

THE VIDEOTAPE RECORDER was invented by a team at Ampex led by Charles Paulson Ginsburg. From the earliest days of television right through to the mid-1950s there was no way of recording television signals. Nearly all television programmes were broadcast live. For some dramas, where for instance a scene outside had to be interpolated, clips of film were resorted to, but where possible this technique was avoided as it was expensive. The need to do almost everything live, including sport and discussion programmes put a great strain on producers and performers alike.

At the receiving end, in the home, there was no way of recording a programme. Viewers had to watch a programme when it was broadcast, or miss it altogether. The invention of the videotape recorder in 1956 marked the beginning of a revolution in television. It became much easier for programme makers to make their programmes ahead of the broadcast, which made editing and a more professional finish possible. From this point on, virtually all programmes were pre-recorded. This gave producers far greater opportunities for quality control and, some might argue, censorship. Occasionally, for example, someone might use bad language in an overheated discussion. If the programme was broadcast live, as on the now-famous occasion when the theatre critic Kenneth Tynan became the first person to use a four-letter word on television, the producer could do nothing about it – except apologize afterwards. If the programme had been taped, it would have been very easy for the producer to edit out the offending few seconds.

An experimental videotape recorder had been demonstrated by the Radio Corporation of America (RCA) in 1953, but the experiment also demonstrated one of the problems. There was so much information encoded into a television signal that on the RCA system four minutes of TV took one and a half miles of tape. This was not a practical system that anyone was going to adopt.

Charles Ginsburg's team overcame the problem encountered by RCA by encoding more of the information across the width of the recording tape. The machine deployed rapidly spinning recording heads that scanned across the width of the tape. As they did so, they laid down a succession of overlapping stripes of recorded information, while the tape itself moved through at the fairly slow speed of fifteen inches per second. With this technique, it was possible to record one hour of television in a ten-inch diameter reel of tape that was two inches wide. This was much more economical in terms of tape used, and in terms of storage space. These early videotape spools were, even so, still quite space-consuming, so it is not surprising that at this stage not every programme was saved in perpetuity by the broadcasting companies; they simply did not have enough storage space to do that.

Ampex sold the first videotape recorder (VTR) for 50,000 dollars in 1956. The first videocassette recorders (VCRs) were sold by Sony in 1971. Video Home System (VHS) tapes in a large, book-sized format were introduced by Panasonic and JVC in 1976. These were very popular for home use for about thirty years. It was and still is possible for viewers to buy blank tapes quite cheaply and tape television programmes they want to watch again, or at a different time. This has given viewers far more control over their time. It has also enabled people to watch favourite programmes over and over again. More and more programmes have become available commercially, introducing far greater flexibility into home entertainment. For educational purposes too, videotape has been a great asset. Teachers have found it enormously useful to be able to use a TV documentary or drama at an appropriate point in a course.

Doubtless videocassette recorders will continue in use, though from the beginning of the twenty-first century there has been increasing competition from a new recording system: DVDs are smaller and lighter, take less storage space, and appear to be more durable.

THE INVENTION OF THE MAN-MADE SATELLITE

(1957)

THE FIRST MAN-made satellite ever to go into orbit round the Earth was launched by the Russians on 4 October 1957. This tiny artificial moon was known as Sputnik-1 and it went into orbit 500 miles above the Earth, taking ninety-five minutes to complete a circuit. It travelled at an astonishing 18,000 miles per hour.

The satellite was a simple device, a metal globe not much bigger than a football with trailing radio antennae. Inside were two radio transmitters that sent out a distinctive intermittent piping sound that could be picked up by ground-based radio stations. The bleep was an extremely clever publicity device, in that it performed no function whatever other than to remind people that it was there. The radio signal was first picked up by Geoffrey Perry, a science teacher at Kettering Grammar School in England; he had a gift for picking up the radio signals from this and many subsequent satellites long before the professionals and became a minor media celebrity on the strength of it.

The Sputnik weighed 185 pounds, which implied that the Russians had developed a fairly powerful launching rocket. The Sputnik was six times heavier than the satellite the Americans were planning to launch in 1958. The Russians had clearly won the first round in the Space Race by getting their weighty satellite into orbit first.

Throughout the Cold War, the Russians were extremely secretive about their planned projects, only announcing successes afterwards, and sometimes concealing failed missions altogether. By contrast, the Americans were very open about their missions, and both the successes and the failures were very public. After the launch of the Sputnik the Russians proudly issued a detailed statement of their achievement, giving the specifications of the satellite. The Americans acknowledged that it was a great achievement.

Dr Joseph Kaplan, who was Chairman of the US National Committee for the International Geophysical Year (1958), said the Russian achievement was 'fantastic'. But it was a great blow to American pride to be beaten in the Space Race in this way. The Pentagon was too stunned to respond immediately, saying only that it needed time to study the Russian report in detail.

Apart from being a technical and political landmark, the first artificial satellite marked the start of an era of what has been called remote sensing. There are now scores of satellites circling the Earth, some of them geo-stationary, rotating with the Earth and therefore staying above the same place on the Earth's surface all the time, but all of them doing something useful. They send back all kinds of information. There is Meteosat, a weather satellite, which every quarter of an hour sends down a photograph of the current cloud patterns; sequences of these photographs are very useful to meteorologists in helping them to understand weather systems and forecast weather. The sequences of photographs from the satellite are often used on television weather forecasts. There is also Landsat, which sends down detailed true-colour photographs of the landscape and can be used as an aid to topographical and vegetation mapping.

Satellites can be used to spot forest fires in the Amazon Basin, and are useful in policing illegal burning. They can even be used to identify what crops farmers are growing. In the European Union, where subsidies are available for growing some crops, there is always the possibility of farmers making false claims about what they are growing on their land. The 'spy in the sky' can be used by EU officials to find out exactly what they are growing. Satellites can be used for more conventional spying too. Expert interpreters are able to identify missile silos, armaments factories, and even divisions of tanks on the move, so the satellite has become a new weapon in modern warfare.

Satellites can also give unparalleled detail of changes in river channels. Mapping a rapidly changing multi-channel river like the

Brahmaputra in India was almost impossible to do by conventional survey methods at ground level. Satellite photographs enable maps of the Brahmaputra's floodplain to be updated instantly which makes understanding the river's hydrology possible. Infra-red photography can give information on the health of vegetation, and give us information on the impact of acid rain on forests in Scandinavia.

The launching of the first satellite was the launching of a new age of data-gathering. Satellite photography has enabled us to find out much more about the world – and how rapidly it is changing.

THE INVENTION OF
THE LASER

(1958)

THE WORD LASER is an acronym; it stands for light amplification by the stimulated emission of radiation. Although lasers did not come into existence until the 1950s, Albert Einstein proposed the theoretical possibility of the process that made lasers possible.

In 1954, Arthur Schawlow and Charles Townes invented the maser. This is another acronym, standing for microwave amplification by stimulated emission of radiation. The technology involved ammonia gas and microwave radiation, and it was very similar to the technology that would produce lasers except that it did not produce visible light. Schawlow and Townes were granted a patent for the maser in 1959; it is used to amplify radio signals and as an extremely sensitive detector for astronomical research.

After developing the maser, Townes and Schawlow proposed the visible laser as the next possibility, using infra-red with or without light from the visible spectrum. Theodore Maiman invented the ruby laser, which is considered the first successful light laser. Gordon Gould invented the light laser at about the same time, and there is disagreement about which of the two men has the prior claim. Gould was the first person to use the word laser, and he may well have been the first to make the first light laser in 1958, but he did not formally apply for a patent until the following year. By then he was too late: his application was turned down and his technology was exploited by other people. Gould did not win the right to his patent until 1977. Other types of laser quickly followed: a gas laser was invented by Ali Javan in 1960, and a semiconductor injection laser by Robert Hall in 1960.

One outstanding property of lasers is that they can be focused down to an incredibly narrow beam. Because of this, lasers have turned out to have all sorts of useful applications. In science they are a great help in spectroscopy. They allow gigabytes of information to be recorded in the microscopic pits of a compact disc or DVD. They can be used to focus relatively low wattage power to such a high intensity that it can be used to cut, burn or vaporize material. They can be used for remote sensing. They have great applications aboard spacecraft. When the Apollo astronauts landed on the Moon, they planted reflector arrays on the Moon's surface. Laser beams are directed through high-powered tele-scopes on the Earth towards the reflector arrays. The time taken for the beams of light to travel to and from the Moon allow us to measure the distance between the Earth and the Moon with much greater accuracy than ever before.

Laser-sighting devices are fitted to military and police rifles to help soldiers and police officers hit their targets. A laser sight is a small visible-light laser fitted to a rifle or even a handgun, lined up so that it emits a beam of light parallel to the barrel. Because of the intense focus of the laser beam, it appears as a small spot even if the target is a long way away. To aim the gun, the user has only to place the red spot on the target and pull the trigger. The new technology is very effective as a deterrent. Criminals who find they have a dot of red light on their chest usually put their hands up and surrender. This idea has been extended to produce a non-lethal laser illuminator; it briefly floods the target with light, which may well be enough to scare an aggressive criminal into running away. If the gun user does not want the target to be aware that laser sighting is being used, an invisible infra-red laser can be used instead. Some countries developed a blinding laser, designed to physi-cally disable enemy personnel, but this is now explicitly banned under the international protocols that outline the rules of war.

Light lasers can also be used defensively. Mobile lasers on the ground can lock on to missiles in flight and destroy them before they reach their targets. A system is under development (mobile tactical high-energy laser) that could be deployed in the field; the idea is that it will be able to track incoming cruise missiles or shells by radar and then destroy them by using a powerful deuterium fluoride laser. Lasers played a huge part in the design of the USA's strategic defence initiative (SDI, commonly known as Star Wars). This was an attempt to design a comprehensive defence system that would use ground-based and satellite-based lasers to knock out incoming inter-continental ballistic missiles. There were many problems here, as the best moment to destroy an ICBM is immediately after take-off, but if that point is many thousands of miles away, the laser would be bent and therefore distorted. This would make aiming the laser very difficult and also reduce its effectiveness. Aiming the laser accurately would be much easier at the last moment, just before it reaches its target, but there would be little to gain from exploding an ICBM, complete with nuclear warhead, over New York to stop it reaching Washington DC.

The same laser-sighting device that is fitted to rifles and handguns can be adapted for use on telescopes to help astronomers to find out precisely which part of the sky they are looking at.

Lasers may be of use in thermonuclear fusion reactors. Short pulses of high-intensity light are fired experimentally at pellets of tritium-deuterium, with the aim of squeezing the pellets and inducing atomic fusion. So far the experiments have been unsuccessful, producing no more power than has been used in generating the lasers, but the work continues.

Lasers can be used as rangefinders, in other words, they can measure exactly how far away a distant object is. This ability can help greatly in aiming, for instance, the gun of a tank. They can be used for tracking, and in combination with radar they can locate aircraft to the nearest three or four feet.

When the laser was first invented, it was envisaged that its main role in the military area would be as a death ray. This idea quickly found its way into comic strips and science fiction, but proved to be impossible to achieve in reality. The idea was that a hand-held device might be designed for ordinary infantrymen to use or a larger laser cannon might be designed to be mounted on top of a vehicle. Perhaps a whole range of death ray weaponry might be produced to replace the existing hardware. But a laser that was powerful enough to injure a person would be far too heavy for an individual soldier to lift. A very high-power laser that could cut through tank armour would be easily misaligned by any knocks or vibrations it experienced, so it would not be at all practicable to deploy it in a battle situation. So, the ray gun of science fiction will probably remain science fiction.

In medicine, lasers are proving to be far more useful. They can be used in very short but powerful pulses of light energy to break apart bladder stones without any need for surgery. They can be used for various kinds of cosmetic surgery, eye surgery, dental surgery.

The laser is one of those technical inventions that at first sight appears unexciting, but turns out to have myriads of uses, from surgery on the retina to traffic speed checks and laser-light displays. We must be thankful that the death ray turned out not to be one of them.

THE INVENTION OF THE PACEMAKER

(1958)

A MAJOR HEART problem is arrhythmia, which means that the heart is not beating properly. This may mean that it beats faster than normal (tachycardia) or slower than normal (bradycardia). An implanted pacemaker can send sequences of correctly timed impulses that will correct both tachycardia and bradycardia. A third type of problem is fibrillation, which means that different parts of the heart are beating in an uncontrolled and unco-ordinated way. Ventricular fibrillation is extremely serious and can result in death within minutes; atrial fibrillation is less serious, but can lead to other problems if left uncorrected.

The pacemaker is one of those stepwise inventions to which several people made contributions at different times. The Australian doctor Edgar Booth gave a demonstration of a portable pacemaker unit in 1931. There is general agreement that the main inventor of the modern pacemaker was Paul Maurice Zoll. He was a Harvard cardiologist who began his research in earnest in 1945. By 1950 he was artificially restarting the heart of a dog. In 1952, he made the critical move to stimulating a human heart. His patient was a sixty-five-year-old man who was suffering from terminal coronary disease and recurring cardiac arrest. Zoll's attempt at external stimulation to keep the man's heart beating was successful; it kept the man alive for another six months.

Paul Zoll's work was not unanimously praised. There were those, even in the medical profession, who thought he was defying the will of God in keeping people alive in this way. A Catholic newspaper was perhaps more enlightened when it wrote that people should not be alarmed by the apparent outlandishness of Zoll's treatment because 'God works in many strange ways.'

What Zoll had discovered was very important. It was possible to resuscitate a patient suffering from cardiac arrest by applying a strong electrical countershock.

By the mid-1950s, cardiac pacemakers were made operating on electrodes stuck to the skin, but these left uncomfortable burns. Electrodes were stuck through the skin, but these simply introduced infections. Another researcher, Ake Senning, made the radical suggestion that the entire pacemaker should be implanted inside the patient's body. This led Rune Elmqvist to design the first pacemaker that could be implanted. It contained a generator that was able to deliver a two-volt impulse. The new transistor technology was a great help in making the pacemaker small enough and light enough to be implanted. A forty-three-year-old patient suffering from life-threatening heart seizures was the recipient of the very first implanted pacemaker in 1958. The patient, Arn Larsson, had been having thirty resuscitations a day. After Ake Senning operated and implanted his pacemaker, Arn Larsson was able to resume a normal active life. Since then, thousands of people who would otherwise have died prematurely have been restored to normal health by pacemakers.

THE INVENTION OF THE MICROCHIP

(1959)

THE AMERICAN ENGINEERS who were responsible for inventing the microchip were Jack Kilby, who worked for Texas Instruments, and Robert Noyce, who worked for Fairchild Semiconductor. The microchip made it possible to install a large amount of circuitry in a very small space. This major innovation – compactness – led the way to the manufacture of a range of miniaturized products, such as electronic wristwatches and electronic clock mechanisms.

Working independently, Kilby successfully encased an integrated circuit within a single silicon wafer. Noyce's contribution to the new technology was to find a way of joining the tiny circuits by printing, which eliminated thousands of man-hours of labour and made mass production possible. These two breakthroughs together drastically reduced the size, weight and cost of electronic components. Robert Noyce went on to found the Intel Corporation, which was to give the microchip memory and logic functions to produce the microprocessor. This in turn would make possible the development of the personal computer.

Today we take the manageable size of our desktop and laptop computers for granted. But before the invention of miniature printed circuits, before the invention of the silicon chip, computers were bulky, heavy and hugely expensive. The revolutionary work of Jack Kilby and Robert Noyce made the democratization of computing possible.

THE INVENTION OF HEART TRANSPLANT SURGERY

(1967)

THE SOUTH AFRICAN surgeon, Christiaan Barnard, performed the first heart transplant operation on 3 December 1967. The operation was a global media sensation as people everywhere recognized that a major breakthrough had been made in medical science. The euphoria diminished significantly when the patient, Louis Washkansky, died less than three weeks after the operation. Evidently replacing the human heart was not as straightforward as it had appeared.

The second heart transplant took place at the National Heart Hospital in London on 3 May 1968. The patient, Frederick West, survived for forty-five days, which was a significant advance and seemed to show that the technique had the possibility of working.

One implication of this new development in transplant surgery was that guidelines were urgently needed to establish when a patient was dead. The American Medical Association adopted a new standard for declaring a potential organ donor dead. It was when two doctors independently declared the patient's death irreversible and beyond hope of resuscitation. The debate continued, as it is by no means obvious when death has finally occurred; a patient's brain may go on functioning for some time after the patient has stopped breathing and after the heart has stopped beating. Eventually the idea of brain death was introduced.

Another implication was that certain patients, such as young traffic accident victims, might be seen as quarries for transplant organs and that surgeons might become more interested in pronouncing death than saving life. There was also the problem of permission. Close relatives of the recently dead are usually in no state to be asked whether they will agree to their loved ones' organs being removed for recycling. The idea of an organ donor card seemed like a better option, and now many of us carry cards saying that we are happy to have our organs reused.

The problem of donors would disappear if artificial hearts could be manufactured. At the University of Utah Medical Center, an artificial heart was successfully implanted into a patient, Barnet Clark, who lived for 112 days afterwards. The artificial heart was designed by Robert Jarvik. Even more controversial was the experimental transplant of a baboon's heart into a two-week-old baby girl at the Loma Linda Medical Centre in California. The baby survived with the baboon's heart for twenty days.

On 17 December 1986 one of the most ambitious organ transplant operations was attempted. The first triple transplant, of heart, lungs and liver, was undertaken at Papworth Hospital, Cambridge, by John Wallwork and Roy Calne. Five years later, surgeons found that they could repair damaged hearts by using muscles from other parts of the body.

The first heart transplant was greeted with great excitement, followed by disappointment when the patient died. The major problem, a repeating problem, is that of rejection. The human body is designed to fight off foreign tissue of any kind; it is part of the body's natural defence against infection. It is possible to feed the transplant patient drugs that will reduce this natural defence, but this in turn reduces the body's immune system and makes it vulnerable to infections. These problems may be overcome in time, but the early promise of organ transplant has not been realized. It is still not a very reliable or safe procedure.

THE INVENTION OF SPACE TRAVEL

(1969)

ROCKET SCIENCE THEORY developed slowly during the first half of the twentieth century, and the solid achievement in rocket technology itself was singularly unimpressive. It was odd, therefore, that there was a general expectation that space travel would soon be happening. In 1951 perhaps the most famous rocket scientist of his time, Wernher von Braun, was saying that a mission to Mars could be accomplished with as few as forty-six rockets, in a round trip that would take three years. Science fiction writers were eagerly anticipating far more ambitious flights that took them to the outer planets. The actual achievement in 1951 fell far short of these perceptions, but there was wild optimism. Von Braun said, 'Man belongs wherever he wants to go – and he'll do plenty well when he gets there.'

The Cold War generated the edge to what became a historic space race between America and the Soviet Union. The Soviets won the first round when Yuri Gagarin became the 'first man in space' on 12 April 1961. Gagarin orbited the Earth in his spacecraft Vostok 1, in a flight that lasted under two hours. Not many months later, the US caught up by putting an American, John Glenn, into orbit in Friendship 7. Then in 1969 came the Apollo 11 flight to the Moon, which was described as the most momentous journey since Columbus's 1492 voyage across the Atlantic.

On 21 July 1969, a man walked on the Moon for the first time. It happened at 3.56 a.m. British Summer Time. The American astronaut Neil Armstrong stepped out of Eagle, the lunar module from Apollo 11, and warily descended a ladder onto the pale grey dusty surface of the Moon. As he planted the first human footprint on the Moon, he said, 'That's one small step for man, one giant leap for mankind.' Shortly afterwards he was joined by his fellow astronaut Buzz Aldrin and the two of them experimented with moving about under the low gravity conditions. The moment was watched on television all over the world, by hundreds of millions of people. Armstrong was intrigued by the texture of the pale dust under foot. 'The surface is like a fine powder. It has a soft beauty all its own, like some desert of the United States.'

The Moon landing came as the climax of the Apollo 11 mission, four days after blasting off from Cape Canaveral (or Cape Kennedy). After an uneventful flight, the Apollo went into orbit round the Moon. Armstrong and Aldrin transferred to the lunar module and started their descent to the Moon. There was a tense final moment, when without much fuel left they had to avoid a boulder-filled crater. The two astronauts stayed on the Moon for less than a day before lifting off to rendezvous with Apollo 11, still orbiting the Moon.

Five days later, scientists on the Earth put the rocks collected on the Moon into quarantine for two months. For the moment frustrated geologists were only allowed to look at them through a window. The quarantine procedure was a precaution, just in case there were organisms of any kind on the Moon that could infect plants or animals on Earth. In fact the rocks turned out to be sterile. They also, when dated, turned out to be very old. They were 4.5 billion years old, and therefore had originated when the solar system as a whole was created. Rocks exposed at the surface of the Earth are much younger, because the Earth is a more active planet and the surface rocks have been recycled many times by landscape processes. The Moon's surface shows every sign of being a very ancient landscape, little changed since its creation, apart from the continuing bombardment by meteorites.

For some reason, millions of Americans were persuaded that they were watching a simulation staged in a film studio, or in fact some desert of the United States, to divert their attention away from the Vietnam War. Conspiracy theories multiplied. Millions more were concerned about the huge sum of money spent on the Moonshot, believing that the money could have been spent more usefully. So, even as the event happened, people were reading a variety of subtexts into it.

The visit to the Moon was man's first step away from the Earth out into the solar system, and undoubtedly more symbolic than useful, as the rock samples brought back by hand could have been collected as easily and far more cheaply by a robot. But it marked a critical stage in man's interest in the cosmos. People speculated about journeys further afield, and talk began about a mission to Mars. The danger and the ever-increasing cost initially ruled that out, but it is an idea that has recently returned.

But after the Apollo mission, the space race petered out. Gradually it became clear that the Soviet Union's space programme had not so much run out of steam as out of cash. The Cold War itself was coming to an end as the Soviet Union could no longer afford to stockpile the expensive nuclear weaponry. Competition shifted gradually into collaboration, with the US joining forces with the Soviet Union, Europe, Japan and Canada to build an International Space Station. The emphasis shifted away from manned flights and after Apollo subsequent expeditions out into the solar system have been undertaken by unmanned vehicles.

Even the outer planets have been flown past now, and the photographs taken during those close encounters have been carefully analysed in order to reconstruct the geology, topography and climate of the outer planets. The most striking thing about them is their sheer diversity. Given their common origin, they are surprisingly different and individual. Another surprise was the unexpected jewel-like beauty of the Earth when seen from the Moon.

The Apollo 11 mission to the Moon may have been a politically motivated stunt to show the Soviets that America was ahead, but it really did demonstrate the sophistication and precision that was pos-sible in modern rocket science. Devices, whether manned or unmanned, could be sent scouting round the solar system to gather data and brought safely back again. It was very much the beginning of a new age, an age of physical reaching-out into the cosmos.

The exploration of the solar system continues. Remote sensing has enabled us to look through the dense cloud layer covering Venus to reveal an incredibly hostile, lead-meltingly hot landscape that we shall never be able to land on. In 2006 a French space probe called COROT was launched in Kazakhstan. COROT is a satellite with a space telescope mounted on it, and by 2009 it will have completed a search for undiscovered rocky planets orbiting other nearby stars. There is the possibility of finding an exoplanet (a planet outside the solar system) which is suitable for human habitation and may already have life on it. Russia plans to launch its Koronas-Foton spacecraft in 2008, to observe the Sun more closely.

NASA meanwhile has thoroughly road-tested the two rovers it sent to explore the surface of Mars, Spirit and Opportunity. They have been working on the Martian surface for three years, lasting far longer than expected. The Mars mission looks increasingly achievable from the technical point of view, though much will depend on whether the Americans will want to fund it. There may also be unforeseen problems in manning a flight that will necessarily take a very long time.

The human body is adapted to a cycle of twenty-four hours. In fact the human body clock runs on a cycle of twenty-four hours and eleven minutes, but this is corrected by our response to sun time. The Martian day is thirty-nine minutes longer, and on a long stay this may prove to be a problem, especially as there will be no familiar corrective. There is no way of knowing whether astronauts will be able to adjust to this. There could be problems on the flights to and from Mars too. Crews on space missions always sleep badly, with astronauts getting on average two hours per night less sleep than they get on Earth. This must have an effect on their alertness and accuracy when undertaking tasks and reading instruments, and could have serious consequences. It is known, for example, that on Earth night-shift workers, who have disrupted sleep patterns, are fifty per cent more likely than ordinary day workers to have a car crash on their way home from work. A crucial factor in the success of the Mars mission, or any other long mission, will be getting enough sleep.

The medium-term prospect of a Mars mission is fraught with danger. The sheer length of the journey is going to create problems for the crew. Missions to other more distant planets would take even longer and be even more dangerous. Spacecraft already use the gravitational pull of successive planets, one at a time, to slingshot them along at high speeds. In the long term, spacecraft may be able to exploit an elaborate network of space freeways that weave invisibly round the solar system. Martin Lo, a NASA scientist, has identified a web of freeways that would allow the exploitation of previously untapped gravitational relationships between planets and their satellites; it is called the interplanetary superhighway. Flying along it would save on fuel and make for greater speed. The superhighway has apparently already been used by the asteroids and comets that roam around the solar system.

Travelling outside the solar system still looks like a science fiction dream. There has been speculation that spaceships might use black holes to leave the universe at one point and come out somewhere else, possibly in a different universe altogether; they might similarly be used as portals for time travel. Most scientists now believe that black holes are so destructive that approaching spacecraft would be torn apart. But that is only speculation, and black holes may not annihilate everything that falls into them. They may have weak sectors that would allow spaceships to fly safely through. The possibility of hyperspace travel is still there. The problem would be returning. Even if it proved to be possible to fly into a black hole and find a wormhole through to another part of the universe, there is no way of knowing where (or when) you would emerge, and you might not be able to get back again. There is also the problem of travelling to your nearest black hole to begin the (even more) hazardous journey to Elsewhere. There is believed to be a black hole at the centre of our own Milky Way galaxy, but that is 26,000 light-years away. Even travelling at near the speed of light, it would take a spaceship 30,000 years to reach it and the spaceship would in any case pass hundreds of other stars and alien solar systems on its way there.

Certainly there is no point in planning normal-route expeditions outside our galaxy. The nearest galaxy to ours is the Canis Major Dwarf Galaxy. Even travelling at the speed of light it would take us 25,000 years to get there – an impossible project. But there is plenty to consider within the Milky Way. There are teams of research astronomers involved in systematic efforts to find life elsewhere in the universe. Their project is called SETI, the Search for Extra-Terrestrial Intelligence, and its focus is naturally our own galaxy. There are enormous numbers of stars within our galaxy, currently estimated at 400 billion, double the number that was believed to exist only a few years ago. It is likely that most of those 400 billion stars have planets, so it is statistically highly probable that life exists in many solar systems scattered through it. In fact widespread life is so probable that we can reasonably expect that there are highly evolved and intelligent life forms on many other planets.

It is probable that there are 10,000 civilizations out there at least as advanced as our own. And that is a conservative estimate. Carl Sagan's estimate was a million civilizations. It is exciting to contemplate what we might learn, what inventions we might acquire, how we might be enriched by contact with just one of those alien civilizations. It is unlikely that they are going to be hostile or aggressive, as to judge from terrestrial models such civilizations are very short-lived. The destructive tend to self-destruct. The lure of ten thousand or a million civilizations is irresistible, the ultimate spur to find more efficient and more revolutionary forms of space travel.

THE INVENTION
OF E-MAIL

(1971)

ELECTRONIC MAIL OR e-mail was invented in 1971 by computer
programmer Raymond Tomlinson. Born in Amsterdam, New York, he
attended the Rensselaer Polytechnic Institute in Troy, New York, where
he graduated in 1963. He then went to the Massachusetts Institute of
Technology to deepen his knowledge of electrical engineering and
received an MSc in 1965.

Tomlinson's great invention is nothing less than a major new form of
communication. Invented in 1971, it accelerated and became very widespread in the 1990s. It works on computers that are linked to each other,
either by the internet or by an intranet system set up by an organization
or company to allow employees to communicate with one another.

Messages are typed in using the computer keyboard, sent electronically, and then received whenever the recipient switches on his or her
computer to check their mail. E-mails can be stored to read later, just
like letters, or they can be read and responded to immediately, so the
response can be almost as fast as phoning.

Electronic mail had its beginnings as far back as 1961, when MIT
demonstrated what it called a Compatible Time-Sharing System, which
envisaged lots of users logging into a central computer from remote dial-
up terminals. The idea was therefore in existence that early, but for a long
time the electronic communication was only between people connected
to the same host computer.

The major leap forward was made by Raymond Tomlinson in 1971.
Tomlinson set up an e-mail system that for the first time enabled people
served by different hosts to communicate with each other. This introduced far more flexibility and made the system inclusive instead of

exclusive. To enable the system to identify the different hosts, Tomlinson introduced the @ sign to separate the user from their machine. This has been used in e-mail addresses ever since. Tomlinson himself sent the first e-mail, but he cannot remember what it was, only that it was 'insignificant'. This is reminiscent of the insignificant content of the first telephone messages. At the time, Tomlinson did not realize how significant his development was. He showed what he was doing to a colleague, and said, 'Don't tell anyone! This isn't what we're supposed to be working on!'

THE INVENTION OF
THE CT SCANNER
(TOMOGRAPHY)

(1971)

THE INVENTION OF the computed tomography scan (CT scan) was a major breakthrough, enabling doctors for the first time to see the detail of soft tissues inside the human body. X-ray photography allowed them to see the outlines of bones, but only the shadows of internal organs.

The CT scan uses a computer to create a sequence of detailed cross-section images of the body. It is as if the body is put through a bacon slicer and the medics are able to examine each slice separately. In this way doctors can pin-point the precise locations within the body, and within a specific organ, that give cause for concern.

The CT scanner was invented in 1971 by Godfrey Hounsfield, a British engineer working for EMI Laboratories. It was, as we have seen so often before, invented independently and simultaneously elsewhere. The other inventor was Allan Cormack, a South African physicist who was working at Tufts University.

The initial system designed by Hounsfield was very slow to operate. It took hours to pick up the data for just one slice and days to assemble a computer image from the data. The system was refined, so that now a CT scanner could supply images of several slices from millions of individual points of data in under one second. It became possible to scan somebody's chest in ten seconds, which was a major step forward in patient comfort and safety.

The CT scan is based on the x-ray principle. Since prolonged exposure to x-rays is dangerous, shortening the procedure significantly increases its safety. The scanner looks like a large washing machine. It has a big square fascia and a round portal through which the patient enters on a kind of conveyor belt. Inside, there is a rotating frame with a tube emitting x-rays on one side and a detector on the opposite side. Each time the tube and detector are rotated through 360 degrees, about a thousand images of one slice through the body are taken with a focus set to a particular thickness. There is also a dedicated computer that translates the points of data into images.

CT scanning is a major refinement of x-ray photography. It allows medics to differentiate between the various soft tissues within the body, and is particularly useful in finding irregularities such as tumours. When it finds tumours, it does so with great precision, determining the exact size, shape and location; this is invaluable in helping doctors to decide on appropriate treatments.

THE INVENTION OF THE CELLPHONE (MOBILE PHONE)

(1973)

ON 3 APRIL 1973 the first public phone call was made on a portable cellular phone. It was Martin Cooper, the cellphone inventor and now chairman and co-founder of ArrayComm Inc, who made the call in New York while he was general manager of Motorola's Communications Systems Division and he phoned his opposite number at a rival company. The moment was the culmination of Cooper's vision for personal wireless communication and it marked a highly significant shift in emphasis. Instead of people needing to get to a phone to make a call, they could make their call wherever they happened to be. Cooper reported that New Yorkers were amazed to see him walking down the street while talking on his new phone. They were amazed because in 1973 there were no cordless phones, let alone cellphones.

There were to be many implications. Car drivers were able to make emergency calls if their vehicles broke down – anywhere. Citizens in difficulties, whether being attacked or robbed or witnessing a crime taking place, were more easily able to call for help.

The new phone was quite unwieldy. The first model was as big as a brick, and weighed two pounds. Cooper started a ten-year process of refining the phone so that it would be acceptable to its potentially huge market. Once Motorola had managed to halve its weight, in 1983, it went into commercial service. At 3,500 dollars each, those first phones were very expensive, but during the next twenty years the cellphones, now more often called mobile phones, became cheaper and cheaper, as well as smaller and lighter. By 1990, there were a million cellphone subscribers in the USA.

Most modern mobile phones are cell-structured. There is a radio link between the handset and the nearest cell site, of which there about 40,000 in the UK. When a handset, for example carried in a car, travels too far from that cell site, a computer system comes into operation and creates a link between the handset and whichever cell site is closest to it. That new link may be on a different channel, yet there is no interruption to the call. The more recent phones are foldable. This is in conscious imitation of the fictional communicators used in the television science fiction series *Star Trek*, although Captain Kirk's communicators operated much like old-fashioned walkie-talkies. There is a strange impulse in us that wants fiction to come true.

The current generation of mobile phones is nudging its way towards internet access. As people come to rely increasingly on internet access for work, entertainment and communication, ways of linking mobile phones to the internet are being explored. With some systems it is now possible to use mobile phones to gain access to e-mails. The phones use low power, just enough to connect them to the closest cell site, so that there is almost no interference among phones using the same radio frequencies.

In the UK there are about fifty million mobile phones in use – almost as many phones as there are people – and that figure has doubled in only six years. The base stations operate mainly under a Global System for Mobile Communications (GSM). This GSM has become the world's foremost and fastest growing telecommunications system, used by one-sixth of the world's population. There are over one billion subscribers, living in over 200 countries, and the number of subscribers is growing at a remarkable rate.

There have been concerns about the possible dangers of excessive exposure to radiofrequency fields. There is a fear that they may increase the risk of brain tumours. The rapid spread in the use of mobile phones shows that people are not afraid of them, but on the other hand smoking was extremely popular in the early twentieth century, regardless of some obvious and some less obvious harmful effects on health. A British government report in 2000 advised that some precautions ought to be taken as the evidence was inconclusive.

The mobile phone has had widespread effects, enabling people to communicate with each other socially, professionally and commercially far more easily. The added facility of being able to leave people text messages on their phones has proved to be a great asset to many people. One surprising feature of mobile phones is the way in which they are transforming life in the less economically developed countries.

Already one-third of Kenyans have mobile phones. In a country where the infrastructure is poorly developed and unreliable, it is helping people to cope with, for example, incidents of stranding by bus breakdowns. It even helps people in the rural areas. Kenyan farmers can use their mobiles to find the price they can expect for their produce in several nearby market centres; they can then choose the most favourable and drive their produce there. It is also possible to receive money by mobile phone, then take the mobile and the code number sent with the transaction to a bank teller, and get the cash over the counter. In a country where many people do not have bank accounts, this innovation is set to transform the economy by helping small businesses to grow. Unexpectedly, the mobile phone is kick-starting the economy of some of the poorer countries of the world – in a way that millions of dollars of international aid donated over the last half-century have failed to do.

THE INVENTION OF THE POST-IT NOTE

(1974)

THE POST-IT NOTE was invented in 1974 by Arthur Fry. Arthur Fry sang in a church choir. To mark the places of the hymns he used slips of paper, just as people have for generations. He became frustrated at the way the slips of paper slid to the floor, causing him to lose the place. He was aware that Spencer Silver, a 3M researcher, had in 1968 developed a weak adhesive when he had been trying to develop a strong one. At the time, no one had been able to see any use for a weak adhesive, but Fry's idea for a temporary sticker for his hymnal was an indicator that a potential real use existed after all. 3M's policy of encouraging innovation meant that Arthur Fry's invention was welcomed. Prototypes were made in 1977 and tried out, and then the product was marketed, as Post-its, all round the world three years later.

The Post-it is a simple piece of stationery consisting of small squares or rectangles of tinted paper with strips of weak adhesive on the back. A generic term for this invention is repositionable note, but Post-it, the 3M trade name, is far easier to say and remember. The adhesive strip is for temporarily attaching the slip of paper to a document for adding comments or reminders, or just as bookmarks. Post-its are made in different sizes to cater for varying needs, and in different pastel colours, which are useful for coding different topics. The original 3M Post-its are a trademark canary yellow, which makes them stand out against regular white stationery.

Post-its might be used by a publisher's commissioning editor, who wants to raise a variety of points arising from the text of a typescript sent in by an author. Rather than writing comments on the typescript itself, the editor can write comments on Post-its, locate them so that they project from the edge of the relevant page, and so mark up the subject for discussion. After the discussion, whether the book is accepted or rejected, the Post-its can be peeled off and discarded.

Post-its are another of those simple but extremely useful inventions that make the lives of those of us who work with paper – and there are a great many of us – a great deal easier.

THE INVENTION OF THE MRI SCANNER

(1977)

THE PHRASE MAGNETIC resonance imaging (MRI) initially had the word nuclear in front of it, but it was dropped for fear of alarming patients; it sounded as if something radioactive might be involved. A long period of research led up to the first use of an MRI scanner on a human being, and when that happened, in July 1977, the event was noted within the medical profession but created no impact in the outside world.

The inventors of the MRI scanner were Raymon Damadian, Larry Minkoff and Michael Goldsmith, and they named their first machine Indomitable. For a long time there were only a handful of scanners in the USA. Now there are thousands. They have also been improved so that images that used to take hours to produce now take only seconds.

The MRI machine is a cube about nine feet in each direction, with a tube running horizontally through the middle. The patient is slid into this bore on his or her back on a table. Radio waves are sent into the body and then received back. The returning signals are converted into pictures by an attached computer. Pictures can be taken of any part of the body and on slices of any specified thinness.

The scanner is a great advance on x-ray photography in that it is capable of great precision. It can be made to focus on, and identify, a cube of tissue only half a millimetre across. Because of the sharpness of focus it offers, the MRI scanner is a powerful diagnostic tool that has allowed great advances in medicine.

THE INVENTION OF THE TRAVELATOR

(1978)

THE TRAVELATOR WAS invented in 1978 in Madrid. Today it is commonly found in large supermarkets, shopping malls as well as airports. The travelator is a horizontal escalator, a simple conveyor belt for transporting people at walking speed, usually about two miles per hour. Travelators are commonly used for taking people and their suitcases and hand luggage from one part of an airport to another, especially where there is a long distance between one terminal and another or between check-in desk and boarding lounge. Travelators are also called moving sidewalks or moving walkways. Whereas the escalator was designed as a moving staircase to take people effortlessly from one floor to another in a building or from one level to another in a tube station, the travelator was designed to move people across large concourses.

The technology of the travelator is very similar to that of the escalator, and it is usually fitted with the same synchronized handrail. The moving floor consists either of metal plates or a continuous rubber belt.

In the 1980s, the French experimented with a high-speed walkway in the Paris Invalides Metro station. Called the TRAX, *Trottoir Roulant Accéléré*, it was too complicated, with excessively elaborate foldable articulated plates, and it was rejected as a technical failure, though the principle of high-speed walkways was not forgotten. A later version was installed at the Gare Montparnasse station in Paris. This whisked people along at 9 mph but, because so many people were falling over, the speed was cut to 6 mph. The high-speed walkway works in exactly the same way as the low-speed version, but there are special procedures for getting on and off. Only people with at least one hand free are allowed on in the first place. On embarking, there is a thirty-feet-long acceleration zone. In that zone, riders stand still on rollers, at the same time gripping the handrail, which pulls them up to the speed at which it is safe to step onto the walkway. At the exit, there is a similar deceleration zone, where the handrail pulls the riders back to a slow enough speed to step off.

A more recent version of this, invented by Anselm Cote, is being tried out at Montparnasse station. It carries people at 7 mph. It is formally called the TRR (*trottoir roulant rapide* or fast rolling pavement), but its users have nicknamed it the TGV (*trottoir a grande vitesse*). This prototype, which is 200 yards long, carries 110,000 passengers per day. Seasoned travelator users stride boldly along it, using it for power walking, while unfamiliar users are more timid; a few fall and hurt themselves and the Paris Metro has had to pay compensation. When they reach the end of the travelator, new passengers appear either relieved or elated. The project manager for the Paris Metro commented that people have to learn how to use the travelator and that the escalator presented similar challenges when first introduced. The particular problems for passengers are the acceleration and deceleration phases at each end.

In 2003 a pair of Finnish inventors, Jorma Mustalahti and Esko Aulanko, proposed a significant refinement to the travelator. Their 2004 US patent application describes 'a plurality of successive conveyors to form a) an acceleration section which includes successive conveyors having constant speeds stepwise increasing in the transport direction, b) a constant-speed section, and c) a deceleration section containing successive conveyors having constant speeds stepwise decreasing in the transport direction.'

This is very close to the travelator I invented in 1972, though mine had several conveyor belts running parallel to each other, not end to end. The outermost belts on each side were to be the slowest moving, running at 3 mph. The next belts inside those ran at 6 mph. The central high-speed belt ran at 9 mph. Boarding passengers would step onto the slow-moving belt. After a few moments of adjustment to its speed, they would move to the medium-speed belt, and then to the high-speed belt. Passengers might stand, walk or even sit on the belts. In time, as with all good inventions, this might be extended, with belts added on each side, to take the fast-moving central belt up to yet higher speeds. Travelators of this type might run not only in weather-proof transparent plastic tubes across cities but longer distances from city to city. Although the speeds envisaged might not be very fast, they would be continuous and reliable. This would be a major contrast to the speed of car travel in cities, which fluctuates enormously: drivers are continually, unpredictably and frustratingly brought to a standstill by traffic lights, roundabouts and congestion queues. It would mean an end to gridlocks, an end to travel stress – and it would be virtually silent.

Exactly this same idea was proposed by the incompetent civil servant Dundridge in Tom Sharpe's comic novel *Blott on the Landscape*, which was published in 1975. It is yet another case of several people thinking of the same idea at about the same time. As far as I know, no one has yet attempted to build a Dundridge-Castleden travelator.

THE INVENTION OF THE DESKTOP COMPUTER

(1980)

THE DESKTOP OR personal computer is another of those inventions that marks the culmination of a series of inventions. The first modern computers, in the 1940s and 1950s, were huge industrial machines that filled entire rooms and operated with agonizing slowness. With time they became smaller and faster until the emergence of a desktop computer that was small enough to fit into people's houses, and literally take up no more space than the top of a desk.

It was the Industrial Revolution that created the spur to make faster error-free calculations for engineering projects. In 1822 Charles Babbage, an English mathematician, built a model of a pioneer computer which he called a difference machine. Babbage's machine, which compiled and printed mathematical tables, was designed to be operated mechanically by turning a handle mounted on top. It was never mass-produced.

The first modern computer was invented by Vannevar Bush, an engineer at MIT, in the 1930s. He built it to help in the solution of mathematical equations needed to solve engineering problems. Bush and his colleagues finally unveiled the differential analyser in 1936, a machine weighing 100 tons and consisting of hundreds of miles of electrical wiring. Clumsy and slow though the machine was, it was said to be 100 times faster at solving equations than a human being. The evolution of the digital computer was stimulated by the Second World War, when it became a matter of urgency for the Allies to break German codes; the Colossus was a computer built by the British specially to break German codes.

The first commercial computer, a rapid calculation machine, appeared in 1951. It was capable of 8,333 additions per second. Although it covered 200 square feet of floor space, this was already much less space than a computer occupied only a few years earlier. And once the computer could be compacted down to three or four square feet of table-top space, it was small enough for people to have in their offices and homes. From 1980 onwards, more and more ordinary people have had their own personal computers, enabling them not only to make rapid calculations but carry out sophisticated word processing.

THE INVENTION OF THE WORLD WIDE WEB

(1990)

THE WORLD WIDE Web was invented in 1990 by Tim Berners-Lee and first became available on the internet in the following year.

Tim Berners-Lee was born in London in 1955 and graduated from Queen's College, Oxford in 1976. He then worked for Plessey Telecommunications Ltd as a software engineer for two years, then as an industrial consultant to D. G. Nash Ltd for a further two years. At this time, while he was a software consultant, he designed an unpublished program which he called Enquire. This program was the forerunner of the World Wide Web. Berners-Lee became the founding director responsible for technical design at Image Computer Systems Ltd (1981–84). Then for ten years he worked at CERN in Geneva. From 1989 onwards he worked on his highly ambitious global hypertext project, which became known as the World Wide Web in 1990.

The World Wide Web is without doubt the greatest invention of our age, and the biggest advance in communications since the invention of television. It consists of a system to organize, link and browse pages on the internet. It is an easy, point-and-click way of navigating and sorting the stored data. The Web has turned the information stored on the internet into a colossal, but very accessible magazine. Addresses on the World Wide Web are instantly recognizable by their opening formula – www, for World Wide Web.

Tim Berners-Lee is a self-effacing man who never had any intention of exploiting his invention commercially. He was motivated by the desire to be useful, and went on expanding the use of the internet as a channel for free expression and collaboration. He has been universally praised for his invention and his willingness to release it for general use. He was knighted in 2003 and later awarded a prize worth £670,000 by the Finnish Millennium Technology Prize award committee. Berners-Lee's invention has significantly enhanced people's ability to obtain information, encouraged new types of social network, opened new avenues for information management and greatly assisted many businesses.

The Web has turned out to be such an easy, cheap and flexible platform for communication that it has sparked an explosion of publishing, both professional and amateur. As a result of its popularity, it contains a colossal amount of information and spans a huge range of topics. The lack of editorial control is both its strength and its weakness; the Web is all-inclusive, a noisy babbling marketplace where the mad, the wicked and the ignorant stand side by side with the sane, the virtuous and the well-informed.

Tim Berners-Lee also designed the URL (universal resource locator) that we are all familiar with – the web address. He also invented HTML, the hypertext markup language. Berners-Lee in effect invented the whole package, an instant global communications system.

THE INVENTION OF THE CLOCKWORK RADIO

(1991)

IN 1991 TREVOR Baylis watched a television news report of the spread of HIV-AIDS across Africa, largely because of ignorance. There was no easy way of communicating crucial information about AIDS and its prevention. Radio would be an effective way to communicate, but most of the 600 million people living in Africa had no electricity to power radios; they might not have access to batteries or be able to afford them if they did.

Baylis thought about the older generation of technology, about old wind-up gramophones and the like. In his workshop he experimented with a carpenter's brace, which was crank-shaped like the winding handle of a gramophone, and found he could drive a small dynamo with it. He could generate enough power to work a transistor radio. Then he added a simple clockwork mechanism which allowed the brace to wind up a spring; then the spring drove the dynamo as it unwound. The clockwork radio was essentially simple in concept, though it took Baylis many hours to refine it so that the radio would run for an hour on a twenty-five second wind. Later, he added a solar cell, so that the winding up was only necessary at night; the radio was to be sun-powered during the day.

It was a perfect piece of 1990s low-technology. Getting investors to take it seriously was another matter. Baylis tried a wide range of organizations and was turned down by all of them. One rejection note read: 'It is very unlikely that UK industry could enter profitably into a licensing agreement with this product. The major customers are Third

World countries which, with severe debts, would not be in a position to pay for this device.'

The media were Trevor Baylis's great allies. The BBC *Tomorrow's World* programme featured his invention in 1994. It was a long-running and very prestigious series; long before, the newly invented silicon chip had been shown, and its implications accurately forecast. Details of the invention were repeated in a similar programme broadcast on a Johannesburg radio station. Hylton Appelbaum, a South African life insurance executive, heard the broadcast and set about funding production. BayGen Power Industries was set up in Cape Town. BayGen made a point of employing disabled people on the assembly line and persuaded various organizations supporting the disabled to join the venture. Technical support was available in Britain from electronics engineers at Bristol University. Production got under way, with 250 disabled people assembling thousands of clockwork radios every month.

Baylis visited the factory. 'As we approached the gate I caught the first glimpse of the words BayGen Power Manufacturing. Seeing the first syllable of my name in letters three feet high was a milestone in my life. It was the culmination of years of struggle. My mind went back to the rejections I had had . . . It brought tears to my eyes.'

Trevor Baylis later received many awards for his invention and became a celebrity. He was born in London in 1937. As a boy he was a great swimmer, but he left school without any formal qualifications. He became an engineering apprentice and studied at technical college, gaining a qualification in mechanical engineering. He later became a stunt man and an underwater escape artist at the Berlin Circus; as an escapologist he was known as Rameses II. This unusual occupation earned him enough to buy a house on Eel Pie Island in the River Thames west of London. Baylis began inventing gadgets to help the disabled, such as foot-operated scissors. He has more than 200 inventions to his name.

In 1997 Trevor Baylis was awarded the Order of the British Empire for services to Africa and also the Presidential Gold Medal by the Institution of Mechanical Engineers. In 1998, for the British Council, he addressed the Conference of Commonwealth Ministers in Botswana. He was also named Exporter of the Year for his outstanding export achievement. In 2000 Baylis invented the electric shoe, which charges the battery of a mobile phone while walking. He undertook a 100-mile walk across the Namib Desert in southern Africa to raise money for the Mines Advisory Group. He wore his electric shoes for this walk and was able to call Richard Branson from the middle of the desert.

But the invention that Trevor Baylis will be especially warmly remembered for is his clockwork radio, which can now be found in virtually every village in the poorest regions all round the world. The clockwork radio has made a huge difference, in enabling people who were once cut off from the outside world to hear national and international news. It enables people to be warned of impending natural disasters such as floods or hurricanes. It gives them useful information about the prevention of disease. It will also, doubtless, do much to widen the aspirations and ambitions of people living in closed and traditional communities. The radio promises to be an agent of major social change.

Trevor Baylis wants to help other people, and that includes inventors. He set up the Trevor Baylis Foundation. Its purpose is to help inventors to decide whether to continue with their ideas or not, and to give practical advice on patents, finance and marketing – the maze of obstacles standing between the idea and the realities of the market place. Trevor Baylis's view is that 'the key to success is to risk thinking unconventional thoughts. Convention is the enemy of progress'.

THE INVENTION OF THE GLOBAL POSITIONING SYSTEM

(1993)

THE FIRST EXPERIMENTAL Global Positioning System satellite was launched in 1978. There are now about twenty-five of them in orbit round the Earth, transmitting signals, allowing GPS receivers to determine the receiver's location. The system was developed by the US Defence Department, and its official name is Navigation Satellite Timing and Ranging Global Positioning System (NAVSTAR GPS). The constellation of satellites is managed by the Space Wing of the US Air Force. The management and maintenance of the system, including replacing ageing satellites, is hugely expensive. The cost is estimated to be in the region of $400 million per year. Even so, civilians are allowed to use GPS free of charge.

The GPS receiver, which looks something like a large mobile phone, calculates its own position by measuring the distance between itself and three GPS satellites. It does this by measuring the time delay between transmitting and receiving each GPS radio signal. Once the distance of the receiver from each of three GPS satellites is known, the location of the receiver is fixed.

The system has become a vital navigation aid, removing the need for any knowledge of the traditional methods for finding latitude or longitude. No geographical expertise of any kind is needed. The GPS receiver tells you your latitude and longitude with great precision, and all you have to do is to look at a map to plot your location on it using those coordinates.

GPS is also very useful for land surveying and map-making. The essence of map-making is to locate a few points in the landscape accurately. For example, if you have the coordinates of the corners of a church and the corners of the churchyard, joining those eight points will create an accurate map of the site.

THE INVENTION OF WINDOWS 95

(1995)

WINDOWS 95 WAS invented by Bill Gates in 1995. William Henry Gates developed an interest in computers while he was a pupil at Lakeside School; he was already programming computers at the age of thirteen. In 1973 Gates went to Harvard where he met Steve Ballmer, who was later to become Microsoft's chief executive officer. While still at Harvard, Bill Gates developed a version of the programming language BASIC, which was to be used for the very first microcomputer, the MITS Altair. He worked on BASIC with Paul Allen, a friend since childhood.

In 1975, at the age of nineteen, Gates founded Microsoft Corporation in collaboration with Paul Allen, after leaving Harvard without graduating. Five years after that they licensed a computer operating system to International Business Machines (IBM) to use in the personal computer industry which was then just getting under way. With the advent of the microchip it was now possible to manufacture computers that were small enough to install in people's homes and offices. The computer system that Bill Gates and Paul Allen created, MS-DOS (Microsoft Disk Operating System), and all the applications systems that dovetail into it have been a phenomenal success.

Right from their earliest days in the business, Gates and Allen were motivated by the belief that the computer would be a valuable tool in every office and home. The Microsoft mission is to improve software continually, to make it easier, more cost-effective, and to make it more enjoyable for people to use computers. Through his Microsoft firm, Gates has been able to maintain his remarkable dominance in the personal computer industry by producing repeated updates that are hugely successful in meeting customer needs – such as Windows 95 and

Windows 98. Microsoft became the world's largest producer of micro-computer software.

Microsoft is now a huge multinational enterprise. It brought in revenues of almost thirty-seven billion dollars in one year recently, and it employs 55,000 people in eighty-five different countries. Adaptation to rapid change and keeping ahead of possible competitors are very important, and the level of investment is very high: over six billion dollars in one year.

By 1986, Bill Gates was a billionaire. He is undoubtedly one of the richest people in the world. He is also, because he in effect controls the internet, one of the most powerful people in the world. He uses his money intelligently and thoughtfully. With his wife he set up the Bill and Melinda Gates Foundation, which has supported global health and learning to the tune of twenty-seven billion dollars.

Microsoft released Windows 95 in August 1995. It was a graphical, user interface-based operating system for computers. It was a significant advance, as it pulled together all the pre-existing MS-DOS and Windows products into one operating system. Instead of the various parts of the operating system being previously available as optional extras, they now became part of a fully integrated system. It was also more powerful, able to cope with file names that were 255 characters long.

Windows 95 did not have a very long life, as it was superseded in 1998 by Windows 98, an upgrade, and in 2001 by Windows XP. Windows XP, standing for 'experience', represents a further upgrade. It has been very popular, bought by more than 400 million users. Windows XP is more stable and efficient than the earlier versions of Windows. The redesigned screen is easier to manage and Microsoft marketed the whole package as more user-friendly than previous versions. Microsoft had another motive for setting it up, which was to combat software piracy; like any commercial enterprise using innovative technology, Microsoft has had to protect itself against piracy. Windows XP in its turn has been followed by Windows Vista, which was released in January 2007.

These later versions of the Windows package, Windows 98, Windows XP and Windows Vista, still use the same basic package and format as Windows 95. By the end of 2001, Microsoft ended its technical support for Windows 95, signalling that the company regarded it as obsolete.

Windows 95 nevertheless remains in widespread use in homes and offices all round the world, simply because many computer owners are unable to afford to upgrade frequently, and given the technical requirements of the later Windows versions it is unlikely that they would be able to upgrade without buying new computers.

Advances in computing are trumpeted so frequently that customers wonder whether perhaps the computer industry is starting to do something for which the car manufacturing industry became notorious: building obsolescence into the product in order to make customers come back and buy again. This business philosophy works well with small and inexpensive items, as William Painter found with his Crown Cork bottle top and King Gillette found with his safety razor and disposable razor blades. But with very expensive items like cars and home computers, it is a marketing strategy that is bound to meet considerable scepticism, resistance and resentment among customers. The rate of innovation and invention in computing is incredibly fast, but the consumers and potential consumers are beginning to suffer from innovation fatigue.

THE INVENTION OF THE 10,000-YEAR CLOCK

(1995)

THE 10,000-YEAR clock was invented in 1995 by the computer scientist Danny Hillis. The motive behind it was to create a monument to long-term thinking. He discussed the idea with friends, then wrote a magazine article about it in 1995. In 1996 a group led by Stewart Brand became the founding board of The Long Now Foundation. One member, Peter Schwartz, proposed 10,000 years as the time frame; that was the length of time that people had had a stable climate and technological progression, the duration (so far) of the current interglacial. There is a manifest need to start, or resume, thinking in deep time.

Danny Hillis himself reflected on the ceiling of the college dining hall at New College in Oxford. In the nineteenth century, when the roof beams needed replacing, carpenters used oak trees that had been planted in 1386 when the dining hall was first built. The fourteenth-century builder had planted those oak trees in anticipation of the time, hundreds of years into the future, when the beams would have to be replaced. Did the nineteenth-century carpenters plant new trees in anticipation of the need to replace the beams in a few hundred years from then? It seems unlikely.

The clock was designed to reflect this longer timescale. The proto-type, called the Clock of the Long Now and built in 1999, is on display at the Science Museum in London, in the Making of the Modern World exhibit. The clock is designed to last 10,000 years. Will it survive? From the evidence around us, the only way for an artefact or building to survive is for it to be made of materials that are very large and worthless, like the pyramids or Stonehenge, or for it to be altogether lost from sight like the Dead Sea Scrolls.

Jonas Salk, who invented the polio vaccine in 1955, could not see the point of Hillis's clock. He said, 'Think about what problem you are trying to solve. What question are you really trying to ask?' Hillis tried to persuade Salk about the need to nurture the future and told him the story about the oak trees. Salk replied, 'Oh, I see. You want to preserve something of yourself, just as I am preserving something of myself by having this conversation with you.' Salk died a few weeks later, which made Hillis remember what he had said: 'Be sure you think carefully about exactly what you want to preserve.'

The requirements of the clock, its design features, were very specific. It has to last 10,000 years and keep the correct time for that long. It has to be easy to maintain and people should be able to work out how it works just by looking at it closely. It should be possible to build it in different sizes. What emerged was an extraordinary-looking clock, as exotic in its way as Harrison's chronometer, and standing eight feet tall. The clock is people-powered. The eventual, much larger version which is expected to stand on a mountain in America, will have a huge weight that people will wind up so that it rises to the top of a tower. The weight will slowly descend on a screw, which will turn and power the clock. The clock is designed to use the noontide heat of the sun to reset itself.

The clock is an unusual invention, perhaps closer to a work of art than a useful device. It is not so much a clock, more a monument to Time itself.

THE INVENTION OF THE DIGITAL VERSATILE DISC (DVD)

(1995)

THE DVD (DIGITAL Versatile Disc) is an optically-read disc that can be used to store data of various kinds, including films with high picture quality and high sound quality. DVDs look very much like compact discs as they have the same weight and the same diameter, 4.7 inches. Although they look similar, they are in fact encoded differently and to a far higher density.

In the 1990s there was an uncomfortable situation where two rival optical disc storage systems were being developed. One was the MultiMedia CD, which was backed by Sony and Philips, the other was the Super Density Disc, which was backed by Time-Warner, Toshiba, Matsushita Electric, Hitachi, JVC and others. The president of IBM, Lou Gerstner, acted as a go-between, hoping to unify the two projects. Many of the people involved could remember the expensive commercial war between Betamax and VHS in the 1980s, and wanted to avoid a repeat of that. The result was that Philips and Sony agreed to adopt the Super Density Disc, asking only for a modification that would increase resistance to scratch damage. The DVD specification was agreed in December 1995.

The DVD format is now routinely used for marketing films and television programmes. From 2003 onwards, DVD-video has become the dominant format for consumer videos. The new format is popular, because DVDs are smaller, more compact, easier to store in the home, and the picture and sound quality are very high indeed. DVD recordables are also available, so that people can record their own choice of radio and television programmes. The higher capacity of DVDs means that when they are used for audio recording they can either hold considerably more music than a CD, with respect to running time, or far higher quality sound. On the other hand the sound quality that is achievable on CDs is already so high that experts wonder whether the super-high sound quality that can be achieved on DVDs will be noticeable to any listeners.

What we are seeing with the invention of the Digital Versatile Disc is a major achievement not only in quality, but in flexibility and economy. It is hard to imagine better technology than the DVD for storing and playing films and television programmes. Minor refinements are certainly possible. From the user's point of view, the loading procedure could be shortened, but otherwise the DVD technology looks unimprovable.

But being unable to see how things might change in the future is the way it has always been. In the seventeenth century, still very much the age of wooden sailing ships, it would have been impossible to foresee that ships two centuries into the future would be made of iron and powered by steam. When Martin Behaim was putting the finishing touches to his great globe in Nuremberg in 1492, showing no New World, Columbus was just making landfall there. His globe was obsolete as soon as it was made.

When Max Planck enrolled at the University of Munich in 1875, he was advised by his professor not to study physics: there was nothing left to find out. A dead subject. His professor's prediction turned out to be dramatically wrong. As the nineteenth century ended, J. J. Thomson discovered the electron, Roentgen discovered x-rays and invented x-ray photography, and Henri Becquerel discovered radioactivity, which led on to the eventual development of nuclear energy; in 1905 Einstein proposed his special theory of relativity. In 1875 it looked as if science and technology had nothing further to achieve. In fact a surge of dramatic discoveries and inventions lay ahead.

When we narrate and evaluate the story of inventions, we can only see backwards: we only know how things came about once they are over. To live through the creative act of inventing or discovering is a turbulent and intensely exciting experience, full of unknowns. The inventor does not know from week to week whether the cherished project is going to work or not. I remember a biographer commenting that there is exactly this problem in biography as a genre. We see a particular shape to the life of, say, Percy Shelley, as a poet who was fated to die young. But Shelley himself didn't know that. He didn't plan to drown that day in 1822 when he sailed out into the Gulf of Spezia in the schooner *Ariel.* Shelley's life must have looked very different to him while he was living it than it does to us looking back at it, across the early death in the boating accident. And so too it is with inventors and their inventions. We know that King Gillette's safety razors turned out to be a triumphant success and that he would become wealthy, but at the age of thirty he thought he was a failure.

There are inventors who went to their graves not knowing how much they had contributed to the common good (or in some cases bad), who simply did not know that they had changed the world. There are also inventors who have been over-praised and over-recognized, like George Stephenson. Many people forget that in 1825 he in effect put the little-known Richard Trevithick's road locomotive on rails: two borrowed ideas. But then, even more people have forgotten that Trevithick took the idea from William Murdock, whose early road locomotive he saw in Redruth in 1786. Everyone has heard of George Stephenson. Virtually no one has heard of William Murdock.

The strange, topsy-turvy world of inventions, with its minefield of patent laws, has nevertheless left us with a world full of remarkable machines, processes and gadgets. Some of them are life-savers, like inoculation and antiseptic surgery, and we rightly value them very highly. But the frivolous inventions help to make our lives more interesting and enjoyable, and they are not to be forgotten.

Philip Astley invented the circus in about 1770. George Crum cooked the first crisps in 1853. LaMarcus Adna Thompson built the first roller coaster in 1865 and George Ferris the first Ferris wheel in 1893. In 1922 the eighteen-year-old Ralph Samuelson from Minnesota invented water-skiing and in 1961 another American, Clayton Jacobsen II, invented the jet ski. Milk chocolate was invented in 1876 by Daniel Peter, bubble gum in 1928 by Walter Diemer and chocolate chip cookies two years later by Ruth Wakefield. Trivial Pursuit was invented in 1979 by Chris Haney and Scott Abbott. Sometimes a hugely serious scientific enquiry can end up producing something trivial and entertaining, like the great Michael Faraday devising equipment for his laboratory experiments and ending up inventing the toy balloon.

With some inventions it is difficult to know whether we should take them seriously or not. An 'all-weather wrinkle-free tie' was patented in 1920 by Jesse Langsdorf. In 1998 someone took out a US patent on a 'kissing shield'. This thin latex membrane stretched over a heart-shaped frame is designed for those who like kissing but fear infection. In 2003 a patent was issued for a beerbrella, a small brightly coloured sunshade that clips onto a beer bottle to keep it cool. Are they serious? Before we give way to laughter, we should reflect for a moment that people through the ages have often found new inventions absurd and either funny or frightening. People laughed at Velcro, vaccination and the clockwork radio, and were frightened by a top hat.

APPENDIX:
INVENTIONS NAMED AFTER THEIR INVENTORS

Agpar scale	Dr Virginia Agpar
Archimedes' screw	Archimedes
Argand lamp	Ami Argand
Armstrong breech-loading gun	William George Armstrong
Bakelite	Leo Baekeland
Beaufort scale	Rear Admiral Sir Francis Beaufort
Belisha beacon	Leslie Hore-Belisha (not inventor, introducer)
Bessemer converter	Henry Bessemer
Biro	Laszlo and Georg Biro
Blacker bombard	Lt. Col. Blacker
Boolean algebra	George Boole
Bowie knife	Jim Bowie
Braille	Louis Braille
Browning firearms	John Browning
Bunsen burner	Robert Bunsen
Clerihew	E. Clerihew Bentley
Coade stone	Eleanor Coade
Colt revolver	Samuel Colt
Daguerrotype	Louis Daguerre
Davy lamp	Humphry Davy
Dewar flask	James Dewar
Diesel engine	Rudolf Diesel
Dolby noise reduction system	Ray Dolby
Doppler radar	Christian Doppler

Ferris wheel	George Washington Gale Ferris Jr
FitzRoy sea area	Vice Admiral Robert Fitzroy (pioneer weather forecaster)
Franklin stove	Benjamin Franklin
Galil assault rifle	Israel Galil
Garand rifle	John Garand
Gatling gun	Richard J. Gatling
Geiger counter	Hans Geiger
Geiger-Muller tube	Hans Geiger and Walther Muller
Gestetner duplicator	David Gestetner
Gillette safety razor	King Camp Gillette
Gore-Tex	Wilbert Gore
Gregorian calendar	Pope Gregory XIII
Gregorian telescope	James Gregory
Guillotine	Dr Joseph Ignace Guillotin (not inventor, proposer)
Hallidie ropeway	Andrew Smith Hallidie
Halligan bar	Hugh Halligan
Hammond organ	Laurens Hammond
Hansom cab	Joseph Hansom
Heimlich manoeuvre	Henry Heimlich
Hoover	William H. Hoover (not inventor, investor)
Hutchinson patent stopper	Charles G. Hutchinson
Jacquard loom	Joseph Marie Jacquard
Jacuzzi	Roy Jacuzzi
Josephson junction	Brian David Josephson
Kalashnikov gun	Mikhail Kalashnikov
Leigh light	Humphry de Verde Leigh
Mae West life jacket	Mae West (not inventor, joke)
Macadam	John McAdam
Mackintosh	Charles Macintosh
Marconi rig (sloop)	Guglielmo Marconi (not inventor, joke)
Mason jar	John Mason
Maxim gun	Hiram Maxim

Melba sauce, Peach Melba	Dame Nelly Melba (not inventor)
Molotov cocktail	Vyacheslav Molotov (not inventor, joke)
Moog synthesizer	Robert Moog
Morse code	Samuel Morse (not inventor, promoter)
Owen gun	Evo Owen
Pavlova	Anna Pavlova (not inventor)
Pasteurization	Louis Pasteur
Petri dish	Julius Richard Petri
Philips screwhead (and screwdriver)	Henry F. Philips (not inventor, promoter)
Pulaski	Ed Pulaski
Prusik knot	Karl Prusik
Raman spectroscopy	C. V. Raman
Remington rifle	Eliphalet Remington
Richter scale	Charles Richter
Rorschach inkblot test	Hermannn Rorschach
Rubik cube	Erno Rubik
Sandwich	4th Earl of Sandwich
Saxophone	Adolphe Saxe
Shrapnel	Henry Shrapnel
Schick test	Bela Schick
Sousaphone	John Philip Sousa
Soyer stove	Alexis Soyer
Spoonerism	William Archibald Spooner
Stark spectroscopy	Johannes Stark
Sten gun	Shepherd, Turpin (and their employer) RSAF Enfield
Stirling engine	Revd Robert Stirling
Strowger switch	Almon Strowger
Talbotype	William Fox Talbot
Thompson sub-machine gun	John T. Thompson
Tesla coil	Nikola Tesla
Theremin	Leon Theremin
Tupperware	Earl Tupper

Uzi sub-machine gun	Uziel Gal
Vernier	Pierre Vernier
Wankel engine	Felix Wankel
Wellington boot	Duke of Wellington
Westinghouse airbrake	George Westinghouse
Winchester repeating rifle	Oliver F. Winchester (not inventor, investor)
Yale lock	Linus Yale
Zamboni	Frank Zamboni
Zeppelin	Ferdinand von Zeppelin